Handbook of Research on Improving Engineering Education With the European Project Semester

Benedita Malheiro
Instituto Superior de Engenharia do Porto, Instituto Politécnico do Porto, Portugal

Pedro Fuentes-Durá
Universitat Politècnica de València, Spain

IGI Global
PUBLISHER of TIMELY KNOWLEDGE

A volume in the Advances in Higher Education and Professional Development (AHEPD) Book Series

Published in the United States of America by
IGI Global
Engineering Science Reference (an imprint of IGI Global)
701 E. Chocolate Avenue
Hershey PA, USA 17033
Tel: 717-533-8845
Fax: 717-533-8661
E-mail: cust@igi-global.com
Web site: http://www.igi-global.com

Copyright © 2022 by IGI Global. All rights reserved. No part of this publication may be reproduced, stored or distributed in any form or by any means, electronic or mechanical, including photocopying, without written permission from the publisher. Product or company names used in this set are for identification purposes only. Inclusion of the names of the products or companies does not indicate a claim of ownership by IGI Global of the trademark or registered trademark.

Library of Congress Cataloging-in-Publication Data

Names: Malheiro, Benedita, editor. | Fuentes-Durá, Pedro, 1968- editor.
Title: Handbook of research on improving engineering education with the European Project Semester / Benedita Malheiro, and Pedro Fuentes-Durá, editors.
Description: Hershey PA : Information Science Reference, [2022] | Includes bibliographical references and index. | Summary: "This book describes European Project Semester (EPS), a one-semester program offered by a network of European universities to students who have completed at least two years of undergraduate engineering studies with the goal to prepare undergraduate engineers to become global actors of change"-- Provided by publisher.
Identifiers: LCCN 2021048750 (print) | LCCN 2021048751 (ebook) | ISBN 9781668423097 (hardcover) | ISBN 9781668423110 (ebook)
Subjects: LCSH: European Project Semester. | Engineering--Study and teaching (Higher)--Europe. | Student-centered learning--Europe. | Project method in teaching--Europe.
Classification: LCC T105 .A53 2022 (print) | LCC T105 (ebook) | DDC 620.0071/14--dc23/eng/20211115
LC record available at https://lccn.loc.gov/2021048750
LC ebook record available at https://lccn.loc.gov/2021048751

This book is published in the IGI Global book series Advances in Higher Education and Professional Development (AHEPD) (ISSN: 2327-6983; eISSN: 2327-6991)

British Cataloguing in Publication Data
A Cataloguing in Publication record for this book is available from the British Library.

All work contributed to this book is new, previously-unpublished material. The views expressed in this book are those of the authors, but not necessarily of the publisher.

For electronic access to this publication, please contact: eresources@igi-global.com.

Advances in Higher Education and Professional Development (AHEPD) Book Series

Jared Keengwe
University of North Dakota, USA

ISSN:2327-6983
EISSN:2327-6991

Mission

As world economies continue to shift and change in response to global financial situations, job markets have begun to demand a more highly-skilled workforce. In many industries a college degree is the minimum requirement and further educational development is expected to advance. With these current trends in mind, the **Advances in Higher Education & Professional Development (AHEPD) Book Series** provides an outlet for researchers and academics to publish their research in these areas and to distribute these works to practitioners and other researchers.

AHEPD encompasses all research dealing with higher education pedagogy, development, and curriculum design, as well as all areas of professional development, regardless of focus.

Coverage

- Adult Education
- Assessment in Higher Education
- Career Training
- Coaching and Mentoring
- Continuing Professional Development
- Governance in Higher Education
- Higher Education Policy
- Pedagogy of Teaching Higher Education
- Vocational Education

IGI Global is currently accepting manuscripts for publication within this series. To submit a proposal for a volume in this series, please contact our Acquisition Editors at Acquisitions@igi-global.com or visit: http://www.igi-global.com/publish/.

The Advances in Higher Education and Professional Development (AHEPD) Book Series (ISSN 2327-6983) is published by IGI Global, 701 E. Chocolate Avenue, Hershey, PA 17033-1240, USA, www.igi-global.com. This series is composed of titles available for purchase individually; each title is edited to be contextually exclusive from any other title within the series. For pricing and ordering information please visit http://www.igi-global.com/book-series/advances-higher-education-professional-development/73681. Postmaster: Send all address changes to above address. © © 2022 IGI Global. All rights, including translation in other languages reserved by the publisher. No part of this series may be reproduced or used in any form or by any means – graphics, electronic, or mechanical, including photocopying, recording, taping, or information and retrieval systems – without written permission from the publisher, except for non commercial, educational use, including classroom teaching purposes. The views expressed in this series are those of the authors, but not necessarily of IGI Global.

Titles in this Series

For a list of additional titles in this series, please visit: http://www.igi-global.com/book-series/advances-higher-education-professional-development/73681

New Considerations and Best Practices for Training Special Education Teachers
Jeremy Bell (University of Louisiana at Monroe, USA)
Information Science Reference • © 2022 • 339pp • H/C (ISBN: 9781799894940) • US $195.00

Digital Distractions in the College Classroom
Abraham Edward Flanigan (Georgia Southern University, USA) and Jackie HeeYoung Kim (Georgia Southern University, USA)
Information Science Reference • © 2022 • 324pp • H/C (ISBN: 9781799892434) • US $195.00

Global Perspectives on Teacher Performance Improvement
Osama Al-Mahdi (Bahrain Teachers College, University of Bahrain, Bahrain) and Ted Purinton (Bahrain Teachers College, University of Bahrain, Bahrain)
Information Science Reference • © 2022 • 302pp • H/C (ISBN: 9781799892786) • US $195.00

Implementing a Virtual Coaching Model for Teacher Professional Development
Suzanne Myers (University of Kansas, USA) Amber Rowland (University of Kansas, USA) and Martha D. Elford (University of Kansas, USA)
Information Science Reference • © 2022 • 258pp • H/C (ISBN: 9781799875222) • US $195.00

Driving Innovation With For-Profit Adult Higher Education Online Institutions
David Stein (The Ohio State University, USA) Hilda R. Glazer (Capella University, USA) and Constance Wanstreet (Franklin University, USA)
Information Science Reference • © 2022 • 317pp • H/C (ISBN: 9781799890980) • US $195.00

Collaborative Approaches to Recruiting, Preparing, and Retaining Teachers for the Field
Maria Peterson-Ahmad (Texas Woman's University, USA) and Vicki L. Luther (Mercer University, USA)
Information Science Reference • © 2022 • 300pp • H/C (ISBN: 9781799890478) • US $195.00

Self-Directed Learning and the Academic Evolution From Pedagogy to Andragogy
Patrick Hughes (Texas Tech University, USA) and Jillian Yarbrough (West Texas A&M University, USA)
Information Science Reference • © 2022 • 276pp • H/C (ISBN: 9781799876618) • US $195.00

IGI Global
PUBLISHER of TIMELY KNOWLEDGE

701 East Chocolate Avenue, Hershey, PA 17033, USA
Tel: 717-533-8845 x100 • Fax: 717-533-8661
E-Mail: cust@igi-global.com • www.igi-global.com

List of Contributors

Table of Contents

Detailed Table of Contents

This chapter aims to introduce the European Project Semester concept and the network of providers together with historical and current data on participants and providers. EPS is a one-semester international exchange program designed for engineering, product designers, and business undergraduates. It embraces student-centered project-based learning and multicultural and transdisciplinary teamwork to help students develop the competencies and skills required for the 21st century. It was created in 1995 in Denmark to prepare future engineers for global challenges, and it combines scientific and technical education, ethical and sustainability-driven problem-solving, intercultural communication, and teamwork. EPS has since been adopted by 19 European higher education institutions, establishing a network of EPS providers across Europe. The network offers international students a wide range of European study locations and cultural insights and, above all, the opportunity to enjoy the EPS learning experience. The number of alumni shows the program's attractiveness, range, and vitality.

This chapter explores the notion of effective communication not only in international contexts but in any situation which brings together people with diverse cultures. The chapter begins with a brief look at some intercultural experiences and then explores the notion of 'culture' and how it can shape our behavior and patterns of thought. From there, it moves on to discuss the 'classical view' of human interpersonal communication—and why it is incorrect—and then describes a model of communication which regards this as a 'construction' or 'negotiation of meaning'. The chapter continues by examining the broad concept of intercultural competence and then looks at some of the key skills, attributes and strategies which can contribute to both effective intercultural and intracultural communication. An appendix includes some example activities designed to develop greater cultural awareness.

In EPS, communication is key. Project teams plan and develop projects involving stakeholders in internationalized contexts. Outcomes are defined in a written report and an oral presentation, both assessed for content and communicative value. The chapter situates technical and professional communication (TPC) in engineering education and identifies communication practices as part of the professional practices of engineers. Main approaches to TPC in engineering are presented, and how TPC is approached from a problem-solving perspective is discussed. The rationale and organization of the TPC course is presented with examples based on a practical methodology that promotes familiarization, practice, reflection, and dialogue. A proposal for integrating TPC in EPS and similar programs is made. From an interdisciplinary view, the chapter reflects on the importance of TPC in engineering education, the needs of globalized professional settings, and the ubiquity of digitalization to discuss how engineering students can effectively engage in international engineering communication.

The purpose of this chapter is to present the experience of applying the design thinking methodology in the development of projects during the European Project Semester. Through an interdisciplinary approach, projects of diverse natures were developed by teams composed of students from different areas of education (mainly engineering and design) and from various nationalities, supported by a team of coaches from distinct disciplines, which demonstrated in practice how the integration between coursework and disciplines is essential and works well in project-based learning. The methodology used, its steps, techniques, and methods applied in the development of projects will be presented, as well as the achieved results.

Engineering aims to improve the quality of daily life on a planet with limited resources. This chapter describes how the European Project Semester offered by the School of Engineering of the Porto Polytechnic contributes to make engineering undergraduates conscious of their environmental and societal impact. The program is a project-based active learning framework where problem-solving is supported by intercultural communication and creativity, ethical and sustainable reasoning, and global contextual analysis. This analysis shows that it fosters scientific, technical, and interpersonal competencies, emphasizing ethical and sustainable design as recommended by several reference institutions through the testimonials left on the project reports and the evolution perceived by the participants during the semester. These findings are supported by a sample of 45 projects conducted by 228 students from 23 countries and 65 degrees. All projects aimed to have a lasting impact on the future engineering practice of participants and, somehow, transform the world positively.

Ethics promotes collective respect, well-being, and development. Its values drive individual, social, and environmental conduct and apply to all human activities. Professional ethics or deontology refers to the moral obligations binding professionals to society in the work sphere. These moral obligations are fundamental in professions that, like engineering, are behind technological innovation, given its current daily and worldwide impact. As such, it is essential to train engineers for holistic problem-solving, that is, driven by ethical, sustainability, technical, and economic constraints. This chapter analyses engineering ethics and how it is addressed within the European Project Semester, an undergraduate one-semester international exchange program, focusing on this aspect of engineering education. European Project Semester relies on multicultural teamwork, project-based learning, ethics, and sustainability-oriented design to motivate and prepare global engineers committed to protecting and improving life on the planet.

Continuous changes in different spheres of human activity rapidly intensifying in the globalized world pose dramatic challenges for individuals and educational institutions whose mission is to prepare their graduates for professional, lifelong success. The changes trigger a continuous discourse on the nature and form of educational paradigms and pivotal values that should be utilized by schools and universities to fulfill their mission. The chapter presents the human-oriented model of engineering focused on empathy, constructed in the group of academics at Lodz University of Technology. This group has defined a methodological approach that can help to guide students and teachers in finding the optimal approach in project-based study programs, most notably in the European Project Semester. Moreover, the chapter includes the results of interviews with students and project mentors and survey results measuring empathy in both stakeholder groups.

Robotics is a multidisciplinary subject that typically involves mechanics, electronics, and computer science concepts. For this reason, robotic projects are particularly well suited to the European Project Semester framework since they allow students with different backgrounds to contribute to the overall team objective in their specific knowledge areas. This chapter briefly presents illustrative examples of robotic projects that have been developed by teams of students participating in the European Project Semester at the School of Engineering of the Polytechnic Institute of Porto. It concludes by presenting and discussing student feedback, namely on the program and the projects developed.

Section 3
Case Studies

The implementation of the European Project Semester at the University Politehnica of Bucharest has outlined the importance of developing soft skills through project-based learning for 21st century engineering learners. This chapter describes the project conducted by a team of four multinational students in 2018. It aimed to develop soft skills (language, written and oral communication, organization and time-management, research and inquiry, self-assessment and reflection, group participation, and leadership) as well as to foster the necessary professional tools for a smooth insertion into the labor market. The development of a 3D-printed robotic arm with IoT capabilities has proved a useful didactic approach in this sense and, at the same time, an efficient way to meet the future economic and technological demands of our changing world. The main goal was to design an affordable robot arm so that people could buy for any application desired.

The European Project Semester program at the University of Applied Science Osnabrück deals mainly with engineering projects in different engineering domains. Developing a product based on complex technical requirements and economic concerns, such as cost-sensible development and marketing and environmentally-friendly design, are goals of a successful project during the one-semester internship at Osnabrück for a multicultural project team. In 2018, a team of four students received the task of developing a speedometer for bicycles with a brake light to enhance the rider's safety, especially in an urban environment. Besides the technical challenges, the team developed a marketing plan for the product. To the students, it was an excellent opportunity to understand the benefits and risks of working in a multicultural team and the necessity of following strict project management guidelines to finish a project as a team successfully.

The issue of design for social innovation is closely linked to citizen participation and the inclusion of the different groups that make up society, particularly the most disadvantaged or at risk of being marginalized. The new product and service design paradigm requires a specific sensitivity in which education and

This case study describes a special edition of the European Project Semester at the course Sustainable Packaging Design and Innovation at the Faculty of Industrial Design Engineering at The Hague University of Applied Sciences (Fall Semester 2017). In this special edition, unique cooperation took place between 12 parties. The parties were three research institutes, six universities, and three companies. Some parties have developed an educational module focused upon sustainable and circular packaging design, including the use of a dedicated tool for life cycle assessment. This module was embedded in the regular EPS. At The Hague University of Applied Sciences, an international class of 16 students worked in four teams on a real-life design assignment. They were offered a wide range of lectures, workshops, pitches, and presentations. The chapter concludes with a review of the followed processes and organizational, managerial, and practical concerns. Although run as a unique edition, all parties discuss to continue this cooperation.

For more than 10 years at the Escuela Técnica Superior de Ingeniería del Diseño, the European Project Semester experience on design and innovation has been led by the authors due to their experience in researching and developing more than 120 products launched in the market. This expertise is key to driving and supervising a student project until the stage of promotion at a professional trade fair. This fact becomes a milestone in the educational context that requires the full involvement of students, the school, and the university facilities. Applying agile methodologies in educational design projects is more than a learning process. It is an experiential activity nearer to professional skills development.

Sustainability is a principal research domain and part of the DNA of the European Project Semester at the Faculty of Design Sciences at the University of Antwerp. This chapter describes how a multidisciplinary EPS team collaborated to design an interactive event, exhibit, and campaign to promote sustainable consumer behavior on a fair at the city campus of the university, with a focus on awareness-raising about fast versus slow fashion. Throughout all development stages, the team improved their work-in-progress during weekly coaching sessions with access to the sustainability-related research group and toolkits. This rich trajectory and the resulting fast/slow fashion fair triggered deep reflections on the team regarding fair fashion as well as taught them to plan a major event with many stakeholders, eco-design for the outdoors, and proactively communicate and develop their public relations and negotiation skills. It also raised awareness about the sustainable fashion cause among the students, staff, and the general public who visited the fair.

Foreword

Talking about the future of engineering students is going beyond just the need to adapt our offer to the needs of the labor market.

We are aware of the potential of our students, and the basis for making this potential reality is to advance in a university with the following characteristics:

- With a system less focused on training for the jobs and trades of today, and capable of educating for the professions and positions of the future.
- With more choice for students and more flexibility in their academic careers.
- With extensive use of technology to personalize learning and increase the capacities of teachers and students.
- With an evaluation system that allows learning and improvement.
- With fewer face-to-face classes and more focus on developing a practical application, research, or innovation and entrepreneurship projects.

The European Project Semester (EPS) program brings together the aforementioned characteristics and helps students discover their strengths and align their training with their best abilities. Students learn as they experience and implement the projects.

I want to thank and congratulate the excellent work and collaboration of the participating professors and professionals and express my gratitude to all the EPS team for the valuable cooperation maintained.

Francisco José Mora Más
Universitat Politècnica de València, Spain

Francisco José Mora Más *is a Telecommunication Engineer from the Universitat Politècnica de Catalunya. Ph.D. in Electronic Engineering, carried out at the European Particle Physics Laboratory (CERN) with a scholarship Doctoral Student. Since 2003 he has been a Full-Professor at the Universitat Politècnica de València in Electronic Engineering. His field of research is the design of data acquisition systems for high energy physics and electronic instrumentation for medical physics. Prof. Mora was Rector of Universitat Politècnica de València from June 2013 to June 2021 and vice-president of Spanish Universities Rector´s Conference (CRUE) from 2018 to 2021.*

Foreword

Engineering education aims to prepare engineering students for their future professional journey, where they will be called on to solve complex challenges affecting individuals, companies, society, and the planet, while considering aspects such as cost, maintenance, and sustainability, to mention a few. Most engineering education institutions still favor professor instead of student-centered learning, focusing primarily on the acquisition of scientific and technical knowledge rather than seamlessly integrating the development of hard with soft transversal skills. Soft skills like design thinking, problem-solving, project management or interpersonal communication and teamwork are essential for engineers to act globally and in close connection with other experts and communities.

As a student-centered capstone design program, the European Project Semester (EPS) exposes students to challenge-based learning, revolving around a project, with special attention to international multidisciplinary teamwork, design, innovation thinking, problem solving and project management to develop a set of desired skills. In fact, this renewed emphasis on design has been identified as the third major shift in 100 years of Engineering Education, this meaning EPS is well aligned with a major trend in engineering education.

The *Handbook of Research on Improving Engineering Education With the European Project Semester* shares best practices in engineering education. This is achieved through close examination and reflection by the European providers as well as from invited colleagues from other longitudes. It describes the adopted learning framework, analyzes how it contributes to the simultaneous development of hard and soft skills, reports on the type of challenges proposed to teams, and delivers a set of team-project cases from the network of providers. Covering topics such as engineering ethics, project management, and sustainable behavior, this book is essential to students in engineering, engineers, engineering educators, educational researchers, academic administration and faculty, and academicians.

This book is the result of an extraordinary supervising and editing effort of Benedita Malheiro and Pedro Fuentes-Durá, who started their academic careers, as engineering educators, in 1989 and 1994, respectively, at the same institutions they are still affiliated with (Polytechnic of Porto – School of Engineering, and Polytechnic University of Valencia, in Portugal and Spain, respectively). This 30-year experience in engineering education is reflected not only in this book but also in the mindsets of the students who benefitted from the European Project Semester. Quoting Paulo Freire, "*Education does not change the world. Education changes people. People transform the world.*" We can only anticipate a better world if and when transformed by these students.

Gustavo R. Alves
Instituto Superior de Engenharia do Porto, Instituto Politécnico do Porto, Portugal

Gustavo R. Alves *graduated in 1991 and obtained an MSc and a PhD degree in Computers and Electrical Engineering in 1995 and 1999, respectively, from the University of Porto, Portugal. He is a professor at the Polytechnic of Porto - School of Engineering, since 1994. He has authored or co-authored 260+ publications, including conference and journal papers with referee process, and book chapters. He co-edited a book, (with Javier García-Zubía, University of Deusto, Spain), about "Using Remote Labs in Education". He was involved in 19 national & international research projects. His research interests include engineering education, remote experimentation, and design for debug & test. He served as program co-chair of the 1st and 2nd International Conferences of the Portuguese Society for Engineering Education (CISPEE2013 and CISPEE2016), of the 3rd Experiment@ (exp.at'15) International Conference, and as general chair of the 11th Remote Engineering and Virtual Instrumentation (REV2014) conference, and of the 3rd Technological Ecosystems for Enhancing Multiculturality (TEEM2015) conference. Dr. Alves currently serves as President of the Portuguese Society for Engineering Education (SPEE), Vice-President of the International Association for Online Engineering (IAOE), and as an Associated Editor for the IEEE Transactions on Learning Technologies and the IEEE Latin American Journal on Learning Technologies (IEEE-RITA).*

Preface

In the mid-nineties, higher education faced many challenges worldwide. How could, for example, the emerging digital technologies be integrated into the existing educational and administrative structures? How should courses and study programs be adapted or re-written in order to satisfy the changing demands of employers and society? How should universities meet the cost of the increasing number of student enrollments and the subsequent need for increased human and capital resources? Would universities be able to continue to fulfill their traditional missions? Would new educational paradigms become necessary?

In the European context, a new trend required urgent shifts in policy if universities were to maintain their crucial role in the development of societal well-being. Increasing globalization and multiculturality meant that promoting mobility was a must. Businesses were becoming location independent, and professional as well as social interactions between people from different ethnic or national backgrounds multiplying. In addition, recognition of the positive effect on employability of soft skills and authentic learning resulted in these being incorporated into curricula. The most significant attempt to modernize higher education was the Bologna Process, with forty-nine participating countries, but many innovative practices at regional and local levels were also born at this time. Beyond transparency, harmonization, and accreditation issues, emerging pedagogical concepts and frameworks such as Project-Based Learning, Design Thinking, Learning by Doing, or Living Lab sprung up everywhere.

One of these innovative practices was the bringing together of a diverse group of students from different fields of study in order to solve engineering problems. In 1995, Arvid Andersen, a professor at the Engineering College of Elsinore in Denmark, started a pioneer pilot project with six students, which, over the next five years, would be consolidated and disseminated. This pilot project became the European Project Semester (EPS), providing a rigorous and intensive learning experience based on student exchange, the carrying-out of a supervised industrial project, and the development of soft skills. Support for promoting student and staff mobility was pretty much universal and, without forgetting the importance of technical knowledge, know-how in soft skills, as well as experience of living and studying abroad, were both becoming increasingly attractive to employers who, more and more, began to recognize that the possession of certain aptitudes and skills not only increased a graduate's employability but also reduced the time he or she needed to become fully integrated into company practices.

Ten years after its inception, EPS had become well-known in innovative educational circles as an especially attractive course and was established in five different countries involving one hundred students per semester. The diversity of cultures, the number of educational institutions, the range of the disciplines involved, and the intended learning outcomes have been growing continuously since then.

By engaging students in long-term team projects across countries, the EPS network offers experiential learning as well as the opportunity to develop exactly those skills in project management and intercultural communication required by graduates today. Continuous changes in many different spheres of human activity, which, moreover, are rapidly intensifying in an ever more globalized world, pose dramatic challenges for individuals and educational institutions whose mission is to prepare graduates for professional, life-long success. This is where the EPS has a global appeal. Today, experts, researchers, and professors from different countries, continents, and universities provide an unforgettable personal experience to five hundred students per year in more than twenty locations.

The genesis of the idea for this book was the desire to create an EPS icon; a symbol to mark 25 years of success. It is inspired by the positive feedback collected from alumni and the vitality of the EPS network. We invited authors of dozens of published papers, as well as others who are content to allow their achievements to go unnoticed, to contribute. The result is a unique volume with compelling examples and reflections. The primary purpose of the book; namely, that of disseminating a set of activities, methods, and singular collaborations, has been supported by forty-two authors from twelve higher education institutions as well as other stakeholders. We are convinced that these chapters can guide and inspire instructors and spark innovation, and that this book will be helpful to managers who want to promote the development of creative thinking. It is not an exhaustive recapitulation of all the EPS projects carried out over the last twenty-five years; that would be exhausting! It is a selected sample of experiences and collaborative teamwork written by the participants themselves.

The book is organized into three different parts. The first part plays the introduction role, explaining the background and the fundamentals of EPS. Its two chapters summarize and present some relevant information, and together they provide a perspective on the benefits of the program and the resources needed to run it. The initial chapter, "European Project Semester," introduces the EPS concept and the network of providers. Facts, figures, and reflections are provided to frame the program's relevance and size. Complementary to this chapter, "United States Participation in the European Project Semester: An Exceptionally Successful Endeavor," describes, as its title indicates, successful exchanges between engineering colleges in the United States and similar institutions in Europe running the EPS, and how the experience has broadened the cultural and educational horizons of the participants. The chapter throws light on some of the advantages of, as well as the barriers to, introducing innovative practices in higher education.

The second part consists of nine chapters exploring the importance of transversal competencies for student development. Each one is devoted in detail to how the EPS manages a soft skill, methodology, or area. The topics addressed are teamwork, project management, intercultural communication, technical and professional communication, design thinking, active learning strategies for sustainable engineering, ethics and how ethical engineering practices are addressed in the EPS, human-oriented engineering practices, and robotics.

The first chapter of this second part is "Teambuilding and Efficient Teamwork," which underlines the importance of team building for effective teamwork and gives suggestions on a few useful tools to help build high-performing teams. After this, "Project Planning, Management, Monitoring, and More" addresses the main concepts and methods involved in the organization and supervision of project activities; thereby providing priceless advice on; for example, establishing correct structures of responsibility, understanding the nature and scope of different kinds of task, or ensuring that the results of the project work are delivered on time. The book continues with two chapters devoted to a fundamental skill of human beings: communication. "Intercultural Communication" explores the notion of effective communication in international contexts and, indeed, in any situation that brings together people with diverse cultures. "Technical and Professional Communication in the European Project Semester" situates technical and professional communication in engineering education and identifies communication practices as part of the professional practices of engineers. The following chapter, "Design Thinking at EPS: An Interdisciplinary and International Experience," presents the experience of applying modern

Design Thinking methodology to the development of projects during the European Project Semester. The steps, techniques, methods, and results achieved are explained and described.

The focus then turns to diversity. Chapters 8, 9, and 10 explore the connection of engineering with contemporary views of sustainability, ethics, and human-oriented practices. "Active Learning Strategies for Sustainable Engineering: The Case of the European Project Semester" describes how the European Project Semester offered by the School of Engineering at Porto Polytechnic contributes to making engineering undergraduates more aware of their environmental and societal impact thanks to a project-based active learning framework where problem-solving is supported by intercultural communication and creativity, ethical and sustainable reasoning and global contextual analysis. "Learning Engineering Ethics: The EPS Contribution" analyses engineering ethics and how it is addressed within the European Project Semester, relying on multicultural teamwork, project-based learning, ethics, and sustainability-oriented design to motivate and prepare global engineers committed to protecting and improving life on the planet. "Towards Human-Oriented Engineering Education: A Model of Empathetic Engineering in the EPS Program at the International Faculty of Engineering, Lodz" presents a human-oriented engineering model focused on empathy and developed by a group of academics at Lodz University of Technology. This group has defined a methodological approach to help students and teachers find the optimal strategy in project-based study programs, most notably in the European Project Semester. Furthermore, this chapter includes interviews with students and project mentors, and survey results measuring empathy in both stakeholder groups. The second part of the book ends with the chapter "Robotics and the European Project Semester." Robotics projects are particularly well suited to the European Project Semester framework because they present a multidisciplinary problem that typically requires mechanics, electronics, and computer science concepts to resolve them. Students from these different backgrounds (and others) can contribute to the overall team objective in their specific knowledge areas. The authors of this chapter briefly present illustrative examples of robotics projects that have been developed by teams of students participating in the EPS at the School of Engineering of the Polytechnic Institute of Porto. They conclude their chapter by presenting and discussing student feedback on the program and on the projects developed.

While the focus in previous chapters has been on student development and preparedness for specific skills and mindsets, the next chapters describe particular cases - and we believe their huge potential for instructors and researchers to engage in learning environments with enthusiastic students and colleagues will be easily seen. Part 3, therefore, is a collection of examples of EPS experiences in different universities, which highlight particular opportunities for cooperation as well as certain megatrends. The series of supervised, student-developed projects we present cover robotics, smart gadgets, social innovation, health, and everyday objects. Of course, it has been a challenge to select the cases to show when we have a database of more than one thousand, and there are many other valuable examples which could be included, however, the editors are confident that the projects we have selected will provide a vital and varied source of motivation for university managers. We hope they stimulate new ways of understanding education, empowering teachers, students and other stakeholders in their common goal; sustainable and lifelong learning. The chapters in Part 3 present various examples of projects involving engineering challenges and design issues that we hope will inspire educators to adopt the EPS approach in their own courses.

"Project-Based Learning With EPS@UPB: Development of a Robotic Arm With IoT Capabilities" describes a project carried out by a multinational team to develop soft skills and foster the development of professional tools by developing a 3D printed Robotic Arm with IoT capabilities. Designing an affordable robotic arm which can be used for many purposes provides the premise for a functional didactic approach. It is, moreover, an efficient way to meet students' social, economic, and technological needs. A further example is the case of Bicycle Speedometer Plus, where a group of students in Osnabrück developed a product constrained by complex technical requirements, an environmentally friendly design, and economic concerns such as cost-sensitive development. The subject was a speedometer for bicycles with the addition of a brake light to enhance rider safety, especially in an urban environment. This project provided an excellent opportunity for the participants to understand the benefits and risks of working in a multicultural team and the necessity of following strict project management guidelines in order to achieve a successful outcome. They also came up with a workable marketing plan.

The chapter "Current Inclusive Design Projects for Social Innovation" gives a sample of the innovation processes underpinning this part of the Valencia EPS program. The chapter underlines the relevance of the interrelationship between open and participatory design methods linked to the issues of accessibility and inclusion in different contexts. The issue of design for social innovation is closely linked to citizen participation and the inclusion of the different groups that make up society, particularly the most disadvantaged or at risk of being marginalized. The new product and service design paradigm requires a specific sensitivity in which education and research have to be combined to produce sustainable innovation. A key aspect in the future of design will be cross-disciplinary knowledge and disciplines more adapted to local needs and specific user groups. Hence, as a matter of constructing the future, the business model must be re-invented in collaboration with all stakeholders.

As described in "The Individual Survival Unit: Transporting Patients in a Protective and Stable Environment in Disaster Scenarios – A Case Study," one can easily understand how students benefit from situated experiential learning projects. An EPS team in Nottingham Trent University investigated and addressed the design and feasibility of a unit that can be used to protect and increase the survival time of an injured individual within a disaster zone. The designed solution provides a stable and protective environment for an individual while avoiding further injury during transportation. To enable the design of an optimal solution, disaster areas were analyzed and identified, providing a focused design problem requiring exploration and resolution. Essential basic life support equipment was researched, listed, and then ranked by importance according to the literature review based upon interviews with experts. The final concept was presented through technical drawings, computer-aided visualization and renders, animations, interface simulation and a real-life situation presentation.

A city is an accessible and inexhaustible resource for considering motivating challenges to those learning any subject, and its nearness and complexity make it ideal for university students. In the chapter "The City as a Laboratory: Concepts, Designs, and Experiments in Higher Education," the authors explain a reflexive type of activity that contributes to exploring the city as a complex environment, collaborating through university activity in the necessary transition towards more sustainable cities. Treating cities as laboratories and as learning and experimental objects has many advantages and plays an essential role in fostering and supporting the transition to circular urban environments. The authors explain this concept and describe the methods and results obtained with two different examples here. The first is a learning experiment on the inclusive city. In contrast, the second discusses the redesigning of the Valencia bus station and includes a preparatory workshop activity. The next chapter, "Embedding the EPS Program in Multi-Party Cooperation at The Hague University of Applied Sciences," describes a special edition of the European Project Semester on the course 'Sustainable Packaging Design & Innovation' at the faculty of Industrial Design Engineering at The Hague University of Applied Sciences. Unique cooperation between three research institutes, six universities, and three companies took place. Some of the partners involved had developed an educational module focused upon sustainable and circular packaging design, including using a dedicated tool for Life Cycle Assessment. This module was then embedded in the regular EPS, an international class of sixteen students working on a real-life design assignment. The chapter identifies the combination of lectures, workshops, pitches, and presentations, and reviews the organizational, managerial, and practical concerns.

The focus then turns to innovation involving engineering students in "Design and Innovation: Furniture for Children." This chapter explores the benefits of running and supervising a student project on children's furniture and then taking the final product for promotion at a professional furniture trade fair. The authors describe the steps involved in implementing the project and the impact the experience had on the students. It is the description of an experiential activity nearer to professional skills thanks to the full involvement of students and staff. Extra details of the learning process are explained.

The final chapter of the book explores the importance of sustainability and learning environments for professional development. "A Fast vs. Slow Fashion Fair on the Global Impact of Local Wardrobe Choices: Promoting Sustainable Behavior at the University of Antwerp" describes how an EPS team collaborated in the design of an interactive event, exhibition, and campaign to promote sustainable consumer behavior at a fair on the university's city campus, with a special focus on awareness-raising about fast versus slow fashion. Throughout all the development stages, the team improved their work-in-progress during weekly coaching sessions with access to the sustainability-related research group and toolkits. This rich trajectory and the resulting Fast/Slow Fashion Fair triggered deep reflections in the members of the team regarding what is and what is not fair (in the sense of 'just') fashion. Students learned to plan a significant event with many stakeholders, as well as eco-design for the outdoors, and how to proactively communicate and develop their public relations and negotiation skills.

For more than a decade, both editors have been committed to spreading the EPS concept worldwide, from China to the USA, from India to Mexico, in order to generate genuinely international environments beyond Europe. They are personally implicated and involved in many actions to improve the learning environments and strengthen the bonds between universities. They are also convinced about the many benefits for the professional development of instructors, which can be found through overcoming the challenges of engaging their students in complex and rich learning environments. A network of higher education institutions (such as those involved in the EPS) can provide invaluable support, understanding, and opportunities.

At the time of writing, it is clear that nothing in the higher education world will be the same as before the pandemic caused by COVID-19. Many organizations have had their learning models disrupted and have had to rapidly adapt to delivering lectures and other study materials through greater use of technology. Within the EPS, face to face experiences remain vitally important. Nevertheless, the program has always emphasized the importance of the skills needed for working in a global virtual world. Now, in the age of avatarism, multicultural and multidisciplinary teams are still possible, and short- and long-term team projects can still become positive virtual learning experiences.

We have been delighted, honored, and excited to receive so many valuable and enthusiastic contributions in putting together this book. There are many people we wish to thank, but these truly heartfelt acknowledgments are handled in a separate section. We end this Preface with the hope that readers of this book will find many valuable ideas and resources for future transformations in higher education.

Benedita Malheiro
Pedro Fuentes-Durá
David Perry

Acknowledgment

The editors would like to acknowledge the support given to this project and, more specifically, to the authors and reviewers that took part in the review process. Without their generous work, this book would not have become a reality.

First, the editors would like to thank each author for the contribution made to this project, investing their time and expertise to share experiences and views in this book.

Second, the editors wish to acknowledge the valuable contributions of the reviewers to the refinement of the submitted chapters. They have helped to improve the quality, coherence, and presentation of chapters. Moreover, we highly appreciate the double task of those authors who also served as reviewers.

Finally, we are obliged to all European Project Semester stakeholders. This joint endeavor would not have been possible without the vision of the founder Arvid Andersen, the gentle guidance of the disseminator Jørgen Hansen, the support of the providing institutions, teaching colleagues and staff, and the sponsoring companies, research labs, and organizations. Last but not least, we wish to express our gratitude to the participating students, the reason for the existence of EPS. They are a source of continuous inspiration, innovation and improvement for the whole network.

Benedita Malheiro
Instituto Superior de Engenharia do Porto, Instituto Politécnico do Porto, Portugal

Pedro Fuentes-Durá
Universitat Politècnica de València, Spain

Section 1
European Project Semester

Chapter 1
European Project Semester

Grazyna Budzinska
Lodz University of Technology, Poland

Jørgen Hansen
Technical University of Denmark, Denmark

Benedita Malheiro
https://orcid.org/0000-0001-9083-4292
Instituto Superior de Engenharia do Porto, Instituto Politécnico do Porto, Portugal

Pedro Fuentes-Durá
Universitat Politècnica de València, Spain

ABSTRACT

This chapter aims to introduce the European Project Semester concept and the network of providers together with historical and current data on participants and providers. EPS is a one-semester international exchange program designed for engineering, product designers, and business undergraduates. It embraces student-centered project-based learning and multicultural and transdisciplinary teamwork to help students develop the competencies and skills required for the 21st century. It was created in 1995 in Denmark to prepare future engineers for global challenges, and it combines scientific and technical education, ethical and sustainability-driven problem-solving, intercultural communication, and teamwork. EPS has since been adopted by 19 European higher education institutions, establishing a network of EPS providers across Europe. The network offers international students a wide range of European study locations and cultural insights and, above all, the opportunity to enjoy the EPS learning experience. The number of alumni shows the program's attractiveness, range, and vitality.

INTRODUCTION

Engineering is a major driver for sustainable socio-economic development (United Nations, 2021). It has helped humankind solve problems and survive crises by responding with viable solutions to emergency situations and has always played a crucial role in socio-economic development while addressing basic

DOI: 10.4018/978-1-6684-2309-7.ch001

Copyright © 2022, IGI Global. Copying or distributing in print or electronic forms without written permission of IGI Global is prohibited.

human needs by connecting social needs with technological applications. That is why the education of engineering professionals is crucial to ensure societal well-being with future technological innovation. Engineering education aims to prepare professionals to address the challenges of the future. This highly demanding goal does not only imply a forecast of future needs, no matter how precarious and unpredictable they can be, but also requires the anticipation of scientific and technological advancement trends. Higher education must face this challenge while including theory-based study programs, addressing industry-oriented practical phases (Dispert et al., 2014) and socio-cultural changes on a global scale. As increased globalization triggers intercultural exchange through the free movement of capital and people, it is inevitable for universities to intensify international exchange in a variety of educational programs, e.g., joint diplomas, international internship schemes, Erasmus+ programs, or international and interdisciplinary projects. Thus, engineering product design is "not an isolated activity" (Andersen, 2001) but rooted in a natural socio-economic environment. As such, it requires high-quality engineering, marketing, and business competencies as well as transversal skills such as collaborating in multicultural and multidisciplinary teams or communicating effectively with a variety of project stakeholders.

The European Project Semester (EPS) attempts to address the challenge by adopting collaborative, experiential, project-based learning (PBL) methodologies. It exposes students to cultural and scientific diversity. To promote learning autonomy, the teams are fully responsible for the project management and execution, solution discovery, selection of materials and components, and assembling and testing prototypes (Malheiro et al., 2019a). EPS has two main goals: (*i*) to motivate young adults to learn engineering; and (*ii*) to provide participants with an international and professional experience (Mulder et al., 2015). To motivate students, EPS adopts teamwork, PBL and proposes real-life problems in close cooperation with various stakeholders, namely, businesses, public institutions, and research centers. Being an international multidisciplinary exchange program, EPS exposes students to the new culture of the host country and higher-education institution. What is more, the students work in teams with peers from different cultural and scientific backgrounds and get insight into the professional environments of associated businesses, institutions, or research centers. The EPS learning framework increases employability prospects of young engineers by simultaneously developing: (*i*) sound scientific and technological competencies; (*ii*) intercultural and cross-discipline communication skills; (*iii*) and a professional attitude.

EPS started in Denmark in 1995, by the hand of Arvid Andersen, and has since spread to nineteen other universities located in twelve European countries: Austria, Belgium, Finland, France, Germany, The Netherlands, Norway, Poland, Portugal, Romania, Spain, and The United Kingdom. EPS has involved hundreds of highly motivated academic, administrative, and support staff from the various providers and thousands of successful international alumni from all continents during its lifespan.

This chapter aims to present an overview of the EPS concept and network of providers. The chapter is organized into the following sections: EPS background, the network of EPS providers, and reflections.

EPS BACKGROUND

While completing studies, young engineers are confronted with constantly globalizing job markets characterized by dynamic and often unpredictable changes. Such changes constitute a test to the curricula of educational institutions. They must equip young graduates with necessary skills to help them successfully build their careers while constructing a better future for all and supporting global communities with progressive solutions. This more holistic approach to engineering calls for addressing the long-term perspective while solving problems, global awareness (Silva et al., 2019), transversal skills, critical thinking, and ethical considerations. Numerous attempts to develop adequate methodologies were initiated long ago by various European universities and were reassured by the Bologna Declaration (Heitmann, 2005). Engineering education has been adopting different forms to enrich educational programs and focus on the global challenges: internships at home and abroad, student mobility exchange programs, capstone projects in collaboration with external institutions, courses exploiting soft skills, collaborative project-based programs. Project-oriented educational schemes conducted in a team setting and tackling real-life problems seem to address a variety of the factors conditioned by the dynamically changing global economies. They facilitate training in technical, personal, and contextual competencies as well as collaborative student-centered research-based learning (de los Rios et al., 2010). Nonetheless, while PBL is generally accepted as an effective method for teaching processes, e.g., problem-solving and decision-making, it tends to be marginalized by the educators themselves, since they lack both the training and the experience (Efstratia, 2014). The EPS, exploring the concept of a project that is conducted in international and multidisciplinary teams, includes a variety of principles that constitute one holistic paradigm and lets each university implement their own specific goals, e.g., close collaboration with companies, focus on engineering design, project-based learning, problem-based learning or design thinking approaches. Figure 1 displays the EPS logo, where the ideas of love of learning, peace, global collaboration, and support are inextricably entwined.

Figure 1. EPS Logo reflects the ideas of love of learning, peace, global collaboration, and support

EPS tries to contribute to the challenges evoked by the dynamically changing environment with an innovative framework (Andersen & Hansen, 2002). The framework of EPS has originated several spin-off programs based on the main principles but focused on a singular core perspective. For instance, the International Design Semester launched at Ingeniørhøjskolen in København (Schjær-Jacobsen et al., 2012) and the International Design Project Semester offered by Universitat Politècnica de Catalunya (Segalàs & Benson, 2012) require students to work closely with companies to come up with engineering solutions that are sustainable from both an engineering and a design perspective. Obviously, the implementation of novel industrial design subject areas or new PBL approaches behooves the creation of new laboratories and workshop facilities. This combination of traditional engineering design disciplines with creative design aims to drive innovation. A recent review of PBL in engineering education indicates that the most common application of projects still happens within the framework of existing courses rather than across courses or at the curriculum level (Chen et al., 2020).

EPS has different but always highly relevant implications for a variety of program stakeholders. At providing institutions, EPS may have various impacts, e.g.: (*i*) ignite impetus for the development of interdisciplinary mindset among students and academics in a new portfolio of engineering courses; (*ii*) enhance the innovative teaching in engineering courses; (*iii*) strengthen a design-oriented approach of engineering courses; (*iv*) enhance student enthusiasm and awareness through more significant contact with senior practitioners; (*v*) improve staff and student understanding of modern engineering design principles while maintaining contemporary professional practice (e.g., manufacturing and construction); and (*vi*) provide experience and an understanding of the value of interdisciplinary working. Participating students have the opportunity to tackle problems outside their disciplines, understand relationships between different fields, develop innovative and creative solutions to open-ended problems, expand personal and interpersonal communication, and foster management, team-working and presentation skills. In addition to providing students with a significant intellectual challenge, interdisciplinary projects help them gain a broader insight into problem-solving and deal with uncertainty, both inherent to the essence of engineering. The providers' experience accumulated through the years proves that both undergraduate and postgraduate engineering students benefit from working in interdisciplinary multicultural teams on complex problems whilst at university. When graduating, they are better prepared for the challenges of future engineering careers.

The program's website displays the EPS goal, concept, and network (European Project Semester, 2022). Figure 2 shows the geographical distribution of the nineteen EPS providers.

Figure 2. Network of EPS providers
(European Project Semester, 2022)

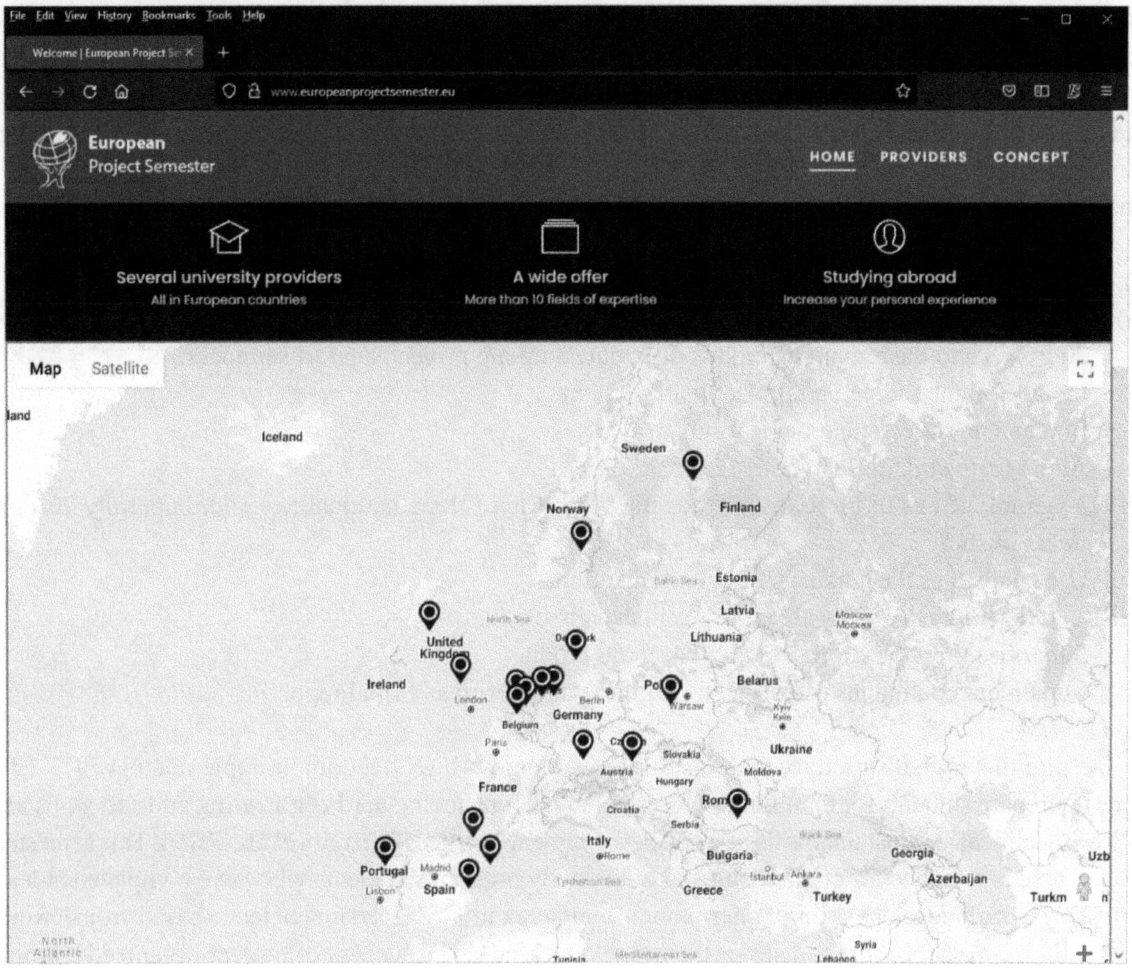

The network of EPS providers covers a large part of Europe. It includes a variety of countries whose cultural and educational specificities contribute to EPS diversity of cultures, educational traditions in engineering sciences, diversity of strategies in problem-solving, economic trends and social values. The high level of diversity creates a favorable environment for learning and developing competencies necessary in globalized markets.

EPS Concept

According to Andersen (2004), the major question in training engineers is "What should we reasonably expect the future engineer to be able to do?". The engineers of tomorrow will need more than technical skills to work in international teams and across borders on solving real-life problems. Teamwork requires a collective effort, a concurrent approach, and specialists with different kinds of expertise (Ponsa et al., 2015).

Modern engineering education calls for a skill base that goes beyond technology and engineering. Today, it requires proficiency in open-ended problem-solving and applications and the ability to collect, analyze and report a large amount of technical, operational, and statistical data. In addition, today's engineers must be able to speak, read and write in good English, work in interdisciplinary and international project groups to solve increasingly complex problems (Skates, 2003). Facing rapid global and local changes, students must learn to learn, which is an increasingly important requirement since knowledge can rapidly become outdated (Hansen, 2012a).

The "10 Golden Rules of EPS" are the fundamental principles behind the program. EPS must:

(i) Last at least 15 weeks.
(ii) Correspond to 30 European Credit Transfer System (ECTS) package, with a project worth at least 20 ECTS and the remaining credits distributed among project support seminars.
(iii) Include teambuilding and project management.
(iv) Implement project-centered learning.
(v) Adopt teamwork exclusively.
(vi) Assemble teams of three to six students from at least three nationalities and, hopefully, distinct fields of study.
(vii) Use English as the language of communication.
(viii) Offer English and local language crash courses.
(ix) Supervise the project process as well as the product.
(x) Implement continuous assessment, including an interim and final report.

Despite this shared structure, each institution still enjoys high freedom in implementing EPS. EPS aims to prepare students for their future professional life, enhancing hard engineering and soft skills and following ethical and sustainable design and development practices (Ferreira et al., 2019). This semester-long course is intended for engineering and other study programs students who have completed at least four semesters. It develops skills of international teamwork in a balanced mix of team-based project work, conventionally taught short courses or themes and tutorials. The success of project-organized learning in the EPS paradigm is measured by participants' ability to handle challenges and situations during the project execution cycle. Students learn to take responsibility for their condition and gain awareness about the importance of lifelong learning. They practice international teamwork, which is more important today than ever since it is essential for a successful engineer who must operate in a multicultural environment. Great attention is paid to goal setting, planning, delegating, cooperating, and respecting the deadlines while navigating towards a common team objective. Hence, the students enrolled in EPS improve their transversal skills and learn subjects distinct from their core training while being mentored or supervised by academic staff (Malheiro et al., 2015a).

The projects are structured in such a way as to expose students to problems of greater complexity and dimension than those faced throughout degree programs. They create an opportunity to work in contact with the so-called real world of work and not only with the academic world (Silva et al., 2018). The project work must be completed as planned in the schedule. The result is presented at an assessment session simulated as a board meeting and in a professionally written report. Self and peer-assessment are undertaken at least twice a semester by each team member, and it plays a crucial role in the student development process and the learning experience. The self and peer assessment consider the quantity and quality of the contribution made by each team member and their participation in teamwork activities.

(Malheiro et al., 2013). The supervisor plays a crucial role in supporting a project team and monitoring its effectiveness. Supervisors use the assessment to monitor teamwork and give constructive feedback and advice (Ashworth, 2010). Project supervision takes different forms at different EPS universities. It may vary depending on the approach in project execution that a specific university or department favors: an academic who guides an EPS project team may play a role of a supervisor having a comparatively higher level of control over the team doings or a role of a facilitator who gives the student high autonomy in decision making and project execution, can also be a mentor or a coach. The variety of approaches to supervision is beneficial for EPS students who can learn in action about different mentoring styles.

Course Structure

Throughout the semester, especially during the first weeks, short, intensive and supportive project courses are provided to students, parallel to project work in international teams. These courses are compulsory and, as a whole, correspond to a total of five to ten ECTS. The project work accounts for the remaining 20 to 25 ECTS depending on the institution. In total, one semester equals 30 ECTS credit points. The courses are designed to support project teams in acquiring the skills and knowledge necessary to maintain the team and execute the project. The range of courses comprises team building and management, communication, intercultural awareness, cross-cultural business behavior, systematic innovation, project management, engineering product design, intellectual property rights in Europe, international marketing, sustainability, European law, English and local languages.

EPS aims at enhancing the employability of students by bridging the gap between higher education and the needs of the work market (Fuentes-Durá et al., 2014). One of its prime objectives is to contribute to the education of engineers who, in the future, must be able to work in interdisciplinary teams, operate internationally, speak foreign languages, cooperate with people of different cultures and communicate well. Therefore, EPS is designed to improve transversal skills. The program formalizes the experience of working together in multicultural and multidisciplinary teams from different countries and different disciplines, facing the need to develop personal competencies, especially the ability to work and communicate within cross-cultural groups.

The projects are assigned to teams of four to six students each, including students from different nationalities and study fields (mainly engineering, product design, media, and business programs). EPS is structured around a central module – a project co-supervised by university supervisors and company mentors – and it includes a set of intensive courses or seminars dedicated to the development of essential professional competencies: communication, sustainability, ethics and deontology, project management, or marketing (Duarte et al., 2020). The program is addressed mainly (but not exclusively) to incoming exchange students so that the class becomes a truly international setting. The aim to expose participants from diverse cultures and scientific backgrounds to teamwork and multidisciplinary problem solving is especially important in capstone projects, transitioning to professional activity (Duarte et al., 2020).

There are several steps or phases in every EPS project (Fuentes-Durá et al., 2014). These are team creation, planning, course definition, research stage, problem definition, project statements, planning, mid-term project development, mid-term project evaluation, project development, project evaluation, and exploitation of results. On some occasions (Malheiro et al., 2015b), projects have adhered to conceptual models as Design Thinking (DT) or Conceive, Design, Implement and Operate (CDIO).

Course definition includes the provision of resources and the selection of students: agreements between universities, level of English proficiency, soft skills orientation, ECTS background, and expectations.

Usually, before they arrive at the host university, students are sent a conceptual statement about each offered project. After the initial welcome session, each project supervisor presents the main outlines and objectives of the proposed project. The projects are deliberately open at this stage as the students themselves will define the project profile and final details. Further discussion with project supervisors follows, usually during 2 or 3 sessions, after which students choose projects according to their interest. The choice is crucial in terms of motivation, involvement, and engagement. Presenting projects at the beginning enables students to decide about the project to follow and form an idea of what the mentor expects and what it will be like to work together. It also enables the supervisors to gauge the team's level of knowledge and interests and make any necessary adjustments to the project brief. The Coordinator of the EPS course decides the composition of the teams based on the student choices and with the help of supervisors and psychometric tests if necessary. Two key points to define the teams are (*i*) all teams should be international (at least three nationalities); and (*ii*) students' backgrounds will enable them to meet the technical demands of the projects due to the necessary subject knowledge they possess.

From the student's point of view, EPS includes a variety of activities. These range from the first reception meeting to selection of projects, icebreaking activities, meetings with supervisors, team building tasks, transversal seminars, team working, submission of reports, networking, visits to local factories, or doing sports such as canoeing or trekking. In general terms, EPS engages students individually and jointly on a particular project based on a new and current situation. The projects require students to solve multidisciplinary problems and encourage ingenuity, open-mindedness, and teamwork. Competences are developed mainly through the students working in certain situations and under certain constraints. They choose an exciting task, a preferred role, collaborate in the definition of goals and the planning, feel part of a team, and get the support of the staff and the pressure of the 360-degree evaluation. The projects are diverse, and a set of options and orientations are available within each particular project. The participants can choose both the project and their particular focus in terms of subject-matter contribution.

EPS aims at creating an engaging learning environment in which each student would find a place to be active and develop their interests. EPS international character, teamwork organized around an involving project rooted in different cultural and economic settings, its flexibility and creative angle, provide an opportunity to involve students representing various backgrounds and preferences and captivate their diverse academic and personal predilections. Global issues and trends reflected in projects themes contribute to the participants' motivation and relevance of outcomes (Segalàs et al., 2011; Armesto et al., 2016). EPS can be a "mind-opener", enabling participants to learn in action, be open to new issues and opportunities, change perspective and be creative, and enrich their experience and increase the potential of future professional success due to academic trips, conferences, seminars, and creative self-reflection. Figure 3 shows the EPS class, 2016, at Universitat Politècnica de València after completing a teambuilding exercise where students were challenged to creatively develop a structure from 16 pieces of paper and a glue stick.

Figure 3. EPS class of 2016 at UPV
(Credit: P. Fuentes-Durá)

The structure was expected to accomplish a set of fixed characteristics such as height, optimal wind resistance. It should be transportable with one finger, aesthetic, sustainable, and include a romantic tale. The students had to collaborate in their project teams, communicate effectively, plan, and ideate creatively while having fun and competing with other teams. Releasing free creativity and having 'serious fun', since the ideas used in the activity could be reinterpreted and used in project teamwork, play a complementary but very important role of an engaging factor that leads to improving communication and collaboration in teams and contributes to openness and sharing, so crucial in collaborative work.

EPS NETWORK

EPS started at The Engineering College of Elsinore in Denmark in 1995 (Hansen, 2010). In 1997 it was transferred to The Engineering College of Copenhagen, now part of the Danish Technical University, where it was carried out every semester till 2013. Meanwhile, several European higher education institutions have embraced the EPS concept. Table 1 lists the country, university, and joining year of the current set of EPS providers. Provider details, including the institutional website, can be found at European Project Semester (2022).

Table 1. EPS providers

Country	University	Year
Netherlands	Avans Hogeschool	2002
Norway	OsloMet - Oslo Metropolitan University, former Hoegskolen i Oslo	2003
Poland	The Technical University of Lodz	2005
Spain	Universitat Politècnica de València	2005
Spain	Universitat Politècnica de Catalunya in Vilanova y la Geltrú	2008
Germany	Kiel University of Applied Sciences	2008
Finland	Novia University of Applied Sciences	2010
France	École Nationale d'Ingénieurs de Tarbes	2010
Belgium	AP University of Applied Sciences, former Artesis Plantijn University College	2011
	University of Antwerp, former Artesis Plantijn University College	
Portugal	Instituto Superior de Engenharia do Porto, Instituto Politécnico do Porto	2011
Netherlands	The Hague University of Applied Sciences	2014
Austria	St. Pölten University of Applied Sciences	2014
Romania	University Politehnica of Bucharest	2014
Germany	Osnabrück University of Applied Sciences	2015
Scotland	Glasgow Caledonian University	2016
England	Nottingham Trent University	2016
Netherlands	Saxion University of Applied Sciences	2016
Germany	Augsburg University of Applied Sciences	2018

Figure 4 displays EPS growth both in the number of providers and in participants. The consistent growth of the network is observed. At the diagram on the left-hand side, the number of EPS providers vs. year is displayed. The development is not linear because, occasionally, a provider stops the EPS activity either temporarily or permanently. The diagram on the right-hand side depicts EPS alumni number vs. year. The observed consistent growth reached 500 students/year in 2016.

Figure 4. Evolution of the EPS network and alumni. Number of EPS providers vs. year (left) and alumni number vs. year (right)

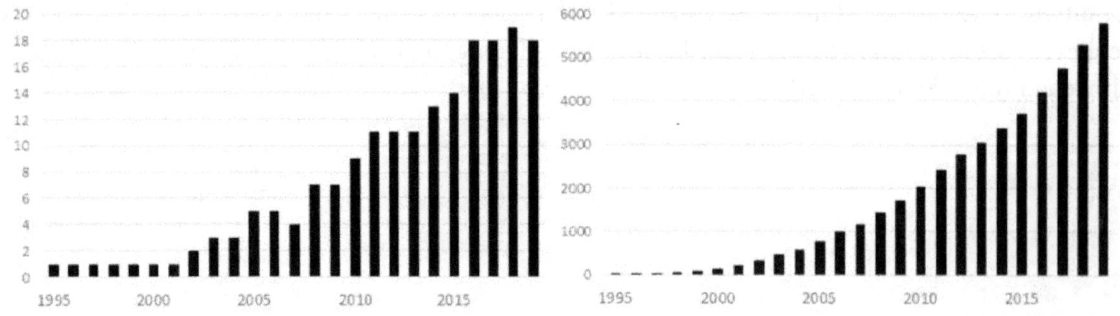

Thanks to successful stories and the academic dissemination of the outcomes (Hansen, 2012b; Abata et al., 2013; Budzinska & Fuentes-Durá, 2014, Duarte et al., 2020), the network has spread over the four corners of Europe and offers a wide range of cultures, implementation flavors, and learning experiences as detailed below in the short presentation of EPS programs at different European universities.

AP University College Antwerp, Belgium

Since 2016, AP has offered different EPS projects in the second semester (from February until June). EPS is based on cooperation between all departments & schools of arts at AP University of Applied Sciences and Arts Antwerp. AP focuses on students, providing them the tools and competencies to realize ideas in a real-world environment in close cooperation with local industry and non-profit organizations. Projects cover different topics related to applying technology to a broad range of contemporary social issues.

Avans University of Applied Sciences, Netherlands

As of 2002, EPS run at the Avans University of Applied Sciences, Faculty of Engineering and Information Technology, in Den Bosch, The Netherlands. The EPS program is organized in close cooperation with the industrial business community of Den Bosch. The students perform real-life projects at companies. The students come from various (mainly European) countries. They represent different study disciplines like Computer Science, Mechanical Engineering, (Technical) Business Engineering, Electrical Engineering, Communication and Multimedia Design, etcetera. All projects have sustainability and multidisciplinary aspects. The EPS program is experienced as very positive and valuable by students and by the participating companies.

École Nationale d'Ingénieurs de Tarbes, France

Starting in 2010, ENIT is the only school in France to offer EPS. ENIT hosts between fifteen and twenty students per semester and sends some of its students to the institutions in the network. ENIT offers this program twice a year (fall and spring sessions). Students follow courses and work collaboratively on industrial projects connecting with cutting-edge technologies thanks to a well-established link with regional industry. Most of the projects have made material engineering and environmental technologies a focus of attention.

Glasgow Caledonian University, Scotland

GCU has been running the EPS program since 2016, and it has brought a wealth of seventy-two international students through the door for that unique GCU and Scottish experience. A successful program, which seeks to develop those softer transferable skills amongst students, giving them not only work skills but skills for life. Students, in return, bring youthful ideas and innovation, particularly those concerning new technologies. The program is centered around industrial projects, which expands students' knowledge and working practices, crucial for employability. Scotland offers a unique cultural experience, is a very inclusive society and GCU is at the center of that with a warm welcome to all overseas students.

Hochschule Augsburg, Germany

In 2018 the EPS program was launched at Augsburg University of Applied Sciences. It is based on the existing Interactive Media study program, an interdisciplinary program run by the Faculties of Design and Computer Sciences. The projects offered are particularly focused on Augmented Reality, User eXperience (UX) Design, and Mobile Apps. They are mainly in-house projects carried out in teams with German and international students. In terms of labs and workshops, students benefit from a well-equipped faculty, ranging from robotics, motion capturing, and UX to traditional printing techniques. Additionally, the Language Centre provides a wide range of language courses and intercultural courses. Augsburg is a lively university town in the southwest of Bavaria, close to the Alps and Munich.

Instituto Superior de Engenharia do Porto, Portugal

EPS at the Instituto Superior de Engenharia do Porto (EPS@ISEP), the school of engineering of the Polytechnic of Porto, was launched in the spring of 2011. Since then, it has welcomed 230 students from twenty-three nationalities and sixty-five engineering, product design, and business degrees, always in the spring semester. The staff comprises local and international teachers, international and counseling officers, and laboratory technicians on a needs basis. EPS@ISEP aims to foster the development of ethical and sustainable design practices, multicultural multidisciplinary teamwork, interdisciplinary problem solving, and a professional attitude. The types of problems addressed cover education, food, health, research, or building, with a special emphasis on smart, innovative solutions for cities, companies, people and planet.

Kiel University of Applied Sciences, Germany

EPS at Kiel University of Applied Sciences in the north of Germany was launched in 2008. Every spring, the interdisciplinary project semester is hosted by the Faculty of Mechanical Engineering. It focuses on important aspects of engineering, IT and software development, mechatronics, business, as well as naval architecture, and mechanical engineering. From 2008 to 2021, 229 students from various European and American partner universities (with a local student in every team) have successfully participated in the program, solving problems, developing new product outlooks and strategies, (re)designing products, creating and improving business plans and management. Almost all projects are offered by local companies and are supervised by both company and university representatives. The university's Centre for Languages and Intercultural Studies offers supporting courses in German and English (and ten other languages).

Lodz University of Technology, Poland

The EPS at Lodz University of Technology (TUL) is offered since 2005. TUL welcomes international students and colleagues both internal and external to the EPS network twice a year. The multicultural environment of International Faculty of Engineering provides a variety of engineering study programs in English and naturally attracts international students worldwide. The program aims to offer high-quality engineering education combined with more generic skills necessary for 21-century engineers who would like to succeed in the globalizing markets. The students work in multicultural teams of four to five members on multidisciplinary projects structured around complex and challenging contemporary problems experienced by companies, institutions, and communities. The projects are frequently linked with university research projects and companies. The students enjoy much autonomy in team and project management. They learn to be independent in research, problem-solving, and decision-making since the mentors play the roles of facilitators rather than supervisors of projects. Their impact on project selection emphasizes the subjectivity of students. Facing the complexity of real-life problems, the EPS students have the right to invite external experts representing different fields and institutions as consultants and advisors, which increases the authenticity and feasibility of the offered solutions. The program provides an opportunity to get acquainted with various modern project learning methodologies and techniques within such approaches as PBL, Design Thinking, Scrum, Agile, industrial design, and others. EPS at TUL prepares students for creating engineering solutions in the context of contemporary global environments with their high dynamics, significant change, complexity and uncertainty.

Nottingham Trent University, England

Nottingham Trent University has hosted the EPS since 2016, and within this time, it has welcomed nearly 100 students from nineteen different partner institutes, both within and outside the European Union. EPS at Nottingham Trent University is delivered within the Product Design subject group and focusses predominantly on 3D design. Students who study with us at NTU come from a wide variety of subject areas, including Engineering, Architecture, Interior Design, Robotics, 3D Design, and many more. The teaching team is usually based around research Professors alongside staff in the Product Design department who work alongside companies to provide live, industry-led briefs that have real-life impact. Briefs change every year depending on research, but in the past have included projects around improving road safety for cyclists, improving product lifetimes, looking at circular economies, and renewable energy harvesting.

Novia University of Applied Sciences, Finland

EPS at Novia UAS was launched in 2010. Since then, it has welcomed 300 engineering, product design, and business students from twenty-five nationalities. The staff comprises local Novia UAS teachers as well as teachers from neighboring universities in Vaasa, and also guest lecturers from the EPS network. EPS at Novia UAS aims to foster the development of multicultural multidisciplinary teamwork, interdisciplinary problem solving, a professional attitude, and ethical and sustainable design practices. The problems addressed cover a wide range of topics focusing on sustainable development within the energy sector, healthcare technology, and food production. In Novia UAS, EPS runs in-house projects initiated by the R&D department or individual teachers and industrial projects.

Osnabrück University of Applied Sciences, Germany

Osnabrück University of Applied Sciences has offered EPS since 2015. Sixty-seven students from nineteen different nationalities have participated since then. Scientific staff supervising the projects is being recruited from professors and laboratory technicians. Local companies and other organizations are being integrated. Project subjects address scientific and technical problems related to the involved laboratories and companies. Robotics in different aspects could be named as one field of the projects. Interdisciplinary subjects associated with the main disciplines at the university, e.g., in agricultural technology, are another exciting branch.

Saxion University of Applied Sciences, Netherlands

In Saxion, EPS enables students to apply the technical knowledge acquired during previous engineering or life science education to real and practical projects since 2016. It allows students to join the job market more successfully, and helps them to enhance their careers. This program has been designed to satisfy society's new demands and train students in a broader range of disciplines, such as intercultural communication, teamwork skills, critical thinking, and languages. Saxion aims to form students capable of dealing with frequent and unexpected changes and developing the skills required for working internationally. An international team of students works on actual and relevant assignments for and with the industry during the semester. The teams are made up of students with different academic backgrounds and are under the guidance of a supervising tutor and specialist. Practical workshops are also included in the project. These focus on project management, personal and professional development, and team building.

OsloMet – Oslo Metropolitan University, Norway

At Oslo Metropolitan University (OsloMet), EPS has been offered since 2003. It has, since 2016, been running both in the spring and fall semesters. Anywhere from eighteen to thirty-eight students attend each semester. As part of the Faculty of Technology, art, and design, it has been truly interdisciplinary from the beginning. In later years, it has been an arena for educational experiments. OsloMet aims to provide projects that are either tied to ongoing research at the university or have an external customer deeply involved in the project. In addition to the focus on problem-solving and professional behavior, OsloMet emphasizes the continuity between semesters. Most current projects are based on previous projects or shall themselves provide a foundation for new iterations in later semesters.

St. Pölten University of Applied Sciences, Austria

Since joining the EPS family in 2014, the St. Pölten University of Applied Sciences has welcomed 124 incoming students from fourteen different countries from within and outside the European Union. The scope of disciplines involved ranges from media technology & production (including film, TV, animation, games, and extended reality) and Internet of Things to Data science & Business analytics. Students interested in participating in the EPS do not apply for a specific pre-defined task. St. Pölten UAS instead prefers to tailor tasks to the individual profiles of accepted students. Projects are carried out in cooperation with the technology sector and the creative industries, which define the requirements for the products to be developed. To fulfill this challenge, students have 24/7 access to vast infrastructure, including cutting-edge labs for electronics, interactive media, TV, post-production, photography, and audio production, as well as a brand-new state-of-the-art Industry 4.0 lab. These EPS projects seek to promote collaborative PBL and strengthen intercultural and interdisciplinary competence.

The Hague University of Applied Sciences, Netherlands

The Hague University of Applied Sciences started EPS in 2014 because the current fast-changing world requires engineers and IT specialists who are prepared to look at problems from different perspectives and who have learned to collaborate with professionals from other disciplines. The EPS is a great way to develop these important skills. The Hague UAS offers five EPS projects at two different faculties.

The Hague UAS has successfully developed EPS projects in the fields of Packaging, Urban engineering, Sustainability, Information Technologies, and Design.

University of Antwerp, Belgium

EPS in Antwerp was launched in 2011 after two years of careful preparation. Since then, it has promoted the personal and professional growth of many participating students and staff from a wide variety of nationalities and study backgrounds, including various engineering specialties, product and multimedia design, business, economic, social, and educational studies. Local and international professors and experts support the students from the first development stage of their first product-and-service-system (PSS) ideas, making the necessary prototypes, various iterative technical tests and user tests until the last stage of the final product-service-system design. The aim of PSS & EPS at the University of Antwerp is to foster the development of the necessary set of skills, self-awareness, and attitude of the EPS students, who are the future professionals in charge of ethical, socially, economically, and economically sound sustainable design projects and practices, to be realized through multicultural interdisciplinary teamwork. Projects address urgent and actual challenges in society, from non-governmental organizations, companies and public organizations, which require an innovative, holistic problem-solving mindset and a system design approach to create solutions across the fields of, amongst others, education, agriculture, fair trade, health care, global development, environmental sciences and social and cultural work, while building on internal and external research, toolkits and expertise.

University Politehnica of Bucharest, Romania

The European Project Semester at University POLITEHNICA of Bucharest was first launched and implemented within the Faculty of Mechanical Engineering and Mechatronics, the Department of Industrial Design, in 2014. Since then, it has been delivered in English throughout the spring semester, addressing students who have completed at least three years of engineering studies. It has already welcomed 206 students from 20 nationalities. Moreover, at UPB, students from different countries and engineering fields work together in multicultural and multidisciplinary groups on hands-on projects addressing product design, mechanics, electronics, aerospace, textiles, artificial intelligence, etc. Through this program, they acquire and develop hard and soft skills (professional communication, leadership, critical thinking, problem-solving, languages, etc.) likely to ensure their smooth insertion into the labor market.

Universitat Politècnica de Catalunya, Spain

EPS at UPC Vilanova (EPS@UPC) was launched in the academic year 2007-08 and has been offered since then once a year, in the spring semester. EPS@UPC has welcomed 430 students from 32 nationalities and from a variety of engineering, product design and business degrees. Projects are proposed by companies, institutions and university research groups and are co-supervised by university lecturers. Project courses are taught by local and invited international teachers and are evaluated by a panel comprising local and international lecturers. From 2011 to 2017, the EPS program was combined with IDSP (International Design Project Semester), focusing on design. The focus of EPS@UPC is on sustainability, problem-solving and innovation, proposing solutions to current problems such as those related to smart cities, technology for health, and the design of sustainable products. EPS project-related courses emphasize on technical professional communication, project management, team building, sustainability, and innovation.

Universitat Politècnica de València, Spain

Universitat Politècnica de València (UPV) hosts EPS since 2005 in the Escuela Técnica Superior de Ingeniería del Diseño. In this period, UPV welcomed nearly 750 students from 37 different countries, both within and outside the European Union. It has since 2016 been running both in the spring and fall semesters. EPS at UPV involves faculty from nine departments and several companies, including multinationals and start-ups. It is an effective support ecosystem for students and faculty, offering new opportunities with added value. Self-reflection and prototyping are the primary learning method practiced by students, faculty, and clients. It involves team brainstorming (over the project and case studies) and explores fresh perspectives. Briefs change every year, but some topics remain similar, like the innovation on design for kids, robotics, and urban social design.

CONCLUSION

Working with an interdisciplinary team of students is a challenging task which calls for intensive interdisciplinary guidance by several experts representing different fields. The experts collaborate with the team mentor in guiding, instructing, and evaluating the learning progress of individual team members (Rohaert et al., 2012). The ongoing collaboration with members representing other disciplines requires a high level of flexibility from the process participants. It challenges their way of thinking and their well-established views and may require students, teachers, and administrative staff alike to leave their comfort zone to be able to understand others and be understood by them. It must also be admitted that a high level of diversity may be a challenge in collaboration among all the stakeholders involved in the process (Segalàs, 2012), may lead to conflicts, and may prolong decision-making processes. Nevertheless, interdisciplinarity and diversity are essential in tackling complex problems, so characteristic for today's precarious reality and, what is crucial, they increase the level of creativity and innovation of a suggested solution. Therefore, EPS intends to explore multidisciplinarity and diversity as an opportunity for students to innovate and generate solutions that can be competitive in the 21-century markets and positively support community needs.

Providing a unique combination of project-based learning, multicultural and multidisciplinary teamwork, ethically driven problem solving, and personal development, EPS constitutes a sustainable engineering paradigm that can successfully address the challenges of today's dynamically changing reality and support students in acquiring competencies that would assist them in building successful careers. The geographical dispersion of the network within Europe contributes to its inclusive and culturally diverse character, which, combined with the multidisciplinarity and interdisciplinarity of projects and strong adaptive potential, produces a rich and powerful learning environment. Moreover, it promotes the sharing of experiences through student and staff exchanges. Sharing good practices and learning from each other triggers continuous improvement and change of themes, projects, and didactic methodologies, resulting in the personal development of students and academics. Participants report personal, technical, and scientific growth. Teaching staff establishes solid and lasting connections leading to joint ventures such as the writing of this book. The network established good practice of meeting yearly at a different partner institution for the EPS Annual Providers Meeting (APM) to share experiences, compare practices, benchmark and strategically discuss plans for the future and develop new ideas (Figure 5).

Since the network has developed through the years, at the APM, participants work in thematic groups, e.g., administration, internal and external communication, marketing, methodology and teaching or strategy. The flexible way of collaboration strives to maintain the uniqueness of all the partner institutions and work towards innovation and improvement. The EPS network attempts to face challenges inevitably connected with its growth and the pressure of external factors. While accepting more universities, the network has grown in quality and quantity, and at the same time, has been changing its structure and the way of operation in an organic way. Having sustainability and quality of education as a goal, EPS is currently confronted with an important strategic question of how to maintain a balance between growth, operational efficiency, flexibility, and adaptability.

Figure 5. EPS Annual Provider Meeting of 2014 hosted by The Hague University of Applied Sciences
(Credit: The Hague University of Applied Sciences)

EPS, its paradigm and philosophy, constitutes a strong voice in a discourse on engineering education today conducted by leading engineering universities worldwide. The public discourse addresses numerous issues; however, the most basic one would be the relation between engineering education with its existing values and paradigms and the challenges provoked by the dynamically transforming reality in its economic, social, communicative and ethical spheres. Numerous universities as Maastricht, Aalborg, Stanford as well as university networks, e.g., CDIO (Malheiro et al., 2015b), articulate concerns regarding required changes in engineering education and look for the solutions in project-based, problem-oriented or, DT schemes implemented in a team setting, more and more often with human-oriented and social angles.

EPS attains the desired educational objectives of implementing problem- and project-based learning, fostering student autonomy and creativity from students engaged in a multicultural and diverse dialogue while simultaneously respecting constraints of higher education institutions and minimizing disruptions to the rest of the engineering curriculum. The CDIO initiative addresses the implementation of sound changes at the engineering curriculum program level, requiring a deep commitment of all involved in the educational process. EPS focuses on the engineering capstone project and internship level. Students must learn to solve real-world problems, develop personal and interpersonal skills and work in teams, and be exposed to cultural, scientific, and technological diversity during their learning period. Initial specification, procurement, and marketing project phases increase students' familiarity with typical entrepreneurship-related tasks, frequently missing in traditional engineering educational practices. EPS can act as a test-bed for project-based learning in institutions, giving the professors a chance to practice and test complex approaches in education and motivating them for more remarkable changes in the overall curriculum and educational methodologies.

Creativity is possible due to students' diversified engineering branches and cultural backgrounds, which are impossible in a traditional engineering degree, and the students' freedom to guide a chosen project. Nevertheless, there is an immense potential for change in increasing diversity of perspectives and the level of inclusion, e.g., engineering disciplines still suffer from low participation of women, which constitutes an exciting challenge for the EPS network to join a public discourse by working out such a learning culture and communication strategy as to become attractive to women students and ignite a positive change.

The future of innovation in higher education will probably include the rise of flexibility and complexity of the curricula through more complex problem-solving competency and project types, as described by Kolmos et al. (2020). In addition, it should focus on Megatrends as smart products and services (Hanzl et al., 2014), design perspective (Verhulst, 2015), or sustainability (Malheiro et al., 2019b), as well as on the necessity to use and optimize face to face and online education according to the latest trends. EPS will be developing, including new providers and partners, experimenting with new learning methodologies, reinforcing the links with industry and society.

EPS, a robust network of higher education institutions, was founded over 25 years ago and had 120 bilateral agreements and numerous annual meetings in its portfolio. The network has expanded to provide an excellent training opportunity to more than 6000 who worked in an international context on megaprojects and developed interdependence, autonomy, and a variety of necessary competencies for their career.

EPS is an engaging educational experience for students and staff. The program constitutes a natural vehicle for the internationalization strategies of higher education institutions. It creates a perfect opportunity for innovation in teaching, research-teaching experiences and links academia with industry.

ACKNOWLEDGMENTS

This research received no specific grant from any funding agency in the public, commercial, or not-for-profit sectors. Nevertheless, the authors thank: (*i*) Arvid Andersen, who ideated and started EPS; (*ii*) all colleagues who have supported and spread the EPS concept around the world; (*iii*) IFE Director Tomasz Saryusz-Wolski for his vision, collaboration with Arvid Andersen and effort to create an environment for EPS sustainable development in Lodz, Poland, and Director Dorota Piotrowska, TUL, for a persistent continuation of the vision; (*iv*) the European Union Erasmus+ mobility program, which has been a significant source of funding for student and staff exchanges; and (*v*) provider universities, which believed in and implemented EPS, for the development of the EPS network, through the people exchange and to knowledge sharing.

REFERENCES

Abata, D. L., Andersen, A., & Krause, W. B. (2013). Transatlantic Interaction with European Project Semester. In *Proceedings of the 2013 American Society for Engineering Education Annual Conference & Exposition* (pp. 23.1264.1-23.1264.13). 10.18260/1-2--22649

Andersen, A. (2001). Implementation of engineering product design using international student teamwork to comply with future needs. *European Journal of Engineering Education*, *26*(2), 179–186. doi:10.1080/03043790110034438

Andersen, A. (2004). Preparing engineering students to work in a global environment to co-operate, to communicate and to compete. *European Journal of Engineering Education*, *29*(4), 549–558. doi:10.1080/03043790410001711243

Andersen, A., & Hansen, J. (2002). Engineers of Tomorrow and Beyond, insight and skills needed to work across borders. In *Proceedings of the 2002 American Society for Engineering Education Annual Conference & Exposition* (pp. 7.500.1-7.500.7). 10.18260/1-2--10461

Armesto, L., Fuentes-Durá, P., & Perry, D. (2016). Low-cost Printable Robots in Education. *Journal of Intelligent & Robotic Systems*, *81*(1), 5–24. doi:10.100710846-015-0199-x

Ashworth, D. W. C. (2010). Project Based Learning in International Teams – Monitoring the Effectiveness of Teamwork. In *Key Engineering Materials* (Vol. 450, pp. 581–584). Trans Tech Publications, Ltd. doi:10.4028/www.scientific.net/KEM.450.581

Budzinska, G., & Fuentes-Durá, P. (2014). Engineering Education for the Future. European Project Semester at Universitat Politècnica de València, Spain, and Lodz University of Technology, Poland. In *Proceedings of the Fourth Valencia Global Conference* (pp.193-201). UPV.

Chen, J., Kolmos, A., & Du, X. (2020). Forms of implementation and challenges of PBL in engineering education: A review of literature. *European Journal of Engineering Education*, *46*(1), 90–115. doi:10.1080/03043797.2020.1718615

de los Ríos, I., Cazorla, A., Díaz-Puente, J. M., & Yagüe, J. L. (2010). Project–based learning in engineering higher education: Two decades of teaching competences in real environments. *Procedia: Social and Behavioral Sciences*, *2*(2), 1368–1378. doi:10.1016/j.sbspro.2010.03.202

Dispert, H., Boudin, C., Morgan, J. A., & McMahon, M. (2014). International study programs: The role of internships, R&D projects, and their evaluation. In *QScience Proceedings, Engineering Leaders World Congress on Engineering Education 2013* (pp. 1–10). 10.5339/qproc.2014.wcee2013.8

Duarte, A. J., Malheiro, B., Arno, E., Perrat, I., Fuentes, P., Silva, M. F., Guedes, P., & Ferreira, P. (2020). Engineering Education for Sustainable Development: The European Project Semester Approach. *IEEE Transactions on Education, 63*(2), 108–117. doi:10.1109/TE.2019.2926944

Efstratia, D. (2014). Experiential Education through Project Based Learning. *Procedia: Social and Behavioral Sciences, 152*, 1256–1260. doi:10.1016/j.sbspro.2014.09.362

European Project Semester. (2022). *Welcome | European Project Semester.* http://www.europeanprojectsemester.eu/

Ferreira, P., Malheiro, B., Silva, M. F., Duarte, A. J., & Guedes, P. (2019). Developing skills in engineering capstone projects with low-cost microcontroller solutions: the EPS@ISEP experience. In *Proceedings of the 11th International Conference on Education and New Learning Technologies* (pp. 5930-5933). IATED Academy. 10.21125/edulearn.2019.1430

Fuentes-Durá, P., Puyuelo Cazorla, M., Gasch Molina, M. I., & Perry, D. (2014). European Project Semester: Good practices for competence acquisition. In *Proceedings of the Fourth Valencia Global Conference* (pp.161-172). UPV.

Hansen, J. (2010). European Project Semester – how engineering students can achieve important competences. In *Proceedings of the 5th International Conference on Engineering Education* (pp. 1296–1298). Academic Press.

Hansen, J. (2012). An International Project/Internship Course. *Proceedings of the 3rd International Conference PRAXIS'2012 Open Discussion Forum.*

Hansen, J. (2012). Practical Elements in Danish Engineering Programmes, Including the European Project Semester. *Industry and Higher Education, 26*(4), 328–335. doi:10.5367/ihe.2012.0108

Hanzl, M., Młynarczyk, A., Tessier, A., Meiriño, M., Miciedes, I., & Noirot, F. (2014). Smart City - a Quest for Innovation within the EPS Framework. REAL CORP 2014 – PLAN IT SMART! Clever Solutions for Smart Cities. In *Proceedings of 19th International Conference on Urban Planning, Regional Development and Information Society* (pp. 563-571). Academic Press.

Heitmann, G. (2005). Challenges of engineering education and curriculum development in the context of the Bologna Process. *European Journal of Engineering Education, 30*(4), 447–458. doi:10.1080/03043790500213136

Kolmos, A., Bertel, L. B., Holgaard, J. E., & Routhe, H. W. (2020). Project Types and Complex Problem-Solving Competencies: Towards a Conceptual Framework. In A. Guerra, A. Kolmos, M. Winther, J. Chen (Eds.). *Educate for the future: PBL, Sustainability and Digitalisation 2020* (pp. 56-65). Aalborg Universitetsforlag. International Research Symposium on PBL.

Malheiro, B., Guedes, P., Silva, M. F., & Ferreira, P. (2019a). Fostering Professional Competencies in Engineering Undergraduates with EPS@ISEP. *Education Sciences, 9*(2), 119. doi:10.3390/educsci9020119

Malheiro, B., Silva, M., Ferreira, P., & Guedes, P. (2015a). CDIO and the European Project Semester: a match for capstone projects? In *Proceedings of the 11th CDIO International Conference* (pp. 1–14). CDIO.

Malheiro, B., Silva, M., Ribeiro, M. C., Guedes, P., & Ferreira, P. (2013). The European Project Semester at ISEP – Learning to Learn Engineering. In *Proceedings of the 1st International Conference of the Portuguese Society for Engineering Education* (pp. 1–8). 10.1109/CISPEE.2013.6701977

Malheiro, B., Silva, M., Ribeiro, M. C., Guedes, P., & Ferreira, P. (2015b). The European Project Semester at ISEP: The Challenge of Educating Global Engineers. *European Journal of Engineering Education, 40*(3), 328–346. doi:10.1080/03043797.2014.960509

Malheiro, B., Silva, M. F., Ferreira, P., & Guedes, P. (2019b). Learning Engineering with EPS@ISEP: Developing Projects for Smart Sustainable Cities. *International Journal of Engineering Pedagogy, 9*(4), 33–49. doi:10.3991/ijep.v9i4.10259

Mulder, K. F., Ferrer, D., Segalas, J., Kordas, O., Nikiforovich, E., & Pereverza, K. (2015). Motivating students and lecturers for education in sustainable development. *International Journal of Sustainability in Higher Education, 16*(3), 385–401. doi:10.1108/IJSHE-03-2014-0033

Ponsa, P., Roman, J., Arno, E., & Pérez, J. (2015). Professional skills in international multidisciplinary teams. *International Journal of Engineering Education, 31*(4), 998–1006.

Rohaert, S., Baelus, C., & Lacko, D. (2012). Project work on wellbeing in multi-disciplinary student teams: a triple testimonial on EPS at Artesis. In *Proceedings of the International Conference on Engineering and Product Design Education* (pp. 1296–1298). The Design Society. https://www.designsociety.org/download-publication/33244/Project+Work+on+Wellbeing+in+Multidisciplinary+Student+Teams%3A+A+Triple+Testimonial+on+EPS+at+Artesis

Schjær-Jacobsen, H., Abou-Hayt, I., Ashworth, D., Jensen, M. P., & Schreiber, M. P. (2012). Industrial design as an innovative element in engineering education. In *Proceedings of the International Conference on Engineering Education* (pp. 281-288). Academic Press.

Segalàs, J. (2012). The EPS Experience at UPC-Barcelona Tech. In L. C. Campos, E. A. T. Dirani, A. L. Manrique, & N. Hattum-Janssen (Eds.), *Project Approaches to Learning in Engineering Education* (pp. 109–124). SensePublishers. doi:10.1007/978-94-6091-958-9_8

Segalàs, J., Esbrí, M. E., & Benson, P. (2011). European project semester: 30 ECTS of PBL in sustainability with multicultural and multidisciplinary bachelor student groups. *Proceedings of 17th International Conference on Engineering Education.*

Segalàs Coral, J., & Benson, P. (2012). International design project semester. In *Proceedings of the 14th International Conference on Engineering and Product Design Education* (pp. 585–590). The Design Society. https://www.designsociety.org/publication/33251/International+Design+Project+Semester

Silva, M., Malheiro, B., Guedes, P., Ferreira, P., Ribeiro, C., Ferreira, F., & Duarte, A. (2017). The EPS@ISEP – European Project Semester Program at ISEP - Implementation results and ideas for improvement. In *Proceedings of the 45th SEFI Annual Conference* (pp. 129-136). SEFI.

Silva, M. F., Malheiro, B., Ferreira, P., & Guedes, P. (2019). Promoting Ethical Practices in Engineering Undergraduates. In *Proceedings of the Fourth International Conference on Robot Ethics and Standards* (pp. 55-63). CLAWAR Association. 10.13180/icres.2019.29-30.07.002

Silva, M. F., Malheiro, B., Ribeiro, C., Guedes, P., Duarte, A. J., & Ferreira, P. (2018). Collaborative Learning with Sustainability-driven Projects: A Summary of the EPS@ISEP Programme. *International Journal of Engineering Pedagogy*, *8*(4), 106–130. doi:10.3991/ijep.v8i4.8260

Skates, G. W. (2003). Interdisciplinary project working in engineering education. *European Journal of Engineering Education*, *28*(2), 187–201. doi:10.1080/0304379031000079022

United Nations Educational, Scientific and Cultural Organization. (2017). *Engineering*. http://www.unesco.org/new/en/natural-sciences/science-technology/engineering/

Verhulst, E., Rohaert, S., & Doorsselaer, K. V. (2015). How to incorporate Sustainable Design in the International European Project Semester programme: Insights from practice. In *Proceedings of the International Conference on Engineering and Product Design Education* (pp. 1-6). The Design Society, Institution of Engineering Designers.

KEY TERMS AND DEFINITIONS

Competence: The capability to do something well. It refers to essential skills that are needed to do a job.

Competencies: The description of a cluster of related abilities, commitments, knowledge, experience, attributes, and skills that enable a person (or an organization) to act effectively in a situation.

Engineering Education: The higher education field dedicated to preparing engineering professionals.

Hard and Soft Skills: Scientific-technical and personal-interpersonal competencies. A broad set of knowledge, abilities, work habits, and character traits that are believed to be important to success in today's world.

Innovation: Renews or updates a domain, product, or service using new processes, techniques, or technologies.

Interdisciplinarity: Analyses, synthesizes, and harmonizes links between disciplines into a coordinated and coherent whole.

Multicultural and Multidisciplinary Teamwork: Work method performed by teams integrating members from different cultures and disciplines typically used to solve interdisciplinary problems.

Multidisciplinarity: Draws on knowledge from different disciplines but stays within their boundaries.

Project-Based Learning: A learning method where students acquire new knowledge and skills autonomously by working together on a challenging real-world problem.

Transdisciplinarity: Integrates the natural, social, and health sciences in a humanities context and transcends their traditional boundaries.

Chapter 2
United States Participation in the European Project Semester: An Exceptionally Successful Endeavor

Duane L. Abata
South Dakota School of Mines and Technology, USA

ABSTRACT

This chapter describes successful interaction between engineering colleges in the United States and similar institutions engaged in the European Project Semester program, EPS, throughout Europe. Dr. Abata was asked to both teach and serve in the administration of EPS. The EPS program described herein was designed to promote student exchange and student interaction between EPS institutions in a multidiscipline environment in Europe. The primary purpose of EPS is to address engineering design in junior and senior years. Several students from the United States participated in this primarily European program for the benefit of broadening their educational and career horizons as well as gain valuable insight into international approaches to engineering design. These students commented that their experience was significant. However, meeting engineering senior design requirements in some cases was problematic due to the interpretation of ABET senior design requirements. This chapter concludes with a brief discussion of the possibility of EPS like-programs in the United States.

INTRODUCTION

European Project Semester, or EPS, is an international exchange program specifically designed to address the design component of engineering students during their junior year. The program is highly successful and involves nineteen Higher Education Institutions in twelve countries throughout Europe.

EPS was conceptualized by Dr. Arvid Andersen, who found a need to expose his engineering students in Denmark to engineering students from other countries. He observed that his engineering students were lacking in international exposure with continental Europe. He then devised a scheme where engineering students from continental Europe would visit his university and work together with Danish students on cultural broadening and engineering design for a semester. Dr. Andersen presented EPS at numerous

DOI: 10.4018/978-1-6684-2309-7.ch002

Copyright © 2022, IGI Global. Copying or distributing in print or electronic forms without written permission of IGI Global is prohibited.

conferences and publications from the late 1990s through 2012 (Andersen, 2000a; Andersen, 2000b; Andersen, 2001; Andersen & Hansen, 2002; Andersen, 2003; Andersen, 2004a; Andersen, 2004b; Andersen 2009; Andersen 2010; Andersen 2011; Andersen, 2012).

Moreover, Dr. Andersen invited this author to teach the subject of Environmental Awareness to EPS students in Copenhagen and later in Oslo, as part of the EPS program. This biannual (spring and autumn) participation lasted for over fifteen years. The intensive two-day presentation covered various aspects in environmental subjects and involved international travel between Copenhagen, Oslo, and the United States. The invitation resulted from this author's engagement in the American Society for Engineering Education, which he served as president for a period, and his knowledge of the subject matter. This experience led to the presentation of the EPS concept at an annual conference of the American Society for Engineering Education (ASEE) (Abata et al., 2013), where it was well received. Additional information regarding knowledge and information of the EPS program in the United States was presented by others also involved in the EPS program (Fuentes-Dura et al., 2015; Shenyang et al., 2020). The concept did not transfer to the American scene for many reasons, some of which are discussed in the following text. However, many students from the United States (U.S.) did participate at numerous EPS sites through an international exchange (Figure 1).

Figure 1. The adventure of living in other countries broadens the vision of EPS students and contributes to their outlook and professionalism throughout their careers

This chapter includes four additional sections. They address the EPS program background, general reflection on the participation of students from the United States in EPS, the challenges of adopting the EPS model in the United States, and the conclusion.

BACKGROUND

The EPS program was started in autumn 1995 with six visiting students at Elsinore Engineering College, in Danish "Helsingør Ingeniørhøjskole", formerly "Helsingør Teknikum". Shortly thereafter, Dr. Andersen moved to Ingeniørhøjskolen i København (IHK) / The Engineering College of Copenhagen located in Ballerup, Denmark, where EPS was restarted autumn 1997 with funding from the Danish government[1]. EPS continued to grow in Denmark, eventually reaching one hundred students each year.

In parallel to this growth, the popularity of the program grew substantially over the following years as other engineering colleges recognized the educational benefits of international exchange (see Figure 2). Too, this international exchange was promoted by the European Union (EU) through the Washington Accord (Washington Accord, 2022), which began to encourage a unified approach to engineering education. Dr. Andersen partnered with Dr. Jorgen Hansen at IHK, who had facilitated the original move from Elsinore to Ballerup. Both were instrumental in the growth of EPS.

Despite its success, EPS in Denmark was terminated in 2013 shortly after the retirement of Dr. Andersen when the administration of IHK was taken over by Danmarks Tekniske Universitet[1] (DTU) / The Technical University of Denmark. Dr. Andersen, along with Dr. Hansen (Hansen, 2010; Hansen, 2012), continued to promote EPS until Dr. Hansen retired in 2014 and Dr. Andersen, a visionary educator and good friend of the author of this chapter, sadly passed away in 2017.

Figure 2. Dr. Arvid Andersen with Dr. Abata overlooking the Grand Canyon in Arizona, 2004

Nevertheless, the EPS program and EPS concept continue throughout Europe. Despite the termination of the program in Denmark, well over three hundred students from forty countries participate in this exchange program each semester at nineteen Higher Educational Institutions, forming a significant network of international engineering education exchange.

The EPS programs are tuition free for participating students which is in alignment with most public colleges and universities throughout Europe. Students must pay for travel and living expenses in the receiving country. Travel expenses are not prohibitive because of the proximity of neighboring countries in Europe and well-developed mass transportation. Living expenses are somewhat comparable. Many grants are available through the Erasmus exchange program of the European Union (EU), in operation since the late 1980s, encouraging students to study at universities in the EU member states for periods of time. It may be safe to add that Erasmus is a sustaining component of EPS as available funding from Erasmus encourages EPS growth and participation. European sites – EPS providers – are also funded by individual governments, the EU, and the participating colleges.

It is important to note that individual EPS programs are not formally coordinated through a central governing body. However, coordination is sound and is achieved through frequent communication and cooperation by site directors. Site directors are mutually committed by the inherent nature and necessity of the overall program. Professional friendships are also a significant driving force that keeps the overall program on track. Annual coordination meetings of the EPS Providers are usually scheduled in autumn at one of the EPS sites. This voluntary meeting rotates around the providers and usually involves more than thirty participants. New concepts, difficulties, and solutions are typical agenda items. As the annual meetings rotate each year, individual site directors became familiar with each of the programs and physical facilities. Moreover, the EPS providers frequently report on the EPS student-centered learning experiences (Rohaert et al., 2012; Segalàs et al., 2012; Budzinska & Fuentes-Durá, 2014; Ponsa et al., 2015; Verhulst et al., 2015; Armesto et al., 2016; Malheiro et al., 2019; Duarte et al., 2020).

Language and Structure

Classes in EPS are taught in English thus eliminating a significant barrier to international exchange. While almost all European students are multilingual, English is recognized as a base language understood by most. Participating students show reasonable proficiency in English during the application process. As mentioned previously, the popularity of the program is such that the program is successfully promoted by word of mouth. Each site does have its own corresponding website that details the application process.

The structure of the program is non-traditional in that students remain as a cohort throughout the semester. Students often remain in one large classroom of reasonable size to accommodate both instruction and design team networking (design lab). Instructors visit the classroom according to a predetermined schedule that varies daily, or students may move to other classrooms as a cohort. In this way, instructors participate in EPS instruction in block intervals allowing each instructor to continue with normally scheduled traditionally taught classes that are not part of the EPS experience. This block scheduling also allows flexibility for visiting instructors that may travel from one EPS site to another. A typical block scheduling scheme is shown in Table 1. The class time devoted to each topic can also be derived from Table 1.

Table 1. Typical block scheduling scheme for EPS

Activities	**Weeks**															
	1	**2**	**3**	**4**	**5**	**6**	**7**	**8**	**9**	**10**	**11**	**12**	**13**	**14**	**15**	**16**
Introduction & Team building																
Project management																
Project planning																
Languages																
Systematic innovation																
EU law & Property rights																
Cross-cultural communication																
Environmental subjects																
International marketing																
Communication skills																
Project review																
Interim report																
Project work																
Examination																
Graduation																

In the model presented in Figure 3, EPS students meet for approximately seven hours per day, five days a week. Three topics are covered during the first week. Thus, students receive approximately 11 to 12 hours of exposure for each subject (Team Building, Systematic Innovation, and Communication Skills).

The semester is a recognized 30-credit unit course under the European Credit Transfer System (ECTS). Project work comprising the bulk of the semester long design experience amounts to between 20 and 25 credit units, and short intensive courses cover the remaining points. It is generally assumed that 2 ECTS credits equal one course credit in the United States following ABET guidelines (ABET, 2022). Thus, the EPS experience is roughly equal to 12 to 13 U.S. credits, or that of a full-time student.

The final grade is determined by the technical content of the report assessed by an individual examination of each student.

Figure 3. Subject matter embedded in EPS program structure

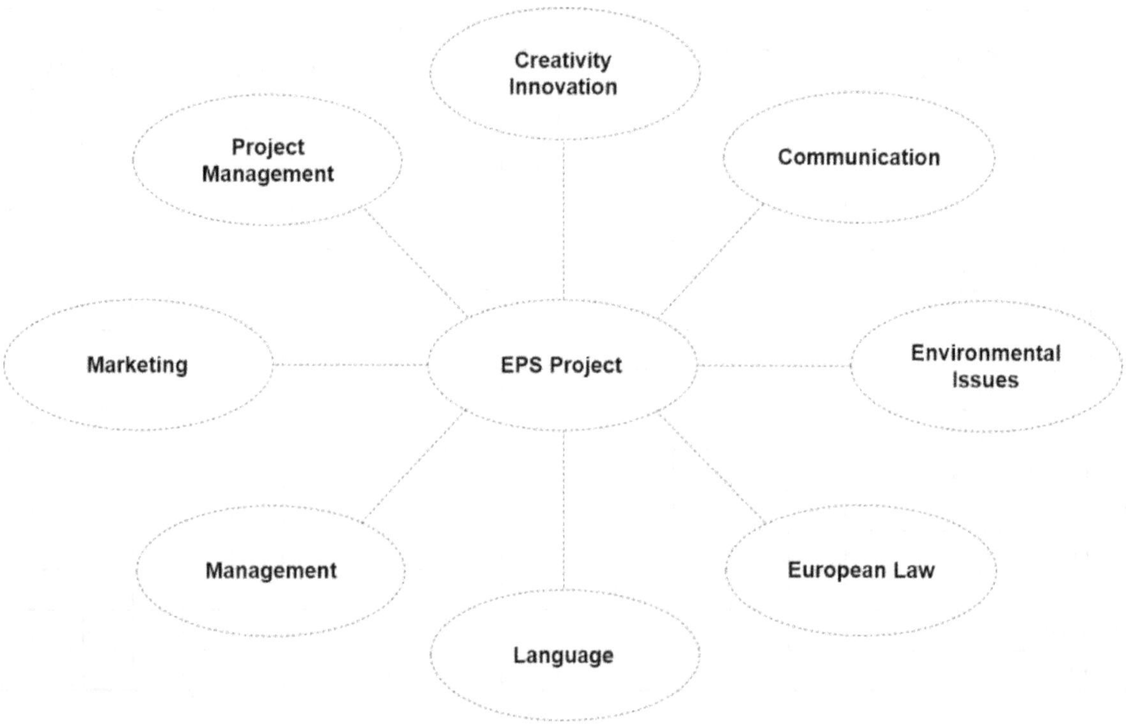

Students receive several grades for the overall EPS experience: one grade for the project, one for the language courses, and one for all the other courses. The final grades are thus both objective and subjective in that attendance and participation are included in the final evaluation. Students are critically evaluated. Those failing to meet the minimum guidelines are either issued "no-credit" for their experience or are asked to return the following semester if sufficient reason is provided for poor performance (illness, language difficulties, lack of background, etc.).

EPS and Experiential Learning

The educational process in the EPS program is best described as experiential learning. Experiential learning is most easily described as the process of acquiring information through the study of a subject without the necessity for direct textbook exposure. In the early 1970s, educational researchers David Kolb and Ron Fry developed the Experiential Learning Model (Kolb & Fry, 1975), which is composed of four elements:

- concrete experience
- observation of and reflection on that experience
- formation of abstract concepts based upon the reflection
- testing the new concepts
- (repeat)

These four elements are the quintessence of a spiral of learning that can begin with any one of the four elements, but typically begins with a concrete experience. The model was named to emphasize links to ideas from Piaget et al. and other writers of the experiential learning paradigm (Beilin, 1992). This model was developed predominantly for use with adult education but has found widespread pedagogical implications in higher education.

There are certain characteristics of a student (or a group of students) that are essential to bring the experiential learning model to a satisfactory end. The learner must:

1. be willing to be actively involved in the experience.
2. be able to reflect on the experience.
3. have and use analytical skills to conceptualize the experience.
4. possess decision-making and problem-solving skills to use the new ideas gained from experience.

The dimensions of experiential learning are analysis, initiative, and immersion, which nicely describe the EPS atmosphere. Once students are identified with a project, they proceed to analyze with the skills they have learned earlier and skills they acquire in the EPS classroom setting. Self-learners having strong academic skills are identified during the application process; thus, initiative is inherent in the EPS student body. It should be noted that most European students are self-learners since semester grades are based not upon hourly exams at regular intervals but rigorous oral and written examinations. Finally, students in EPS are completely immersed; the one-room learning environment and interaction with each other on an analytical, cultural, and social basis is evidence of this immersion experience. Language and cultural barriers are overcome with understanding, thorough communication, and interpersonal reward (Itin, 1999; Merrian et al., 2007; Hutton, 1980). Though both methods (experiential and traditional classroom) aim at instilling new knowledge in the learner, academic learning does so through more abstract, classroom-based techniques, whereas experiential learning actively involves the learner in a concrete experience.

GENERAL OBSERVATIONS AND CHALLENGES

The author of this chapter, Dr. Abata, met Dr. Andersen at an annual meeting of the ASEE many years ago. Dr. Andersen explained the concept and asked if Dr. Abata could assist with recruiting students from the United States. This brief meeting was the beginning of Dr. Abata's participation in EPS and a wonderful friendship with Dr. Andersen which lasted nearly twenty years. Dr. Abata served as a guest lecturer in the EPS program at two sites for more than fifteen years (Oslo and Copenhagen). He also acted as U.S. liaison with EPS site directors during this period and attended many of the annual meetings held at various EPS locations. This was a time of significant growth of the program.

In his academic and administrative roles during this period, Dr. Abata worked with students and faculty members to encourage student participation from the United States, resulting in a small but steady flow of student exchange. Memorandums of Understanding between universities and colleges in the United States and each of the EPS sites are necessary because the EPS program is tuition free funded by the European constituency and engineering programs in the United States are tuition and fee based. These formal memoranda establish a one-to-one reciprocal student exchange allowing tuition transparency for participating students. On a mutually beneficial note, these memoranda have encouraged many students from Europe to attend a semester of study in the United States.

An EPS class at one of the sites typically ranges from about twenty to thirty students. Not all sites offer EPS in both semesters. In an EPS class of thirty students, one can expect to find representation from ten to fifteen countries. EPS directors have found that selecting two or three students from a country increases the success and satisfaction rate since strange surroundings (different cultures) are better understood and accepted with a supporting colleague. Attention is also placed on student background and major area of study. Potential design projects are established before the admission process; thus, student selection is also based upon the student selecting a satisfactory design experience that will meet his or her academic and interest needs.

The typical EPS class then is composed of students from a variety of engineering study areas and different cultural and linguistic backgrounds with the common goal of engineering design (Figure 4).

Students in teams of three to six or slightly more are assigned a design project based upon their interests identified during the application process. EPS students are generally very mature and eager to learn, and, because of the international setting they themselves are immersed in, they are very respectful of international lecturers and visitors. There is an expected struggle with expression as students converse in a language other than their mother tongue. U.S. students learn quickly that patience and compassion are part of the EPS experience. U.S. students also learn rather quickly that their multilingual shortcoming is a handicap. During a lecture a low level of chatter is experienced as students ask fellow colleagues for help with interpreting the lecturer's words, and page through dictionaries to assist with translation. Despite the maturity level of the students, attention span is somewhat limited, so it is essential that lectures are punctuated with numerous breaks, discussion, and possibly brief "hands-on" assignments.

Figure 4. A typical EPS class
(Spain, ca 2014)

One of the most important yet easily missed observations is that students from different cultures think differently. The approach to a design problem (a problem having a design solution instead of a single numerical answer) is taken from different perspectives. Solutions too are obviously diverse and representative of the background of a student; together in teams this results in what can be termed "collective creativity" of an interdisciplinary and multicultural capstone experience.

The social aspect of a multicultural group of eager college age students is yet another valuable component in the EPS experience. Social gatherings and interactions are not formal nor part of the program structure. They appear spontaneously and with great enthusiasm. It is not uncommon for groups of five to ten students to travel together on the weekends to historic or picturesque sites, or events that pleasantly saturate the European continent. Travel by train is often the choice of these weekend experiences. Lodging may be in the parent's home of one of the students. Friendships formed during the EPS experience (now in its twenty-fifth year) are frequently life-lasting.

Exit interviews of U.S. students have been routinely conducted by the primary author of this paper. Interview questions, presented to the student in an informal manner, typically are:

- What specifically did you learn about the design process?
- What other academic knowledge and skills have you experienced?
- What did you learn about team building and team interaction?
- What credit transfer do you expect or have already received by your home institution?
- Finally, did you enjoy your EPS experience?

About forty students from the United States were interviewed through the years from three institutions. It is noted that the period covered was seventeen years from 2001 to 2017. For the most part students from the United States were in their junior year of study, did not have prior experience with international study, and were not fluent in any language other than their mother tongue of English. The exact profile of each student including their cultural background or where they attended the EPS program was not recorded although most attended EPS programs in Denmark, Oslo, Germany, or Spain. Prior cultural experience is almost impossible to determine since the United States is considered a 'melting pot' of cultural backgrounds, mostly from western European countries. EPS is, of course funded by the countries of Europe. Thus, students from the United States were invited by reciprocity agreements between the sending and receiving institutions and their own personal funds without any scholarships or funds from the European Union. The response to this survey was overwhelmingly positive. However, some students commented that credit transfer is problematic (discussed below). Probably the best expressions of support come from testimonials of students which, in some instances, were documented:

Let me start by saying that I cherish this experience as much as I do my four years at Notre Dame. I have learned more about myself, the world and life in the last five months than I have in my previous 24 years of living. I came to the program believing that with my credentials and experience, I would be able to easily succeed in this setting. I have never been so wrong in my life. My error was in that while I have experience working with Americans of various ages and egos, we are all American and therefore predominantly of the same mindset. When I started working within a multi-national team, I found myself having to reinvent my work habits, my communication skills and my choice of words. It was one huge and complete paradigm shift. (mechanical engineering graduate from Notre Dame).

Although choice of words and subject vary, the above testimonial is typical of many gathered over the years. There appears almost a re-birth of thought and enlightenment that occur within the student. This positive experience is one that is carried back and shared with other students unable or not interested in participating. The experience (and investment) is multiplied many times over.

Barriers Encountered by U.S. Students

The EPS experience is a good one but not for everyone. There are several barriers encountered by U.S. students that are both self-generated (or self-imposed) and institutionally imposed. Some of the more common barriers are expressed below.

Funding

The cost of EPS to U.S. students may be prohibitive. While maintenance costs (living expenses) are about the same, except in the Scandinavian countries where such expenses are a bit higher, transportation costs to and from EPS are high. A typical airline ticket may cost, in today's currency, over a thousand dollars. While this investment is not prohibitive in the overall scheme of one's education, it falls at a time when many U.S. students struggle with ever rising tuition and fees at U.S. institutions. There are additional costs associated with initial transportation and settlement in the new environment. EPS programs also require an upfront 5000 USD deposit which ensures the receiving EPS site reimbursement for unexpected difficulties such as early return, health issues, etc. This deposit is held in a security account and returned at the end of the exchange. At some sites, this deposit remains in the student's or parent's account with appropriate documentation from a U.S. monetary institution.

Visas

While visa requirements were somewhat lax in the early stages of the EPS program, it has now become essential that all U.S. students have travel visas because the semester length is longer than that of tourist status. Visas have become a challenge. In many cases, students must apply for a visa, travel in-person to an embassy (or consulate) for an interview that may last no longer than five to ten minutes, and return to their home institution to wait for approval which may take as long as several months. Passports are, of course, required before the visa application, and a passport application may require a month or more for processing. (It is worth mentioning that the recent COVID-19 pandemic has essentially prohibited any participation in EPS. This is hopefully a temporary interruption and EPS will resume as the pandemic is resolved.)

Tuition in the United States and EPS

The funding model for engineering education in the U.S. is considerably different from that in Europe. Higher education (B.S. degree and higher) in the U.S. is either public or private. Public institutions are funded at the state level, endowments, and by student tuition with very few exceptions (e.g., U.S. Military Academy at West Point, U.S. Naval Academy, etc.) by the federal government. Indirect charges levied with on-campus research also may significantly contribute to the institution's general fund. Private institutions rely on student tuition and endowments and hence tuition is generally much higher. Thus, student tuition is much higher than that of European institutions. EPS is offered tuition free or almost tuition free to European students.

Reciprocity is often offered as a solution to offset this tuition difference. That is, U.S. students participating in EPS will pay the usual tuition amount to their own institution, and this amount is then used for the tuition of a student from Europe attending a U.S. institution. This is generally done by either formal or informal agreement on an institution-to-institution basis. Arrangements rely on individual coordinators, that is, individuals who negotiate the agreements. In short, U.S students seeking opportunities with EPS schools often have limited options unless the EPS school is willing to offer an EPS opportunity at their own expense.

Fear of the Unknown and Isolationism

Students interested in the international experience are self-driven. Yet there is an inherent fear that this investment may not work, or adjustment may be difficult or insurmountable. This is particularly true of engineering students which by nature are cautious and, for the most part, lack risk-taking skills. Language is part of this fear. English is obviously the spoken language throughout the United States, and because of the physical size of the country, the general population does not have exposure to other languages or the need to communicate in other languages. There is some exposure to Hispanic (Historically Spanish) languages because of the proximity of Mexico and immigration from Mexico and South America, but the general population relies on Hispanic-speaking people to learn and communicate in English. Some of the traditional European languages such as German, French, and Spanish are taught in most high schools across the country; however, exposure is limited to a year or two of classroom learning. Immersion in a different language is rare. Although EPS is taught in English, which is the mother tongue of U.S. students, they still are somewhat reluctant to travel to a foreign country where immersion outside of the EPS classroom may be encountered. For the most part, this is self-imposed as many Europeans in larger cities have some knowledge of the English language. U.S. students with some language skills in German, Spanish, or French, are less intimidated. Quite obviously, the adventurous student is more likely to seek an international experience, particularly with encouragement from members of the faculty or from a parent.

One of the primary purposes of EPS and other study abroad opportunities is to gain exposure to other cultures and languages. Nevertheless, many U.S. students are reluctant to leave their comfort zone because of inbred isolationism due to the size of the United States compared to other European countries. As an illustrative point, driving across the United States with reasonable breaks for meals and overnight lodging can take up to two weeks, whereas driving across a country on the European continent requires a day or so. Thus, U.S. students and the country in general suffer from isolationism, that is, the ability to remain apart from the affairs or interests of other countries because of the ability of self-containment. It is somewhat ironic that this isolationism was one of the underlying goals behind the creation of the EPS program initially. The innovator behind EPS, Dr. Anderson, started EPS because he witnessed Danish students having the same reluctance to travel to other countries.

ABET Influence/Impact

The impact of engineering accreditation in the U.S. has an enormous impact on students seeking EPS opportunities since the content in engineering programs is largely and indirectly controlled through this nationwide accreditation process.

The main (and only) accreditation agency for engineering education in the United States is ABET, formerly the Accreditation Board of Engineering and Technology where the name ABET originates as an acronym (ABET, 2022). ABET is the recognized U.S. accreditor of college and university programs in applied and natural science, computing, engineering, and engineering technology and is, in practicality, the only accreditation process for engineering programs in the U.S.[2] Employers rely on ABET accreditation as the accepted standard for engineering education. Almost all engineering programs in the U.S. are ABET accredited (ABET, 2022).

ABET has a surprisingly small administrative structure with headquarters in Baltimore, Maryland. Rather ABET is composed of several hundred unpaid volunteers from industry and academics chosen by each of the various professional engineering societies to represent specific disciplines in engineering. For instance, the American Society for Mechanical Engineers has several volunteers that represent the society who serve as ABET Evaluators for mechanical engineering programs that seek accreditation. The accreditation process is highly structured, although there have been improvements to the process in past years that allowed greater program flexibility.

Each program pursuing or renewing accreditation must be reviewed by an ABET Evaluator during an on-campus visit about every six years. A successful accreditation results in a maximum six-year period following an accreditation visit. The ABET Evaluator examines, amongst other things, student records during the accreditation visit. Since it would be impossible for a reviewer to examine all student records, a random sample of student records is requested by the Evaluator and then reviewed in a highly structured process. University (or college) administrators are rightly concerned that a U.S. student who has participated in EPS is included in this random sample meets all program requirements. Since EPS is non-traditionally structured, credit transfer is a qualitative judgment of the ABET evaluator, who most likely is not familiar with EPS. Thus, administrators are overly cautious with what is acceptable credit transfer. This does not mean that a U.S. student cannot participate in EPS but rather what the student can use as course content and credit for graduation requirements. Tuition is charged almost exclusively on a credit-by-credit basis so the student may be paying tuition for non-credit courses (courses that cannot be counted toward the baccalaureate degree).

Senior Design Requirement

The accreditation agency, ABET, in the United States, requires a "capstone design experience" during the senior year, the fourth year of study for engineering students. The words "senior year" are underlined because EPS recruits and involves students during their junior year or the third year of study. This discrepancy was and is the source of credit transfer for U.S. students. This design experience must involve the "cumulation of engineering knowledge" gained at the end of the academic experience (ABET, 2019; ABET, 2022). Generally, this is a semester long design course involving student teaming in a design project much like that found in the EPS experience. Multidisciplinary teaming is encouraged, although not required. Construction of the design project is not required although some U.S. colleges encourage this activity. Acceptance of the EPS design experience as a substitute for this ABET requirement depends upon the decision of members of the faculty of a college sometimes on a case-by-case basis. At some colleges this was not an issue. At other colleges, this was a significant challenge even though the experience at EPS may have been actually superior to that encountered at a home institution because of the interdisciplinary and multicultural nature of the EPS activity. Some faculty members simply stated that the EPS experience was not rigorous enough and did not involve significant engineering analysis (even though in some cases it did!). In many cases, a student had to demonstrate that the project met ABET guidelines when he/she returned to campus. Students often could not transfer this experience and were thus required to redo this design experience at their home campus. Since tuition is charged on a per-credit basis in most U.S. institutions, this resulted in additional expense, duplication of activity, and a loss of time. Students faced with this predicament oftentimes benefited from a "champion" on their home campus that would help them with their argument both in content (substance) and politically as well. On a final note, some EPS schools were lax with the design experience in which case the challenges by college faculty in U.S. colleges may very well have been justified. Another argument encountered which contributed to this skepticism was that the EPS experience occurred during the junior year rather than the senior year of study.

In some cases, U.S. students were not resistant to the consequences discussed in the previous paragraph because they held the belief that a semester abroad at a European college far outweighed the requirements to repeat some, or almost all, of the material at their home institution whether it be justified or not.

Organizers of EPS programs sometimes were aware of these issues when defining design projects and assigning teams involving students from the U.S. to alleviate or reduce this challenge at the student's home institution. This may or may not be possible because projects are sometimes hard to define or difficult to identify during the planning stages of the EPS semester.

Credit Hour Transfer

This was, and is, one of the most difficult barriers with U.S. students participating in EPS. Credit earned by students participating in EPS is based upon the European Credit Transfer and Accumulation System (ECTS). Students earn 30 ECTS. Exchange agreements between participating countries in EPS must agree to this 30 ECTS policy except for students from the United States. This exception is made to encourage U.S. participation but may be problematic since many here in the U.S. do not or are not willing to understand the ECTS system. Roughly 2 ECTS equals one credit hour in the United States. Through the memorandum of agreement with their institution, a U.S. student must pay the usual tuition directly to their home institution to allow this one-to-one exchange. Generally, credit transfer approval does not take place until the student returns with transcripts in hand. In many instances, a student may receive only a small number of transfer credits (i.e., a student pays a full 12 credit tuition but may receive only three to six transfer credits). This credit transfer issue is embedded in: (*i*) the student's ability to convince his/her advisor and department head; and (*ii*) the innate fear that an ABET program evaluator will recognize and discount this credit transfer during the ABET review process. This latter point is sufficient for administrators to error always on the side of caution. This is particularly true at institutions that are conservative in nature, even though international exposure is clearly enunciated and encouraged in ABET criteria (ABET, 2019). This has an added negative effect in that certain faculty members, particularly at conservative institutions, may discourage international exchange programs and dismiss such programs as a waste of time. Members of the faculty in the United States often look for content rather than experience. Recall (described earlier in this chapter) that the EPS program is not traditionally structured with regular classes and traditional course numbers.

The EPS experience poses additional challenges because of the nature of the material covered, which is integrated throughout the semester rather than taken in neat little silos as in the U.S. This proved to be problematic when transferring credit, although the integrated course package may be vastly superior to siloed packages. This cohort format of both the student group and classroom setting is not employed in the U.S., and it is difficult for members of the faculty in the U.S. to understand it as a high-quality experience in their own rather parochial definition, most likely because of the momentum of the accreditation agency ABET.

Therefore, the credits earned for the EPS experience were often discounted or not counted at all. In almost all cases, U.S. students were pleased to have participated in this international experience and therefore felt that the experience was reward enough for their participation in the program.

Housing

Students traveling abroad face the problem of finding suitable housing for the semester. Since all students enrolled in an EPS program are, by definition, international, all EPS sites have systems in place to streamline or alleviate this difficulty. In most instances, housing is found by EPS assistance ahead of the exchange. This is one of the major benefits EPS has over many other semester exchange programs.

Design Project Selection

During the application process a student identifies three projects of interest based upon a paragraph description. Teams are organized at least one month before the start of EPS based upon student interest. Students receive an email informing them of their project and project team before they arrive on campus and are asked to discuss the project with a professor or supervisor at their home university. In this way, the student is aware of the amount of work required to receive proper credit by his/her home university. Soon after arriving at the EPS site, students are strongly encouraged to further discuss their projects and teams with their home supervisors so that projects can be further refined to meet home university requirements. At this critical time, students also meet with their design teams and their project supervisors to ensure proper subject material, depth, and breadth are reached for home university requirements. While this process is detailed and EPS site directors try their best to place the student on a relevant design project, the design project assignment may or may not reflect the actual needs of the student. Students too may not be fully knowledgeable of the design depth requirement at their home institution. In some instances, design projects are not convincingly sufficient to meet home institution requirements. Advanced advising at the home institution is essential, but considering the many steps and barriers involved in the application process, this is often overlooked.

One Semester Design Experience

EPS is a semester long design experience. Many capstone design requirements in the U.S. are two semesters in length. There is always the question of whether this one-semester experience will count toward the capstone design requirement in the U.S.

Academic Calendar Differences

There are significant differences in the U. S. and European academic calendars, and these calendars differ from institution to institution. European calendars begin earlier in the autumn semester but, like U.S. calendars, run to the Christmas holiday period. European calendars may begin earlier or later in the spring semester and may run to early June.

While it seems that start and end dates should be a minor consideration, one must also consider student employment during the summer. While summer employment is a concern for all students, it is a major concern for U.S. students because of the much higher cost of tuition in the U.S. the end of spring term in the U.S. has trended toward ending early at most colleges and universities so that students can seek summer employment in early May since the competition for summer employment is quite strong. U.S. students thus perceive a disadvantage in seeking summer employment when they participate in EPS programs that end in early to mid-June. U.S. students seeking summer employment are thus compromised and summer employment may be essential to meet tuition, fee, and living expenses for the coming academic year.

EPS MODEL IN THE UNITED STATES

The United States, while superficially viewed as a homogenous culture, a melting pot, it is anything but that (U.S. Census Bureau, 1995). Although citizens are tied together by a common government and a common language, there are significant regional differences, as any astute traveler will attest. The culture of the urban environment of the East Coast differs significantly from the rural Midwest Plains. The progressive culture of the California coastline is vastly different from the conservative cultures of the West or South. Engineering students are extremely mobile during their initial careers, oftentimes leaving a small engineering college in rural America for the hustle and bustle of a large urban city. Many students have difficulty with adjustment, and many leave their initial career choices after only a few years.

It would seem reasonable and logical that the EPS model would work nicely in the United States where such an exchange program, formally structured, would add significantly to their academic experience. Students in South Dakota might spend a semester (or year) at an institution in New York and vice versa. Several years ago, this concept was proposed to the Deans Engineering Council of ASEE[3] and received reasonable support. However, there are significant barriers to such a venture, and these barriers are probably insurmountable.

Funding of the Model

Most public engineering colleges and universities in Europe are funded by their respective central governments. There is an altruistic desire to make education available to all at little or no cost (reflected in higher taxation rates). The funding model in the U.S. is much different. Of course, public engineering colleges and universities in the U.S. are funded partially by an individual state and partially by tuition and fees, oftentimes with general funds augmented by indirect charges on research grants and contracts. Some public institutions are also partially funded by large endowments. Funding for a like-EPS program is necessary to provide infrastructure (faculty release from normal teaching duties, classroom facilities, and necessary administrative tasks). There are only a few opportunities for federal funding from such agencies as the U.S. Department of Education (DOEd) or the National Science Foundation (NSF). DOEd provides limited opportunities to augment novel approaches to engineering education, e.g., the Fund for the Improvement of Postsecondary Education, and these opportunities are very specific. NSF provides opportunities in research, including pedagogical research as mandated by their federal charter[4]. Funding for such an undertaking as a like-EPS program would require a ten-year commitment which is not likely in the yearly fiscal cycle. Industrial funding could be a possibility, but such coordination and commitment would be a tremendous undertaking and highly improbable.

Enrollment Competition

Competition for enrollment in academic programs, including engineering, poses yet another challenge. Public (and private) universities rely heavily on enrollment numbers for tuition and fee income which is a significant component of the general fund of a university. While an exchange program as that described would essentially be transparent, there is a perceived risk that students may not return to their home institutions resulting in loss of income. It is an inherent characteristic of competition to avoid any damage, perceived or otherwise. This difficulty could only be overcome when university presidents (or chancellors) embrace this effort and communicate their convictions to the administration and faculty.

The Conservative Nature of ABET

While one of the underlying themes of ABET 2000 was to inspire innovative techniques in engineering education, our accreditation body ABET inherently promotes traditional approaches to engineering education. The structure of the ABET process is driven by professional societies and industrial representatives. The accreditation process occurs after the academic program is in place, not before. Since accreditation is highly valued (rightly so). Program evaluators are selected from both industry and academia. Program evaluators may not view a novel and non-traditional program such as that described herein the best of light. Consequently, engineering colleges and universities are highly reluctant to venture into the unknown without adequate incentive. Rather, the conservative approach is more the norm.

At this conjuncture, it is relatively safe to say that unless there is a very strong incentive to formulate and put into place a reasonable like-EPS program in the United States, such a venture is unlikely to happen or evolve into the rich networked program offered by European Project Semester. While this relatively negative statement may seem as though this novel approach should be abandoned, nothing should be further from such an educational outcome. There are always possibilities of creative and acceptable approaches.

CONCLUSION

This paper briefly describes the formation of a novel and successful program now in-place and fully operational in Europe called the European Project Semester, or EPS. This program had meager beginnings at one small engineering program in Denmark and now encompasses several engineering colleges in European countries with an annual enrollment of over 600 students. EPS encourages student exchange from all European countries, Brazil, China, Japan, South Korea, and the United States. While there has been interest and participation by U.S. students, almost all have found their EPS experience extremely positive. There are several hurdles that U.S. students must accept and overcome; these hurdles are not insurmountable but may require an investment of both time and money. Nevertheless, energetic and venturous U.S. students have an opportunity with EPS that will enhance their career interactions on the global playing field of today's economy and enlighten their understanding of the many cultures and people that make up the international engineering community. While this EPS program could be adapted to the United States to bring together differences in regional cultures and creative problem solving, it is unlikely to occur under the present circumstances due to funding constraints, the individualist nature of the American physique, and the conservative nature of the accreditation process. Perhaps greater participation in the EPS program by students from the United States could be achieved by: (*i*) addressing the perceived rigor of senior design required by U.S. students; (*ii*) developing a more succinct funding path either through industry involvement or government; and (*iii*) acceptance of the widespread benefits of the international academic experience within the United States by academia, employers, and/or state and federal governments. Generalizations are always dangerous however it is perhaps fair to say that the overall view of international education by the general population of the United States tends toward inward looking rather than culturally broadening. This is most likely due to the size of the population (now exceeding 300 million), immense area of the country (9.8 million km^2), multi-cultural and diverse background of the population composed of people from almost every country in the world, attitudes of isolation because of physical location and incorrectly perceived self-sufficiency and of course the pollical nature of the human psyche.

REFERENCES

Abata, D. L., Andersen, A., & Krause, W. (2013). Transatlantic Interaction with European Project Semester. *Proceedings of the 2013 American Society for Engineering Education Annual Conference.* 10.18260/1-2--22649

ABET. (2019). *Criteria for Accrediting Engineering Programs 2020-2021.* https://www.abet.org/accreditation/accreditation-criteria/criteria-for-accrediting-engineering-programs-2020-2021/

ABET. (2022). *ABET.* https://www.abet.org/

Andersen, A. (2000a). International Teamwork and Communication in Engineering Education (Discussion paper). *Workshop on Internationalisation, SEFI 28th Annual Conference.*

Andersen, A. (2001). Implementation of Engineering Product Design Using International Student Teamwork—To Comply with Future Needs. *European Journal of Engineering Education, 26*(2), 179–186. doi:10.1080/03043790110034438

Andersen, A. (2003). Collaborating Learning in Interdisciplinary and Intercultural Teams. *Proceedings of the Third International Conference on Engineering and Computer Education*, 1-3. https://copec.eu/congresses/icece2003/proc/pdf/T131.pdf

Andersen, A. (2004a). Preparing Engineering Students to Work in a Global Environment to Co-operate, to Communicate and to Compete. *European Journal of Engineering Education, 29*(4), 549–558. doi:10.1080/03043790410001711243

Andersen, A. (2004b). Student Based Learning in A Multicultural Environment. *Proceedings of the 2004 American Society for Engineering Education Annual Conference.* 10.18260/1-2--13893

Andersen, A. (2009). Keynote on Project Management and Teamwork. In *Proceedings of the First Ibero-American Symposium of Project Approaches in Engineering Education* (pp. 10-10). University of Minho.

Andersen, A. (2012). The European Project Semester: A Useful Teaching Method in Engineering Education. In Project Approaches to Learning in Engineering Education. SensePublishers. doi:10.1007/978-94-6091-958-9_3

Andersen, A., & Hansen, J. (2002). Engineers of Tomorrow and Beyond: Knowledge, Insight and Skills Needed to Work Across Borders. *Proceedings of the 2002 American Society for Engineering Education Annual Conference*. 10.18260/1-2--10461

Andersen, A. J. (2000b). Assessment Techniques used in Multidisciplinary and Cross-Cultural Student Teamwork. *Proceedings of the 2000 American Society for Engineering Education Annual Conference*. 10.18260/1-2--8174

Armesto, L., Fuentes-Durá, P., & Perry, D. (2016). Low-cost Printable Robots in Education. *Journal of Intelligent & Robotic Systems*, *81*(1), 5–24. doi:10.100710846-015-0199-x

Beilin, H. (1992). Piaget's enduring contribution to developmental psychology. *Developmental Psychology*, *28*(2), 191–204. doi:10.1037/0012-1649.28.2.191

Budzinska, G., & Fuentes-Durá, P. (2014). Engineering Education for the Future. European Project Semester at Universitat Politècnica de València, Spain, and Lodz University of Technology, Poland. In *Proceedings of the Fourth Valencia Global Conference* (pp. 193-201). UPV.

Duarte, A. J., Malheiro, B., Arno, E., Perrat, I., Fuentes, P., Silva, M. F., Guedes, P., & Ferreira, P. (2019). Engineering Education for Sustainable Development: The European Project Semester Approach. *IEEE Transactions on Education*, *63*(2), 108–117. doi:10.1109/TE.2019.2926944

European Project Semester. (2022). *Concept | European Project Semester*. http://www.europeanprojectsemester.eu/concept

Hansen, J. (2010). European Project Semester – how Engineering Students Can Achieve Important Competences. In *Proceedings of the 5th International Conference on Engineering Education* (pp. 1296–1298). ICEE. https://www.ineer.org/Events/ICEE2010/papers/WS11/Paper_1107_1296.pdf

Hansen, J. (2012). Practical Elements in Danish Engineering Programmes Including the European Project Semester. *Industry and Higher Education*, *26*(4), 329–336. doi:10.5367/ihe.2012.0108

Hutton, M. (1980). Learning from action: a conceptual framework. In S. Warner Weil & M. McGill (Eds.), *Making Sense of Experiential Learning* (pp. 50–59). SRHE/Open University Press.

Itin, C. M. (1999). Reasserting the Philosophy of Experiential Education as a Vehicle for Change in the 21st Century. *Journal of Experiential Education*, *22*(2), 91–98. doi:10.1177/105382599902200206

Kolb, D. A., & Fry, R. (1975). Toward an applied theory of experiential learning. In C. Cooper (Ed.), *Theories of Group Process*. John Wiley.

Malheiro, B., Guedes, P., Silva, M. F., & Ferreira, P. (2019). Fostering Professional Competencies in Engineering Undergraduates with EPS@ISEP. *Education Sciences*, *9*(2), 119. doi:10.3390/educsci9020119

Ponsa, P., Roman, J., Arno, E., & Pérez, J. (2015). Professional skills in international multidisciplinary teams. *International Journal of Engineering Education*, *31*(4), 998–1006.

Rohaert, S., Baelus, C., & Lacko, D. (2012). Project work on wellbeing in multi-disciplinary student teams: a triple testimonial on EPS at Artesis. In *Proceedings of the International Conference on Engineering and Product Design Education* (pp. 1296–1298). The Design Society, Institution of Engineering Designers.

Segalàs, J. (2012). The EPS Experience at UPC-Barcelona Tech. In L. C. Campos, E. A. T. Dirani, A. L. Manrique, & N. Hattum-Janssen (Eds.), *Project Approaches to Learning in Engineering Education* (pp. 109–124). SensePublishers. doi:10.1007/978-94-6091-958-9_8

U.S. Census Bureau. (1995). *Celebrating Our Nation's Diversity*. CBET 95/01.

Verhulst, E., Rohaert, S., & Doorsselaer, K. V. (2015). How to incorporate Sustainable Design in the International European Project Semester programme: Insights from practice. In *Proceedings of the International Conference on Engineering and Product Design Education* (pp. 1-6). The Design Society, Institution of Engineering Designers.

Washington Accord. (2022). *International Engineering Alliance*. International Engineering Alliance. https://www.ieagreements.org/accords/washington/

KEY TERMS AND DEFINITIONS

ABET: The accreditation agency for engineering education in the United States.

Accreditation: Official approval of college and university programs, in order to maintain satisfactory standards.

European Project Semester: A one-semester international exchange program, based on experiential learning and international and interdisciplinary teamwork, aimed to address the design component of engineering education.

Experiential Learning: The process of acquiring information through the study of a subject without the necessity for direct textbook exposure.

Interdisciplinary: Related to teams, experiences, and projects where, beyond the standard boundaries, several disciplines contribute in a coordinated and coherent whole.

International and Interdisciplinary Teamwork: A joint work experience by teams of students from distinct scientific and cultural backgrounds.

International Student Exchange: Student mobility programs between higher education institutions from distinct countries that allow students to study abroad for a predefined period – a year or semester – with recognition of their studies.

One Semester Design Experience: A one-semester dedicated to the design of an engineering project intended for engineering undergraduates.

Senior Year: The final year of higher education studies before graduation.

Skills: Broad set of knowledge, abilities, work habits, and character traits that are believed to be important to success in today's world.

ENDNOTES

1 As of January 1, 2013, IHK has merged with DTU (Technical University of Denmark). The Engineering College is actually located in Ballerup, which is about a fifteen-minute train ride from Copenhagen city center.

2 The accreditation of programs mentioned occurs mainly in the United States but also internationally. As of October 2020, 4307 programs are accredited, distributed to over 846 institutions in 41 countries.

3 Engineering Deans Council, American Society for Engineering Education, a representative body of engineering deans throughout the country that meet regularly at the ASEE Annual Conference and semi-annually at the Engineering Deans Institute, another ASEE conference event.

4 Private communication with both the Office of International Science and Engineering and with the Division of Engineering Education and Centers at the National Science Foundation.

Section 2
Skills and Themes

Chapter 3
Teambuilding and Efficient Teamwork

Roger Nylund
Novia University of Applied Sciences, Finland

ABSTRACT

Teamwork is a preferred way of working in today's working life. Teambuilding helps both individuals and teams identify their strengths, weaknesses, and thereby, opportunities and challenges. This chapter presents some central findings on efficient teams and teambuilding and gives suggestions on a few useful tools to help build high-performing teams. Building trust within a team is paramount. Understanding one's own contribution and the value of others, different contributions are also important. A suggested way of setting up a teambuilding workshop for students is outlined. Furthermore, this chapter highlights the advantages of putting effort into teambuilding activities at the beginning of a project.

INTRODUCTION

Teamwork is encouraged and required in many workplaces today. This is partly due to the fact that the content of work has become too complex to be grasped by single individuals (Varney, 2021). Since working in teams has been found to be necessary, teambuilding activities have become a part of starting projects in all kinds of work environments and increasingly also in education. Now, since teamwork is so important for increasing organizational productivity, "everyone is jumping onto the bandwagon to ensure team success" (Wheelan et al., 2020). The globalization of work has also brought the need for cross-cultural understanding and abilities to work with people from different backgrounds, both professional and cultural. Teams will not become high performing by themselves. In a study, out of 700 teams within companies, only 17% were rated as high performing (Wheelan et al., 2020). Thus, there is clearly a need for education on teamwork in higher education preparing young professionals for a career in working life. The objectives of this chapter are to give insight into a small part of teambuilding and teamwork theory and to give examples of how to utilize these theories in higher education.

DOI: 10.4018/978-1-6684-2309-7.ch003

Copyright © 2022, IGI Global. Copying or distributing in print or electronic forms without written permission of IGI Global is prohibited.

This chapter is organized in four additional sections: a background section on teambuilding; a section dedicated to building efficient teams; a section focused on teambuilding in higher education; and the conclusion.

BACKGROUND

Teambuilding can be achieved in many ways. In some companies, it has traditionally meant going away to a remote place to eat, drink and relax – in Finland to enjoy a sauna – and just be together. This is a good start, but so much more can be achieved. Today in many organizations, human resource managers are encouraging, and often also facilitating, activities with the purpose of building a basis for efficient teamwork. What then are the main prerequisites for efficient teams? A few suggestions have been given. One traditional suggestion is to look at the individuals, find the best experts in their fields, and put together a team of top performers. This is rather common in sports. Sometimes this method is successful, but often less so. Another approach based on the individuals is to try to find out how well the personalities involved in a team seem to fit together. For mapping personalities, there are personality indicators such as Big5 (openpsychometrics.org, 2019), Myers-Briggs (The Myers-Briggs Company, 2021), and many more. Team role indicators, such as Belbin's team role indicator or team-role descriptions such as de Bono's thinking hats, are also widely used. In a large study undertaken by the Google corporation, they found out that these things do matter, but that there were other things even more important for creating top-performing teams (Duhigg, 2016). From the teambuilding point of view, there is one overarching concept that describes the goals with such activities, that is, *shared mental models*. Mental models are organized knowledge structures that allow people to predict and explain the behavior of people and the world around them (Mathieu et al., 2000). In multicultural teams, this is a challenge. The mental models differ at the start. The faster the mental models become known, understood, aligned and shared, the sooner a team can perform. The types of mental models that can be influenced in teambuilding are models of work, teams and teamwork. To put it simply, when team members learn to know each other's understanding of these issues, they become more predictable to each other. When a team puts effort into creating a team contract, they find out their differences in their mental models about working together in a team. Still, they can also start finding common grounds for team interaction. It helps them to identify potential challenges and strengths in the team.

BUILDING EFFICIENT TEAMS

In this chapter, a short overview of key findings on teamwork and teambuilding will be given. Furthermore, a suggestion for how to undertake a teambuilding workshop for students over two or three days, based on many years of experience from working with student-teams, will be given. This sub-chapter about theoretical findings is structured in the same way as the description of the suggested workshop that follows. It all starts with self-knowledge and self-awareness because, in order to work well with and understand others, one needs first to understand one's role and contribution. Then it continues towards teams and efficient teamwork.

In human resource management practices, such as recruitment and personnel development, different personality profiling tools are frequently used. This is a prevailing practice, even though personality psychology as such is a disputed field of study. All profiling tools, such as personality indicators or team-role indicators are only meaningful in organizational environments to the extent that they can help to *predict future behavior*. If a profiling tool lacks predictive power, the outcome can only temporarily satisfy our endless curiosity about who we are and how we are perceived by others. The Big5 personality indicator and the Belbin team-role indicator have both gained popularity due to their perceived predictive power (Aritzeta et al., 2007). This is also why these two are chosen to be presented here, rather than some of the many other tools available. This does still not constitute a statement about the validity or usefulness of other tools. Surely there is an abundance of useful tools to choose from. However, in the author's opinion, one central criterion for choosing assessment tools is that they are based on solid scientific findings. One such tool is the Belbin test already mentioned.

Belbin Team-Roles

Belbin's team role theory can be used in teambuilding activities. Belbin identified nine team roles, out of which an individual tends to be drawn to two or three roles. The idea is that each person has two or three preferred roles, one of which will be adopted when working in a team. The role one adopts is based on the person's own personality, preferences and situational factors, such as what other roles are present in the team or how comfortable the person feels. Belbin's team-role inventory is not a psychometric test measuring personality, but rather it measures individual *behavior* when working in a team (Belbin.com, 2014). When used by professionals, the process of identifying one's team role consists of self-evaluation and an evaluation by an observer who knows the evaluated person. Quite often, though, only self-evaluation is used (Aritzeta et al., 2007). Belbin identifies nine team roles that an individual tends to take when working in a team. These roles are described briefly below.

The *Plant* is the creative innovator who comes up with new ideas and approaches. They thrive on praise, but criticism is especially hard for them to deal with. *Monitor-Evaluators* are best at analyzing and evaluating ideas that other people (often Plants) come up with. These people are shrewd and objective, and they carefully weigh the pros and cons of all the options before coming to a decision. Monitor-Evaluators are critical thinkers and very strategic in their approach. *Specialists* are people who have specialized knowledge that is needed to get the job done. They pride themselves on their skills and abilities, and they work to maintain their professional status. *Coordinators* are the ones who take on the traditional team leader role and have also been referred to as the chairmen. They are often excellent listeners. *Resource Investigators* are innovative and curious. They explore available options, develop contacts, and negotiate for resources on behalf of the team. *Team Workers* are the people who provide support and make sure that people within the team are working together effectively. These people fill the role of negotiators within the team and they are flexible, diplomatic and perceptive. *Completer-Finishers* are the people who see that projects are completed thoroughly. They ensure that there have been no errors or omissions and pay attention to the smallest details. *Implementers* are the people who get things done. They turn the team's ideas and concepts into practical actions and plans. *Shapers* are people who challenge the team to improve. They are dynamic and usually extroverted people who enjoy stimulating others, questioning norms, and finding the best approaches for solving problems. The first three roles presented above are thought-oriented roles, the following three are people-oriented and the last three are action-oriented (Belbin Associates, 2014). Getting the students to think about teamwork within this framework has at least three potential advantages.

Firstly, it gives an awareness that there *are* different roles to take when working in a team. Without such reflections, engineering students tend to be *very task-oriented* and *less process-oriented,* and even less oriented towards the *social aspects* and roles. This can lead to a performance far below the potential of the team. Especially when working with undefined problems or challenges, social interaction within the team becomes important. Moreover, this is the type of problems engineers will face in their professional life. Belbin shows the importance of identifying the team member's team roles and finding a balance between these roles within a team. The Belbin team-role inventory provides the means to do so.

Secondly, a team-role test, Belbin or some other, helps the students to identify their *own* strengths and contribution to the team. It can be of importance, especially for those with less visible roles, to see that their role actually matters and is needed in building a well-functioning team. The team-role descriptions also highlight potential drawbacks in the different team roles.

Thirdly, in the best case, it also helps team members to *understand other* team members' *different* contributions and their value for the successful completion of the project. This can be valuable for those with more dominant roles, helping them to realize that they actually need those others with less visible roles to successfully complete their task or project. However, individuals are not static entities with given features that stay completely the same over time and in different situations. A team situation brings forth plenty of dynamic developments. This dynamic nature of teams has been studied ever since the Hawthorne studies in the 1930s, and our understanding of the phenomena at play still develops (Buchanan & Huczynski, 2019).

The Dynamic Nature of Teams

Studies that focus on the dynamic nature of a team are looking for *similarities in behavior in different stages* of a team. They are based in the fields of group psychology, social psychology, and sociology. Tuckman's famous list of stages, forming, storming, norming, performing and adjourning, is one of the most well-known models in group psychology (Bonebright, 2009). Even though it has been used far beyond its limitations[1], and there are grounds for questioning the linear progress of team dynamics, there is one feature that, in the author's opinion, seems relevant in any team process. That is the *storming* face. This is when there is a struggle for leadership, disputes about ways of working and values to uphold in the team. This stage is needed for building *trust.* Many studies show the significance of trust as a basis for efficient teamwork (Duhigg, 2016; Brahm & Kunze, 2012; Sheng et al., 2010). However, it is impossible to trust a facade. In the initial stage of a team, team members tend to politely assess the situation and not reveal very much of themselves (Tuckman 1965). They keep up a facade. Thus, the storming face is needed, simply for the participants to *become visible* to each other. When something is important enough for an individual to fight for, it reveals the *core values* of that individual to the team. In the storming face, in conflicts over priorities and goals, team-member personalities, values, and goals become visible to the others. That gives a basis for building trust within a team. One can only trust somebody who is known to an extent that makes the other's future behavior at least somewhat predictable. Getting to know each other's values and priorities does not necessarily mean streamlining of values and goals, on the contrary. Still, by bringing them "on the table", they can be handled. This should be relevant regardless of whether the team process turns out to be linear or non-linear.

The Importance of Building Trust within a Team

Building trust includes a process of getting familiar with the values involved. Trust is easily established between persons with similar sets of values. The concept of values is discussed more in the Ethics chapter of this book, but the point to be made here is that they play a very *practical* role in our lives, guiding our actions (otherwise they are not values but wishful thinking). However, values are often to be found on a sub-conscious level from where they need to be brought forth and reflected upon. One way of "finding" the values is the struggle in the above-mentioned storming face. Another, more straightforward way to find them is simply to use a value-inventory tool as described in the next sub-chapter. Apart from having common values, the perceived team support is another thing that has been shown to build up trust and team commitment (Sheng et al., 2010). Therefore, teambuilding activities should enhance a supportive mindset within the team rather than a competitive mindset. The competitive mindset could rather be channeled to compete with other teams or organizations. An EPS project is well suited for becoming an eye-opener for the students regarding the value of helping each other. They come from different universities. They do not compete but depend on each other in the strive towards a successful project result. Supporting each other means letting go of the Malthusian trap, which is the mindset that one's gain must be someone else's loss. In a team, helping each other creates win-win situations. Fostering a "giver" mindset makes the team as a whole more successful. There is massive evidence from organizational research in different fields showing that helping behavior makes teams and organizations successful. Helping behavior depends heavily on norms, but norms are hard to detect even if they are the most powerful forms of social control over human action (Grant & Patil, 2012, p. 547). Every small step towards understanding the benefits from helping each other and also creating group norms based on a supportive mindset is of great value for the project, but more so for the student's future success in working life. In conclusion, self-awareness, understanding of the team roles, and positive group norms are important building blocks for a successful team. This fits quite well into the empirical findings from the extensive study undertaken within the Google corporation (Duhigg, 2016).

The Aristotle Study

In the Aristotle study, made within Google corporation (Duhigg, 2016), they found a few common features in their top-performing teams. The most important was not to have the most brilliant individuals in a team. It didn't seem to be the perfect match of team roles either. What turned out to be important was the group norms. What the top-performing teams at Google had in common was, first and foremost, a sense of *psychological safety*. Each member of a team felt safe to speak their minds without getting ridiculed. The atmosphere was supportive and it was safe to fail in trying something new. This corroborates much of Adam Grant's studies about helping behavior at work (Grant, 2014).

The second thing the top-performing teams had in common was *dependability*. They knew they could depend on each other to do their part of the job and deliver on time. In a way, this shows that personal features and personality do matter. The third feature in the top-performing teams was *structure and clarity*. The goals and roles of each member of the team should be crystal clear to every team member. They should also have knowledge on how to do their part of the job. This is obviously not very often a given in teamwork. The team has to work it out by themselves, together. The fourth feature of top-performing teams was that they found their jobs *meaningful*. Motivational theory (Ryan & Deci, 2000) and resilience studies (Staab, 2017) both identify a sense of meaningfulness as being of central importance for motivation and resilience. Later studies also connect a sense of meaningfulness to occupational health and well-being (Nylund, 2019). Then again, it should be quite obvious; who would like to work with something they find meaningless. Still, many people struggle to find meaningfulness in what they are doing at work and in their studies. A team coach can be helpful in this process. The fifth thing that top-performing teams had in common at Google was a sense of making a difference, i.e., that their work had an *impact*. These two, a sense of meaningfulness and impact, are clearly and closely connected. If the work has an impact, it is probably also perceived as meaningful.

The Importance of Good Communication

Although a short chapter like this, by necessity, will be lacking many important things, one aspect of good teamwork cannot be bypassed. That is communication. To build trust, to get to know each other, to find out about each other's values and team roles; all of this requires open communication. When working in multicultural and multidisciplinary teams, as is the case within EPS, the communication challenges need to be addressed. We misunderstand each other in so many ways because of cultural differences (Hofstede et al., 2010; Trompenaars & Hampden-Turner, 2020). However, in the EPS context, there is a mandatory course in English and Cross-Cultural Communication, and in this book there are two separate chapters concerning that. Thus, the importance is just mentioned here as a reminder that it can and should also be addressed in teambuilding sessions.

Before moving to the practical part, one more comment is needed. The above referred scientific evidence is biased. The short-term advantages of leading by fear or other power-oriented ways of working together are not advocated at all. That is a deliberate choice of the author, although these practices are very common. In the author's understanding, young students rather need solid scientific evidence about the usefulness and advantages of cooperation, helping each other, listening respectfully to each other and creating a safe work environment. That is beneficial for the individuals who do it and for the teams where they work (Grant, 2014). In the teambuilding workshop, which is always part of the European Project Semester, these findings can be shared and illustrated through different activities and group tasks.

TEAMBUILDING FOR STUDENTS IN HIGHER EDUCATION

Teambuilding can be used to speed up the processes of getting to know each other, bonding and finding an understanding of team dynamics in the beginning of a study program, module or project. Teambuilding within EPS is often realized as a workshop over two or three days in the beginning of the project semester. In the following, a loose structure and suggested content for such a course are outlined, together with some thoughts on possible connections to learning goals. All of this is presented based on the above-mentioned theories. Please note that this chapter is not a handbook, it is rather meant as food for thought for potential teambuilding coaches. It builds on ideas shared with the author by very generous people like Pedro Fuentes and David Swetnam (personal communication) and on more than a decade of experience from trying out different approaches for teambuilding with many groups of students.

The idea is to start from self-knowledge and self-awareness, then go further to awareness of the others, the task, and opportunities, and finally, to get an idea of what constitutes a high performing team. Figure 1 illustrates the suggested structuring of the teambuilding sessions.

Figure 1. Illustration of the suggested way of structuring teambuilding sessions
(Author's own figure)

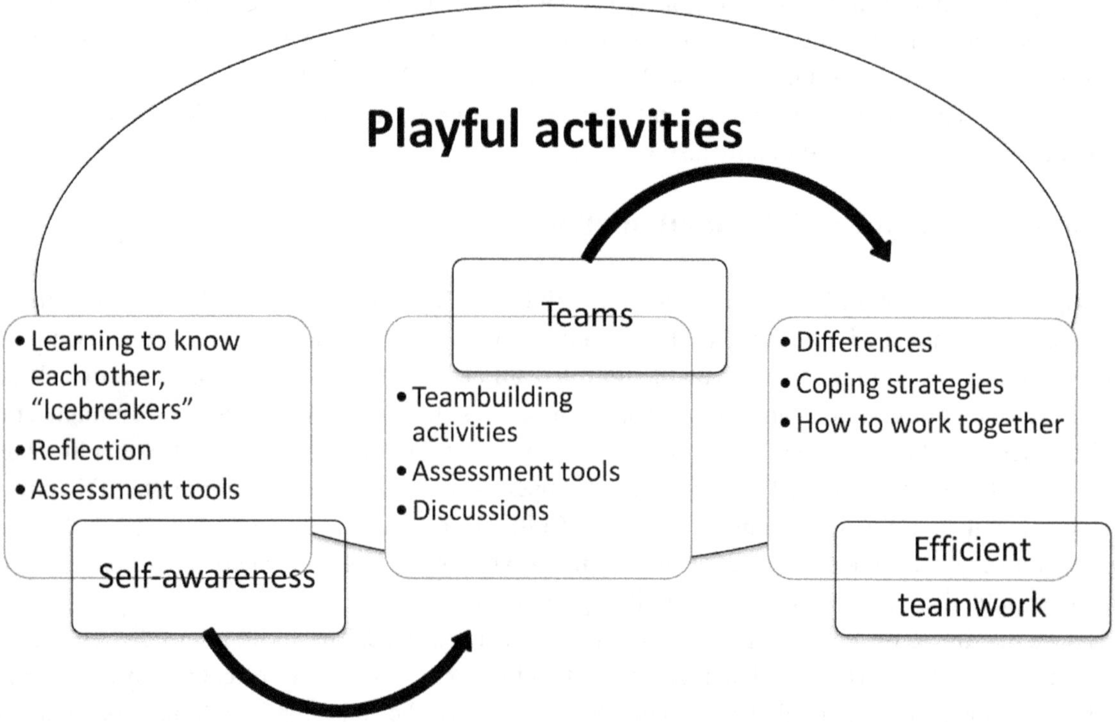

Icebreakers

Self-awareness and efficient teamwork are strongly intertwined. Since the teambuilding workshops are held in the beginning of an EPS project semester, bringing together students from different fields and different countries, some *icebreakers* are needed at the start. It is important to create a safe and friendly environment and to get the students to relax from the start. Various icebreaking activities are easy to find online or in the literature (e.g., Peragine & Hudgins, 2017). This kind of activities can be the content of the first morning sessions. In the author's experience, it is good to mix highly activating tasks with more reflective ones. This way both the "doers" and the "thinkers" get a feeling of inclusion and the students avoid getting exhausted or bored. This is the balance for the instructor to keep track on throughout the workshop. Teambuilding is not just another set of lectures, it should engage and challenge the students both socially, physically and intellectually. It is surprising how different groups of twenty to thirty students can be. One could expect such a mixture of nationalities, gender and cultures would somehow even out much of the differences between groups as a whole. Still, this is not the author's experience. In each group, there are a few individuals who tend to define the behavior of the group as a whole. Thus, it is a good idea to choose the icebreaker activities in a manner that will also help the instructor to get to know the group in terms of activity-level, dominating individuals, more withdrawn students, etc. Oftentimes, the challenge for the instructor is to limit how much space the most outspoken students are given to give the shy ones a fair chance to be seen.

Starting Up the Sessions

A starting game where the students are asked to line up on a line (if the floor has no lines to use, it is easy to make the line with tape on the floor) can give the coach an idea of the stage of group formation the team is in. When the students have lined up, they are asked to sort themselves in height order without stepping off the line. After that, in order of date of birth. For both of these, there are random things to discuss, like the fact that persons born at the beginning of the year get so many more chances to develop their talents than those born at the end of the year. At the age in childhood when "talents" in all fields are found, the advantage of being almost one year older than your competitors is huge. That is probably one of the reasons why for example a majority of the ice-hockey players in the National Hockey League are born at the beginning of the year. This very unfair random thing that we cannot affect can be used as an example of the necessity to be open for other options and not lock oneself to one single dream in life. Neither should one blame oneself for factors outside of one's control. Studies show that young people get a lot of stress and anxiety from being told they can become anything if they just try hard enough. In some cases, this is simply not true. One of the most successful women in business today, Sheryl Sandberg, says: "Life is never perfect. We all live some form of Option B" (Sandberg & Grant, 2017). Life is not ruined if you cannot become a pilot because you wear glasses. There are other goals to pursue. This discussion can also be used for highlighting the importance of focusing on things in your influence instead of focusing on things that are important for you but outside your influence (Covey, 2020). To connect most of the activities or games to some findings about working well as a team has shown to be a better way to get the messages to stick, than traditional lecturing (Heath & Heath, 2010). Of course, every coach has to choose his or her own activities and what – if anything – to highlight with each activity. Some activities should probably be selected just because they are fun and activating. Every task or activity that makes the students interact can fulfill teambuilding goals. In this game, one can quickly see how relaxed and comfortable the students are with each other in the way they make their way past each other on the line. Some give a hug; others try to turn the back and not touch the other student. The coach should pay attention to whether some of the students feel uncomfortable with physical contact. As a group, students tend to get more comfortable with physical contact in a while. If many students try to avoid contact in this exercise, the coach might conclude that more icebreakers of different kinds are needed. The program for teambuilding workshops must be adjusted to the dynamics in each group. For groups that already know each other well, there is much less need for icebreakers and the activities can be more focused on aspects of working well together.

Multitasking Session

The first-afternoon session could consist of one bigger group task. This very useful idea was adopted from following the teambuilding sessions held by Pedro Fuentes in UPV, Valencia, many years ago. The students are given 8 to 10 tasks to finish in, for example, 90 minutes. These tasks should be mixed, preferably to highlight some of the Belbin team roles or at least the differences between analytical "head persons", socially-oriented "heart persons" and action-oriented "hands persons" (Gazibara, 2013). The given time, if 90 minutes, is long enough to provide the students with the initial impression that they have all the time in the world. The students work as project teams and afterward there can be much to reflect on. Below is a suggested list of questions to raise:

- Did you feel any urgency at the beginning of the 90 minutes? Usually, not at all.
- What about in the last ten to five minutes? Usually hectic activity, almost panic. Time management can be discussed.
- Did you start by planning and distributing tasks? 90 minutes/10 = 9 minutes per task, but if the work is divided, for example, into two sub-groups, the team will have 18 minutes per task. Usually, there was no planning, which gives the opportunity to discuss the importance of planning the work and to have some tactics for the execution.
- Did anyone take the lead in your group? Leadership roles can be a great asset in a team, but also a burden. It can sometimes be good to raise the question of whether the one who spontaneously takes the lead also is the most suitable leader. There are probably no definite answers to be given here, but the question is intriguing. At least, it is not a given that the person spontaneously taking the lead is the most suitable for that.
- Did you listen to each other's ideas? Listening is the most important – and least to be found – communication skill. There is much to discuss about the advantages of being a good listener.
- Were there any conflicts? Why/why not? Usually not, but this can give a chance to discuss Tuckman's team dynamics and the "storming" that may be ahead for the team.
- Who took on the analytical tasks, and who jumped onto the physical activities? This indicates what direction one's team role is, social, analytical, or action.

Usually, the students have prior experience from teamwork, so it can be fruitful to ask them to share these experiences, what they would like to bring to this team, what they hope to avoid. A team observation questionnaire can be handed out to help the students in reflecting on different aspects of their cooperation.

Belbin Team Roles

In order to get the students to reflect on their own team roles and behavior, Belbin team-role inventory can be used. A brief introduction of Belbin team-role theory and discussion about roles present and absent in each team can help students understand possible challenges they face in their teams, at the time being or in the near future. There are indicative free versions of the Belbin-test to be found on the Internet. Better self-awareness can emerge by identifying one's preferred roles, contributions to the team, and allowable weaknesses. Analyzing the combination of roles within each team is also recommended. In such a discussion, the members can learn to predict some of the strengths of each team, but also get indications on challenges likely to occur. Things to look for can be missing roles and overlapping roles. Overlapping roles is usually a smaller problem than missing roles. As a team coach, one learns to use this tool better over time by following the dynamics and progress in teams where potential challenges were identified in the test. It might be recommendable not to tell the students too much about the challenges indicated by the Belbin test, in order to avoid self-fulfilling prophecies.

Personality Profiling

If there is time, personality indicators, such as Big5 (openpsychometrics.org, 2019), can also be used for the students to get food for thoughts about themselves and their contribution to the team. The introduction of such tools can also be motivated by the fact that the students are likely to be profiled in recruitment processes soon. A word of warning may be appropriate here. Personality indicators, used by amateurs, can be dangerous. Therefore, coaches who intend to use one of these should, at least, do two things. Study basics in personality psychology and about the features of the tests they intend to use, and use them with caution, telling the students that this is only an *indication* of what features they might have. It is also good for the coach to keep in mind that perceived meaningfulness is important for most people (Sinek, 2011). This also applies to different tasks given. Sometimes it is good to have the students do things first and then afterward help them to reflect on the meaning of the exercise. Sometimes it is better to explain the idea before the exercise. If a personality indicator is used, it is recommendable to give an introduction of the basis and background of such a tool beforehand. If personalities are explored, the question about situational factors would also be necessary to discuss. Do we have a fixed personal identity, or do we rather act according to the situation? For many EPS students, the exchange semester is their first extended period of time away from their own culture. This can be a confusing experience, raising questions about one's own identity and culture. The situational factors can be strongly at work in such a situation.

Temperaments in Teams

Another recommended activity in the range of self-awareness is to study different temperaments in pairs or groups for ten to fifteen minutes and then present them in five minutes per pair to the rest of the students focusing on three things:

- Short description of the temperament; some are more or less self-evident, others are not.
- Advantages of having a person with this temperament in a team.
- Disadvantages of having a person with this temperament in a team.

After each presentation, there can be a discussion about experiences and reflections on the temperament presented. For each temperament, the students can then be asked to "rate themselves on a scale from 1 to 10 in order to identify and become aware of their own dispositions. The temperaments to discuss can be chosen by the coach, but the following list has shown to be useful: Activity, sensitivity, adaptability, distractibility, steadfastness, rhythm, accessibility and mood (author's translation from Toivakka & Maasola, 2011). For example, the disposition "sensitivity" can be a possibility to make the students aware that there can be highly sensitive persons in their team. To be a highly sensitive person can be enriching, but team members need to understand that they sense things so much stronger and also different from the majority. Other temperaments can be starting points for other kinds of discussions since they all have a bearing on working in teams.

Values and Norming

As a bridge to creating efficient teams, a value-inventory can be used, for example, the "Life Values Inventory" (Brown & Crace, 1996). The students grade 42 values from "almost never guides my behavior" to "almost always guides my behavior" on a scale from 1 to 5. Then the scores are grouped, identifying values for achievement, belonging, concern for the environment, concern for others, creativity, financial prosperity, health and activity, humility, independence, loyalty, privacy, responsibility, scientific understanding, and spirituality. The scoring gives an indication of the person's value hierarchy. This is always a difficult task, but it helps the students to, at least preliminarily, sort out their values. Obviously, one has to put them in a context to get realistic results, but this exercise still helps to navigate and to make values visible. The next step is to compare the individual values within a team and to agree, as a team, on three to four core values. Finally, the group should define norms, protecting their core values. Often the students suggest norms such as:

- Helping and supporting each other.
- To honestly tell when they have problems with their part of the task and to ask for help.
- Listening to each other's opinions and respecting differences.
- Delivering in time and coming to meetings in time.

Students tend spontaneously to identify important things that researchers have spent years to find out. In a very limited way, they are corroborating the findings about efficient teamwork mentioned in the previous part of this chapter. This value-exercise can be extended to a task of making *team contracts* or a *team-canvas* including also very practical agreements such as:

- Where and when to work.
- Communication within the team and with other interested parties (team coach, customer, etc.).
- Rules for the use of electronic devices when working together.
- Rules, with consequences, for coming late or being absent.

In Tuckman's terms, this is to kick-start the norming stage. Norming is relatively easy as long as it is just paperwork. The difficult part is to live up to the norms agreed upon. In the author's experience, this is where the storming often begins. Thus, in Tuckman's model, the sequential order of the stages a team goes through might not be the most important takeaway, but the realization that these typical stages tend to occur – in some order. Often the team is going back and forth between the stages. Tuckman's model, forming, storming, norming, and performing is also good to discuss. It can be a good idea to highlight that, in the storming phase, *conflicts* can be necessary to build trust within the team. Suppose the coach for the teambuilding sessions is not part of the local team of EPS coaches. In that case, it is good to get the local team coaches on board to follow the continuance of the norming process because it is definitely not anything that can be completed in one afternoon.

Suppose the groups come up with norms about respecting and listening to each other. In that case, that can give an excellent opportunity to run some games highlighting the importance of careful listening – or to have a discussion on what it means, how to create an atmosphere where this will be realized, etc. Listening has many aspects. In a respectful team, one listens to understand rather than to find ways to win a debate. Of course, when the team sorts out and evaluates different solutions for their project, they need to be critical of the suggestions on the table, but also this can be done respectfully. It is good to remind the students to make constructive use of our human features to be very good at finding weaknesses in other people's arguments and very poor in finding weaknesses in one's own reasoning (Mercier & Sperber, 2018). Another aspect of listening in multicultural teams is to be aware of the constant risk for misunderstandings based on differing cultures and communication behavior. There are many funny games to play about how easily misunderstandings or misinterpretations can occur.

Other Activities

One aspect of working together efficiently is to know when to divide the work and when to co-operate as a team. A very straightforward illustration of this is a game called "Lost at Sea" or alternatively "Crash-landing on the Moon". Both can be found on the Internet and consist of a description of the situation and a list of 15 items for the students to rank in order of importance, first individually, then as a team. Afterward, the ranking is compared to expert opinions. The takeaways from this exercise are among others:

- Highlighting that teamwork gives better results than individual reasoning for analytical tasks about unfamiliar issues. Usually, only a few students have better individual scores than group scores.
- The importance of strategic thinking, should one choose items helping the team to stay alive as long as possible or with the first priority to be found (lost at sea) or based on some other priorities.
- The importance of doing things systematically. It can, for example, be a good idea to group the items into five most important and five least important, leaving five items in the middle category. Then it is easier to rank the items within the three groups. Indeed, there are other systematic approaches to apply.
- Finding out something about the team's decision-making procedures. Some teams simply cannot agree on how to proceed while others apply democracy, meritocracy or simply follow the loudest.
- It gives an opportunity to discuss deadlines and the pain connected with decision-making. The games are about surviving in an urgent situation. Sometimes, decisions have to be taken quickly based on incomplete information and knowledge. This is very uncomfortable, even painful, for the analytical, cautious person.

Another nice and energizing exercise is to build with Lego. Every team is given an identical set of about 20 to 30 pieces, both in size and color. The instructor hides a complicated model built from one of these sets of pieces in one end of the room. The teams are then to build an exact copy of the model. Only one person is allowed to go and see the model and then come back to the table to instruct the team on how to build. They are allowed an unlimited number of runs to check the model. The teams can be given the freedom to change messengers from time to time, but only one can be away from the table at the time. The first team to build an exact copy wins. Suggested takeaways from this activity:

- The importance of Work Breakdown Structure in project management, students tend to try to bring far too much information at the time. They neglect the fact that each piece contains three pieces of information, size, color and position. The realistic strategy would then be to bring the information of two or three pieces at the time since our memory capacity for random information is around seven pieces of information. The team should break down the task into small enough subtasks in order to succeed.
- Change management; when a team realizes a mistake, they usually, given the competitive situation, mess up completely because they don't have a strategy for how to handle the changes.
- Once again, work division or working together? The best strategy here is probably actually to have only one builder. The same goes for the messenger. Usually, teams relying on only one messenger are more successful than teams sending different members to check the model.
- The importance of clear communication and tactics. The messenger must decide on whether to build layer by layer or one side at a time.

CONCLUSION AND DISCUSSION

The activities suggested above are only examples. Every teacher should find activities that resonate with his/her own personality, goals and ambitions. The big challenge for a coach in teambuilding – if the ambition is to raise awareness of how to build efficient teams – is to mix activities with information and theoretical reflections in a way that engages the students. In the author's experience, each group of students is surprisingly different. This means that the coach needs to be prepared with a bigger repertoire of activities than he or she expects to need in the time given. That gives the possibility to improvise based on the character of the group. The whole event should preferably exemplify the good atmosphere, inclusion, enthusiasm and engagement expected in teamwork. That is why teambuilding is not just playing around for the coach but a constant observation and reflection on how the students act and work together. The best results tend to come from situations where the coach can catch some idea from the group and work further from there. Co-creating the workshop is, of course, more challenging for the instructor but often more rewarding for the students since it gives a sense of inclusion. In this way, and on a personal level, teambuilding is more demanding than lecturing on any given topic. To engage the group, the coach needs to be engaged as well. There are several potential benefits of successful teambuilding. It can help the group to bond, build trust and perform. Thus, it is well worth the time and effort when done with a "serious fun" mindset.

REFERENCES

Aritzeta, A., Swailes, S., & Senior, B. (2007). Belbin's Team Role Model: Development, Validity and Applications for Team Building. *Journal of Management Studies, 44*(1).

Belbin, R. M. (2010). *Team roles at Work*. Elsevier Science.

Belbin Associates. (2014). *Method, Reliability & Validity, Statistics and Research, A Comprehensive Review of Belbin Team Roles*. https://www.belbin.com

Bonebright, D. A. (2010). 40 years of storming: A historical review of Tuckman's model of small group development. *Human Resource Development International, 13*(1), 111–120. doi:10.1080/13678861003589099

Brahm, T., & Kunze, F. (2012). The role of trust climate in virtual teams. *Journal of Managerial Psychology, 27*(6), 595–614. doi:10.1108/02683941211252446

Covey, S. R. (2020). *7 Habits of Highly Effective People: Revised and Updated*. Simon Schuster.

Duhigg, C. (2016). *What Google Learned from its Quest to Build the Perfect Team*. https://www.nytimes.com/2016/02/28/

Gazibara, S. (2013). Head, Heart and Hands Learning – A challenge for contemporary education. *Journal of Education Culture and Society, 2013*(1), 71–82. doi:10.15503/jecs20131.71.82

Grant, A. (2014). *Give and Take, Why Helping Others Drives our Success*. Penguin Books.

Huczynski, A. A., & Buchanan, D. A. (2019). *Organizational Behaviour*. Pearsons.

Heath, C., & Heath, D. (2010). *Made to Stick, Why Some Ideas Survive and Others Die*. Random House Publishing.

Hofstede, G., Hofstede, G. J., & Minkov, M. (2010). *Cultures and Organizations: Software of the Mind* (3rd ed.). McGraw-Hill Professional.

Mathieau, J. E., Heffner, T. S., Goodwin, G. F., Salas, E., & Cannon-Bowers, J. A. (2000). The Influence of Shared Mental Models on Team Process and Performance. *The Journal of Applied Psychology*, *85*(2), 273–283. doi:10.1037/0021-9010.85.2.273 PMID:10783543

Nylund, R. (2019). *Unga ingenjörers psykologiska kontrakt inom teknologi-industrin*. Acta Wasaensia 433, University of Vaasa.

openpsychometrics.org. (2019). *Big Five Personality Test*. https://openpsychometrics.org/tests/IPIP-BFFM/

Peragine, J., & Hudgins, G. (2017). *365 Low or No Cost Workplace Teambuilding Activities*. Atlantic Publishing Group.

Ryan, R. M., & Deci, E. L. (2000). Self-Determination Theory and the Facilitation of Intrinsic Motivation, Social Development, and Well-Being. *The American Psychologist*, *55*(1), 68–78. doi:10.1037/0003-066X.55.1.68 PMID:11392867

Sandberg, S., & Grant, A. (2017). *Option B: Facing Adversity, Building Resilience, and Finding Joy*. Barnes & Noble.

Sheng, C.-W., Tian, Y.-F., & Chen, M.-C. (2010). Relationships Among Teamwork Behavior, Trust, Perceived Team Support, and Team Commitment. *Social Behavior and Personality*, *38*(10), 1297–1306. doi:10.2224bp.2010.38.10.1297

Sinek, S. (2011). *Start with Why*. Penguin Books.

Sperber, D., & Mercier, H. (2018). *The Enigma of Reason*. Penguin Books.

Staab, H. (2017). *Das Salutogenesekonzept nach Aaron Antonovsky*. GRIN Verlag.

The Myers-Briggs Company. (2021), *Myers-Briggs Type Indicator® (MBTI®) | Official Myers Briggs Personality Test*. https://www.themyersbriggs.com/en-US/Products-and-Services/Myers-Briggs

Toivakka, S., & Maasola, M. (2011). *Itsetunto kohdalleen, harjoituksia itsetuntemuksen ja vuorovaikutustaitojen oppimiseen*. PS-Kustannus.

Trompenaars, F., & Hampden-Turner, C. (2020). *Riding on the Waves of Culture: Understanding Diversity in Global Business* (4th ed.). McGraw-Hill Professional.

Tuckman, B. W. (1965). Developmental sequence in small groups. *Psychological Bulletin*, *63*(6), 384–399. doi:10.1037/h0022100 PMID:14314073

Varney, S. (2021). *Leadership in Complexity and Change*. Walter de Gruyter GmbH. doi:10.1515/9783110713343

Wheelan, S. A., Åkerlund, M., & Jacobsson, C. (2020). *Creating Effective Teams – Guide for Members and Leaders* (5th ed.). Sage.

KEY TERMS AND DEFINITIONS

High Performing Team: A group of goal-focused individuals with specialized expertise and complementary skills who collaborate, innovate, and produce consistently superior results.

Mental Models: Organized knowledge structures that allow people to predict and explain the behavior of people and the world around them.

Personality: The unique combination of qualities in a person that makes that person different, as shown, systematically, by the way the person feels, thinks, and behaves.

Team: A group of people who act together to achieve a shared goal.

Team Role Theory: A theory about the different roles played by members of teams that suggests that individuals with a particular personality are most comfortable in a specific role.

Teambuilding: The process of encouraging group members to work well together and the activities that help a group work together more effectively as a team. It is commonly used as a two-word noun, i.e., team building, and as a hyphenated word when referring to team-building activities, exercises, and skills. The authors found it more appropriate to use a single word when teambuilding is the focus.

Teamwork: The activity of working together in a group with other people, especially when this is successful.

ENDNOTE

1 Tuckman made his studies among therapy groups. Still, the stages he identified have later shown to be relevant also in other kind of group dynamics.

Chapter 4
Project Planning, Management, Monitoring, and More

Pedro Fuentes-Durá
Universitat Politècnica de València, Spain

Philippe Fillatreau
École Nationale d'Ingénieurs de Tarbes, Université Fédérale de Toulouse, France

David William Charles Ashworth
Technical University of Denmark, Denmark

ABSTRACT

Project management is the structuring of a project in terms of visible and manageable activities, which are the building blocks for executing the project across its lifecycle. In particular, one of the critical phases of the lifecycle of a project is project planning. Project planning must include the main strategy for a working method within the project team. It should structure the tasks and describe the tools for monitoring the execution and controlling the project's progress. On the one hand, this chapter describes the essential steps of a project. On the other hand, it presents the most used perspectives and methods such as PMBOK, PRINCE, Agile, Scrum, and supporting technologies such as Asana, BaseCamp, Trello, Slack, or Jira.

INTRODUCTION

The purpose of this chapter is to provide a general overview of project management (PM) with some reference to how this is applied in an EPS course. EPS students participate in PM courses and practical sessions and are guided through supportive supervision by senior staff to train PM skills along with the program. However, PM is such a vast field that here we can only provide an outline of what it is and a set of highly appropriate references and resources. The introduction of the chapter presents some terminology and essential concepts. The following section briefly describes the basic elements of a project, and the subsequent expands on these descriptions and elements. After that, a section deals with the all-important topic of planning and is followed by a section describing the importance of proper closure. The last two

DOI: 10.4018/978-1-6684-2309-7.ch004

Copyright © 2022, IGI Global. Copying or distributing in print or electronic forms without written permission of IGI Global is prohibited.

sections, respectively, describe some of the most common PM methodologies and available supporting software. The chapter ends with a short conclusion.

Types of Projects

A project is a piece of planned work or an activity or a set of tasks that are finished over a period of time and intended to achieve a particular, specific, unique purpose (PMI, 2017). Effective project management is essential to the profitability of any company. The values from a solid PM relate to efficiency, legitimacy, power and control, and stakeholder satisfaction (Eskerod & Riis, 2009). Obviously, there are some necessary preconditions (both technical and human factors) in order to harvest the values.

In our context, the term "Project" will imply working together as a team towards a common objective, using the skills of individual team members. We can see a project as (Maylor, 2010):

1. Task resolution, using a specific working method with specific "rules-of-the-game".
2. An organizational working method for the realization of ideas or for solving a complex problem.
3. An organized, provisional relationship of collaboration for solving a task or creating an innovative solution.

In the EPS framework, a project is a set of tasks which, through controlling and applying assigned resources, must fulfill a specific objective within a specific time limit and by a specific international team of students with specific skills. Often, an industrial partner acts as a project client and cooperates with the development. Below are a few examples of some of the EPS projects which have been carried out over the years and which are described in this book:

- The Individual Survival Unit Transporting patients in a protective and stable environment in disaster scenarios.
- A Fast versus Slow Fashion Fair on the global impact of local wardrobe choices.
- Development of a Robotic Arm with IoT Capabilities.
- Renovation of Valencia Bus station.
- Speedometer for a bicycle with some enhanced features.

As these examples show, a project working approach can be applied to many different types of tasks or problems. Although these may be classified in different ways, we may say that there are seven main types. These are shown, with examples, in Table 1. Each type exhibits diverse attributes and different levels of complexity and uncertainty and requires specific methodologies.

Table 1. Illustrative examples of some types of projects

Type of Project	Examples
New product development	A pump, an automobile engine, an UAV, etc.
System development	New IT solution in an organization.
Organization development	New organizational structure or new corporate identity.
Technical construction work	A technical plant, a house, a building, a bridge, etc.
Planning/Clearing-up Tasks	Supplying a fundamental appraisal of factors influencing vehicle traffic in a specific area over a certain time.
Arrangements	A world tour of a symphony orchestra or the planning and running of an exhibition.
Research work	development of a new chemical or mapping with satellite data.

Before tackling their major projects, EPS students are often introduced to the skills, tools, research methods, software, etc., that they may need by working on a mini-project over 2-5 days. This simulation exercise helps to train individuals, fine-tune teams, and enables students to become familiar with their team members and working methods. Some examples of mini-projects are devising the launch of a product to better night's sleep, developing an App providing details of music festivals, putting together a team and equipment to participate in a car rally, or devising a new chocolate bar for a specific market. This last example has been the most used and the most successful. Students are given the following information:

There are many different types of chocolate products on the market. Some are available globally and, as a result, well known and easily recognizable. Others are equally well known, but in a more constrained geographical area – for example, Europe; or even more restrictively, such as in a single country. Sometimes, the same product may have alternative names, for instance, Snicker© and Marathon©.

So, what factors influence the design of a new confectionery product? What is the target market? What would be the product's ingredients and the marketing strategy, including the packaging?

If you research confectionary products, you will find there are many variables – the products may be pure chocolate but could be a mix of dark and milk; they may contain candy, fruit, nuts, biscuit, cereal, etc. The percentage of chocolate and other ingredients vary – what are these percentages, and why are there so many different options? How does this affect the price? How do the above factors influence manufacturing?

Each team is then given a target group: children aged from 8 to 14, construction site workers, adults on a diet and for whom chocolate is forbidden, environmental activists, for example. After working on their chocolate bars, the teams present their ideas, initial research, and calculations. Feedback on the proposals is given to enrich the activity and spark innovation.

Project Selection

Projects are initiated from the identification of a need for a new product, service, or system. This identification can arise from different sources like a user, employee, outside organization, etc. For a project to go ahead, it must be justified - there must be a need, a level of importance, and feasible solutions should exist to make the project viable. Various criteria will need to be satisfied considering alternative solutions, estimated cost, and returns for each alternative. As these are estimates, a risk assessment is also a necessary requirement. A well-structured selection and evaluation process with all relevant parties involved increases the probability of management approval for the project to go ahead. Table 2 shows a typical feasibility evaluation procedure.

Table 2. Project feasibility evaluation procedure

Start	Who does it
Identification of a need	Client, personnel, management
Is the need important? (Stop if the answer is no)	Management (marketing)
Technical evaluation	Management (engineering)
Is the project feasible? (Stop if the answer is no)	Management (engineering, finance)
Development of alternatives	Engineering, R&D
Cost-Benefit analysis	Engineering, marketing, finance
Selection of the best alternative	Management
Defining the project	Management

The feasibility of a project is not limited to its technical aspects but is also strongly related to its financial feasibility. Key parameters related to the context of the client, the supplier, and the technical, economic, and regulatory environment should be taken into account. Figure 1 presents a decision diagram for a project's Go / No Go procedure from the early stages. Note that in the real economic/industrial world, the project may be made public, and interested parties invited to tender.

Figure 1. Go / No Go decision diagram from invitation to tender phase on
(Credit: Philippe Fillatreau)

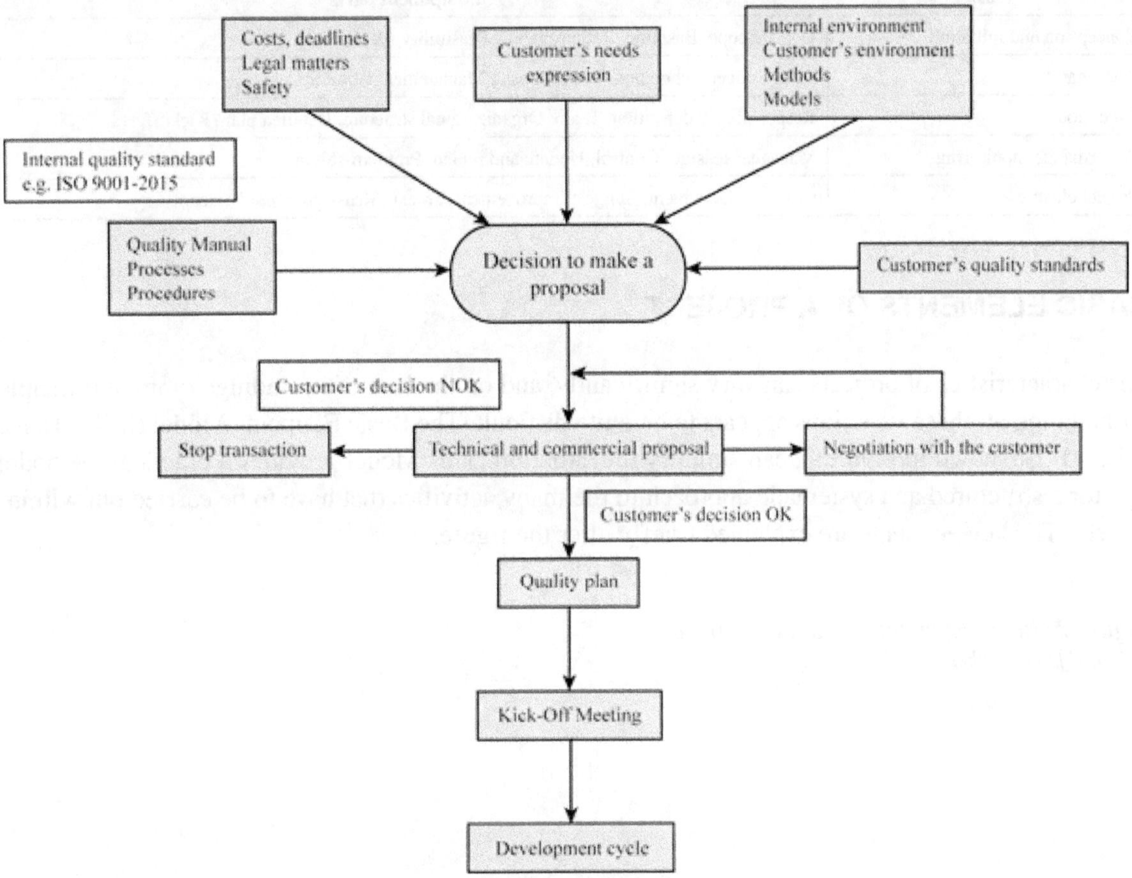

Project Life Cycle

PM is a concept for managing a complex undertaking and the organization required to achieve it. Any project passes through several phases during its life cycle. The phases may vary in size and complexity. They can occur over varying time periods and can be analyzed in different ways. Table 3 shows a typical set of five phases (developed by the Project Management Institute) that make up a project life cycle. As can be seen from Table 3, the five phases of PM are (*i*) conception and initiation, (*ii*) planning, (*iii*) execution, (*iv*) performance/monitoring, and (*v*) project closure.

The Project Management Institute (PMI), which was founded in 1969, is the world's largest non-profit membership association for the PM profession. It sets the certification standard for project, program, and portfolio management and offers training and certifications. The gold standard of certification (there are others available) from the association is the Project Management Professional (PMP)(R) certification.

The ISO 21500:2012 (ISO, 2012) provides guidance for PM. It can be used by any organization, including public, private, or community organizations, and for any type of project, irrespective of complexity, size, or duration. This standard provides a high-level description of the concepts and processes considered to form good practice in PM and is commonly used for EPS projects.

Table 3. Project phases and their component parts

Phase	Component parts
Conception and initiation	Goals, Scope, Baseline, Requirements, Feasibility, Desirability
Planning	Plan, Budget, Scheduled, Bid proposal, Management Commitment
Execution	Responsibility definition, Team, Organizational structure, Detailed plan, Kickoff
Performance/monitoring	Manage, Measure, Control, Update and replan, Problem-solving
Project closure	Closeout, Document, Suggest improvement, Transit, Reassign, Dissolve team

BASIC ELEMENTS OF A PROJECT

The characteristics of projects can vary significantly, and on the face of it, finding common principles for handling all these variations appears to be quite difficult. The Basic Elements Model (Mikkelsen & Riis, 2017), shown in Figure 2, can simplify the situation. This Model provides a practical methodology for a structured and systematic approach to the many activities that have to be carried out within a project. The key elements are explained briefly after the figure.

Figure 2. Basic Elements Model of a project
(adapted from Mikkelsen & Riis, 2017)

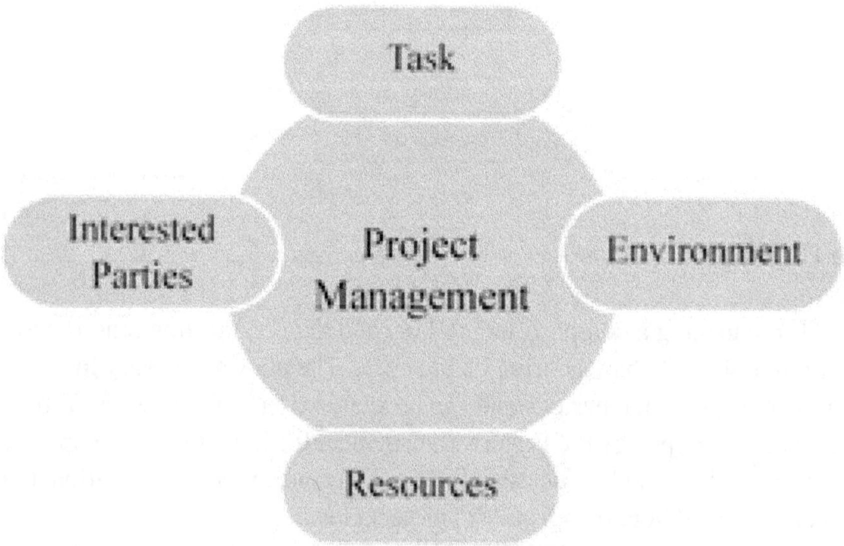

Identifying the **task** is the starting point for any project. The big question here is how to define the task because task definition is very dependent on the task itself; it may be simple or complex and require one semester (as in EPS) or three years. Essentially, the definition of the task becomes the main aim of the project, and from that, an additional set of tasks will be defined in more detail. Thus, the task definition is the intended outcome (the aim) of the project. It should be defined as a single sentence and expressed

clearly and concisely. For example, "to design an outdoor theatre on the university campus" is not as good a task definition as 'to design an outdoor theatre on the university campus which is adaptable to different kinds of events (music concerts, film, theatre, poetry readings, dance, etc.)".
The EPS projects discussed in this book assume that any task has the following characteristics:

1. The task is complex.
2. It requires contributions from people with different skills and backgrounds across the organization.
3. It is unique because a similar task has never occurred before (it is not routine). In real industrial life (as in the EPS), a project is always specific, and no two projects are ever the same. Even if the list of objectives and the customer are the same, environmental factors will probably differ. Indeed, especially if some time has passed between two apparently similar projects, many things may have changed: technology, quality standards, market competition, human resources such as changes in personnel and the corresponding skills or level of expertise, etc.

For some projects (e.g., in research work), the aims and objectives are very difficult to define, and only a broader vision of the result is possible initially. In such cases, it is often necessary to establish some intermediate aims, which can be adjusted if required during the project. However, in general terms, without a specific definition of an aim, the desired outcome remains unknown and hence impossible to work towards. Also, the lack of a specific aim makes it difficult to conclude whether a satisfactory outcome has been reached.

Apart from the main aim, it may also be necessary to establish subsidiary (or secondary) aims. These are the objectives, i.e., the subsidiary aims, that will need to be fulfilled in order to achieve the main aim. The aims and objectives for the task must be well defined, well understood, accepted, and supported by everybody in the project group.

There will always be interdependence between the task and **the resources**. The resources may cover an extensive range of elements, including knowledge, skills, people, buildings, machinery, equipment, available budget, and so on. The ideal situation is to define the task initially and then hold the necessary resources for achieving the objectives during the stipulated time limit.

There is a need to identify positively and negatively **interested parties** in any project. Positively interested parties (who may be individuals, companies, organizations, and so on) are those who are in favor of the project and maybe ready to contribute to its success by providing resources. On the other hand, negatively interested parties wish to prevent the task from being achieved. The fossil fuel industry, for example, would prefer to see very slow adoption of sustainable energy policies. Sugar manufacturers are concerned about increasing the amount of sugar in many foods. Environmental groups or local inhabitants may be against the destruction of a natural area for building an airport. Collectively, these people are known as the **stakeholders** of the project. Pinto (2007) defines stakeholders as:

Persons and organizations such as customers, sponsors, performing organizations and the public, that are actively involved in the project, or whose interests may be positively or negatively affected by execution or completion of the project. They may also exert influence over the project and its deliverables.

For the success of a project, it is therefore essential to examine and analyze who are the positively and negatively interested parties and use this knowledge to establish an operational base for generating ideas.

The project **environment** (or context) refers to where the finished project will exist in the future and the internal and external factors affecting the project. Some examples of environmental factors are the marketplace, the current level of technology, competing products or systems, laws and regulations, the economy, the political situation in the region or the country, the company culture and working practices, or the availability of resources.

DEFINING THE SCOPE OF A PROJECT

Knowledge about a project grows with time, obviously. However, the project team must focus on searching for all accessible information right at the start of the project. This is mainly because the consequences of decisions made in the early stages are more important in the sense that they could have knock-on effects. It is also because the project scope should be dimensioned technically and financially at the beginning so that the objectives are clearly defined and known by the project team and all stakeholders (especially the client). Thanks to this, the project team knows the available budget to ensure that costs are kept within the limits set.

A comprehensive and realistic analysis of the recent and present environment and context is required to know where activities can start. It is essential to define the parts of the project and what is not part of the project. The project team members must know where the project work should stop; in other words, where an influence on the project does not exist. Trying to influence external and unchangeable parameters is a waste of time.

Indeed, quality ISO 8402:1994 standard (ISO, 1994) defines quality as "*The totality of features and characteristics of a product or service that bear on its ability to satisfy stated or implied needs*". It is up to the subcontractor or supplier to ask whatever should be asked to make the expression of the need clear. The supplier is the expert on whatever it is (product or service) the client wants to subcontract and should therefore know how to clarify, exactly, what is required. The supplier should also bear in mind that some needs may seem obvious to the client and may therefore not have been expressed.

In addition, it is indispensable to make the project goals clear. Obviously, items such as delivery times, quality requirements, minimizing costs, and team satisfaction should be included in the objectives, but also the limitations and exclusions of the project. These essential elements help to avoid many misunderstandings with the client.

The Basic Elements Model creates a good base for arranging a more operational structure. Each Basic Element is looked at, one by one, and the activities are identified to obtain this operational structure. Which of the Basic Elements from the previous section becomes the starting point will depend on the nature of the project.

Defining the project's scope basically consists of making the context and objectives clear and dimensioning the project. This phase should provide a clear framework and a rough global organization to be used later as input to the project planning phase. It is essential to review and validate this phase with the client before proceeding to the project planning phase.

The main aspects of being tackled in the scope definition phase are:

- to define the objectives (The task)
- to define the stakeholders in the project (including the representatives of the client(s))
- to define the resources and environment

- to define the main deliverables
- to define the main milestones and their schedule
- to define the project requirements, limits, and exclusions
- to analyze risks and define related actions.

The Task

The task must be described in a way that makes it possible to measure the result of the project work concerning the desired objective. Otherwise, success might not be demonstrable, which is important for every project. As explained above, this should be done by identifying the aim of the project and the outcome the project seeks to achieve. However, even if the task description makes clear what is wanted, it will probably never be in a form that can be used for the practical organization of activities. For example, the main task is "to develop a chocolate bar for environmental activists". In order to achieve this, the main task has to be broken down into smaller sub-tasks.

The theory of setting objectives suggests that a project's objectives should be SMART (Zwikael, 2014) — Specific, Measurable, Attainable, Relevant, and Time targeted. SMART goals help focus attention towards goal-relevant activities, drive and control effort energy, encourage persistence and lead to strategy development (Boyne & al., 2007). The explanation of the five letters of the SMART acronym as defined in Doran (1981) is:

- Specific: all projects are specific, as emphasized in previous sections; the definition of the goals should be as specific as possible and not generic or vague.
- Measurable: goals should be measurable to allow stakeholders of a project to determine clearly whether the goals have been achieved or not when performing project acceptance.
- Attainable: goals should be attainable. This does not mean that the goals should be easy to reach. For the students' interests in an academic context or to be appropriate and suitable to a company in a very competitive context, goals should be, of course, challenging.
- Relevant: it is crucial that the goals defined are relevant at all levels of an organization (individual, team, or company) and mainly for the client.
- Time targeted: goals should be time targeted to help reach client satisfaction by on-time delivery, avoid having to pay delay penalties, or maintain a good level of efficiency in the development teams (be they employees or students).

It should be mentioned here that the meaning of the SMART acronym is not standardized. In many websites, but only rarely in research papers (Aghera, 2018), we can find Specific, Measurable, Attractive, Realistic, and Timely (or Time-bound). 'Attractive' here means that the objectives themselves provide a high level of motivation, involvement, and efficiency and will result not only in the satisfaction of the customer but also in the satisfaction of the development team, and more generally, of all stakeholders. Realistic: corresponds mainly to "Attainable" in the definition of the SMART acronym presented above.

Sub-tasks will give a more down-to-earth, practical understanding of what the project is all about and will help to obtain explicit, manageable activities to which the necessary resources can be allocated. Sub-tasks will also provide a basis for the later planning, performance, and follow-up. They can help the team to formulate the subsidiary aims described above. Splitting up the task into smaller activities will give a mix of different types of sub-tasks. Initially, sub-tasks may be listed in random order, while diagrams can be used to identify and show each separate activity (or event or process). Using charts allows activity sequencing and parallel activities to be identified. The ordering of sub-tasks can be done in many different ways, depending on the nature of the project and the decisions made by the project team. This kind of structuring can be especially useful to show:

- The functions of the sub-tasks.
- The relationships of the sub-tasks within the broader system.
- The geographical or physical location of the sub-tasks.
- The implications of the sub-tasks concerning their organization and execution.

The Stakeholders and Interested Parties

The stakeholders are the organizations and human resources directly running the project (this includes a clear definition of the client) and all interested parties during the project and potentially beyond. In all cases, interested parties are vital for successfully completing the project because they must directly or indirectly support the necessary resources. It is essential to make sure there is a deeply rooted interest in the matter from their side. Without this, the project work will be difficult and probably also impossible to carry out through the duration of the project. Moreover, it could be the case that any negatively interested parties will act against the project. In this situation, efforts should be made to secure a group that can see more advantages than disadvantages to fulfilling the aim.

The Resources

The resources may be human (personnel with the required knowledge and skills for the project in question), material (such as buildings, machinery, technical equipment), or financial (money, funding from different sources). When identifying and determining resources, it is vital to be realistic. There might be two fundamentally different situations:

1. The necessary resources for the task will be allocated to the project in a flexible way.
2. The available resources are already allocated and cannot be changed.

The question could be, "Which one of these two situations should dictate how the project will be carried out?". Obviously, they are fundamentally different, and it is necessary to be realistic in situations where the task and resources are related to each other. Can we get the resources necessary to solve the task, or must the task be limited or redefined to fit the available resources?

The Environment

The environment can be broken down into several specific areas that affect or are affected by the project. What, for example, is the political environment in which the project exists? How could the project be affected by changes in political power or influence? What is expected in terms of bribery and corruption? Will there be conflicts between different political factions? These factors may affect how the project management team needs to engage with certain stakeholders.

The geographical location of the project is another element that could have a bearing on, for example, the availability of utilities and human and material resources. Extremes of local weather may have to be considered, or there might be topographical constraints affecting access to the project site to take into account.

Further aspects include any local regulatory and legal requirements, the cultural and religious environment in which the project exists, and how this may affect things such as accommodation of workers, local holidays, working hours, or the technological tools, skills, and experience available which could influence the speed and efficiency of the project. One more crucial environmental influence is the market conditions. Market conditions tend to be dynamic and, depending on the prevailing economic environment, can be beneficial or detrimental to the performance of a project.

From these few examples, it should be clear how important environmental and contextual factors can be in the success or failure of project management. Normally, after considering the interested parties and contextual factors and forming conclusions about which activities these might affect, two elements are left: the task itself and the resources available or necessary for the task.

The Main Deliverables, the Main Milestones, and their Schedules

Once the stakeholders' needs and resources are made clear, the main deliverables should be listed and used as inputs to the project organization and planning. Project managers need to understand their organization's mission and strategy to make appropriate decisions and adjustments (Gray & Larson, 2021).

A first timetable for the project can then be defined by associating a deadline to each deliverable defined at this point. This association allows defining a rough first schedule for the project.

Risk Analysis

Many risk analysis techniques can be found in the literature (Kerzner, 2017). All of them aim to identify and assess potential risks by taking into account information such as the probability that the risk will happen and the impact on the project if it does.
Risk analysis may be used at all steps of project management in different ways:

- to define preventive and corrective actions
- to take the analysis of the risks into account when estimating development time and costs
- to take the analysis of the risks into account in order to establish priorities in the requirements of the project, in the deliverables to be produced, and in the tasks to be performed by defining different plans (A, B, etc.), for example.

Project Process, Monitoring, and Costing

Projects are started, planned, and executed to achieve a certain solution due to work carried out. If the objective is, for instance, to have a product manufactured, this objective is not fulfilled by merely offering a collection of drawings, after which the product might (or might not) be built later. Reporting with arguments for or against building the product will not fulfill the objective either. Shtub et al. (1994) emphasized organizational design as well as the types of data and systems needed for successful decision-making. They were stressing integrative concepts rather than isolated methodologies.

Monitoring and controlling project work is the process of tracking, reviewing, and reporting progress. Thanks to this, teams can meet the performance objectives defined in the project management plan. A key benefit of this process is that it allows stakeholders to understand the project's current state, the steps taken, and the budget, schedule, and scope forecasts. In the form of graphs or charts, different tools are chosen to suit the nature of the project and the item to be monitored. Graphics should be visual and easy to understand as well as effective in showing performance and deviations accurately. Commonly used charts are schematic block diagrams, bar charts, histograms, and Gantt charts (Gantt, 1919; Vanhoucke, 2018). Standard indicators are cost and schedule variances or performance indexes (PMI, 2017), and information related to monitoring indicators can be found in the PM literature. Monitoring should be carried out at every level, ranging from the global aim to the smallest sub-task, and frequent periodic meetings should take place to assess the state-of-the-art of all these aspects. In addition, it is vital that records are kept of all formal meetings – decisions taken, changes to the plan, tasks achieved, etc.

Estimating the cost of a project requires a lot of skill and experience and is crucial to a company's success. Costs can be of very different natures and could be related, for example, to human resources, materials, travel expenses, infrastructure, dissemination, or the environment. Usually, a team effort is required to calculate realistic costs.

A variety of costing methods exist. The choice depends on the nature of the project. One approach is to compare actual and estimated costs with previous contracts (historical database). Verbal quotations (where they occur) must always be confirmed in writing. Lengthy projects, especially, may see changes in labor and materials costs. Companies should take great care in accepting fixed-term cost contracts. Generally, a clause in the contract to allow for such changes is good practice. Within this consideration are also the consequences of the overrunning of the project in terms of cost (delay) penalties. Below is a brief outline of some costing methods often applied in project management situations:

1. Generic Costing: This method is ideal for contracts with modular elements (variable structures), where projects vary in size and configuration. Computer database packages allow quick and easy updating and comparison of costs and are ideal for post-project audits. The benefits are the simplicity for standard proposals of configurable projects, the comparison with similar projects, and the support of databases.
2. Creative Costing: This method is used where projects are unique, and each stage will be costed from the source. As there are no standard proposals, the creation of new estimates becomes necessary. In practice, most projects contain both of the above–standard and non-standard configurable elements.
3. Cost Coding and Estimating: This method is suitable for manufactured or modular products and is commonly used as an estimating tool. The code is built up in levels. The aim is to produce a logical breakdown, similar to a Bill Of Materials (BOM) used with Materials Requirements Planning (MRP) systems.
4. Check Lists and Estimating Forms: This method can be used with any project. Typical cost items that need to be considered are Foreign currency exchange (rates and charges), Safety costs, Contingency factors, Packing, freight and insurance, Warranty, customer support, training, handbooks, installation, etc.
5. Cash Flow Analysis: Cash flow is always critical, especially on large projects involving staged payments. Spreadsheets are ideal for this form of control, although strictly speaking, it is an ac-

countancy function. Establishing a payment schedule for each project (with a given amount to be paid by the client at each delivery and associated milestone) may help monitor cash flow.

PLANNING

The structuring of a project in terms of visible and manageable activities provides the building blocks for planning the project's lifespan. The planning procedure must clarify the following issues: the main strategy for a working method within the project team, the structure of the tasks and sub-tasks within their activity fields, and the choice of tools for monitoring and controlling the progress of the project.

It is vital to distinguish between a detailed working program and a strategy for the overall project. Detailed planning may come after the strategy of reaching the goal has been launched, but never before. One of the main requirements for project work is a fixed completion date. Time scales and deadlines are therefore scheduled to meet these targets.

Planning is an iterative process, using the first step to obtain approximations, the second step to refine the values, and as the project progresses, further reviews of all or some of the steps may be necessary to make adjustments. The authors suggest the following steps for planning a project:

1. Outline the main strategy based on the basic elements.
2. Structure the activity fields. Subtasks are arranged in work packages and assembled in a logical order to give the Work Breakdown Structure (WBS). The WBS aids the development of the project plan by showing all the steps from the top down. The WBS displays and defines the product to be developed or produced by hardware, software, support, or service elements and relates the work activity elements to each other and also to the end product. If required, the team can then apply a hierarchical cost code system that will provide a basis for budget preparation and control.

The WBS is a key process, yet it can be difficult. It consists in developing a tree structure starting from a first level composed of the main deliverables/activities. Then, each of these is decomposed into sub-deliverables or sub-activities until a level at which the sub-deliverable or activity can be managed or developed by a clearly identified set of team members. After that, the tasks required to achieve each sub-deliverable can be clearly identified and listed.

Other breakdown structures (functional, organizational, product, resource, risk) can help manage a project. The ISO 21511:2018 standard (ISO, 2018) provides relevant terms and definitions, concepts, characteristics, benefits, uses, integration, and relationships related to WBS. Figure 3 shows an example of the first level of a WBS developed by EPS students. The project was to design a house that used positive energy.

Figure 3. First level of a WBS for an EPS project devoted to designing a house using positive energy (Vandelvelde et al.,2018)

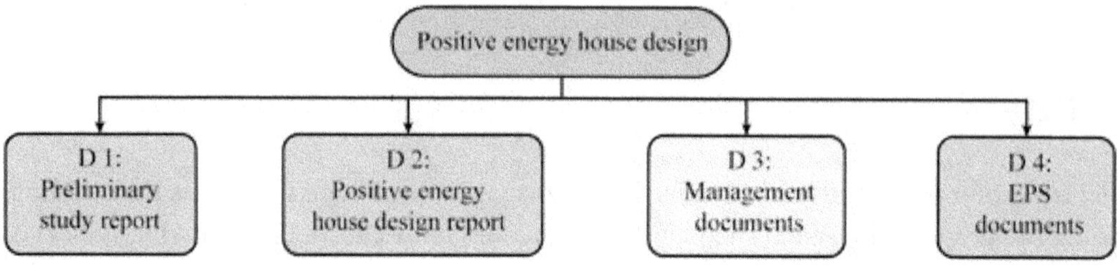

Figure 4 shows the complete D1 branch of the WBS for the same project. In the Figure, 'D' = Deliverable, while 'T' = Task. We can see that four deliverables (in the form of various documents) have been identified together with eight tasks (Vandervelde et al., 2018).

Figure 4. The D1 branch of the WBS shown in Figure 3 (Vandelvelde et al.,2018)

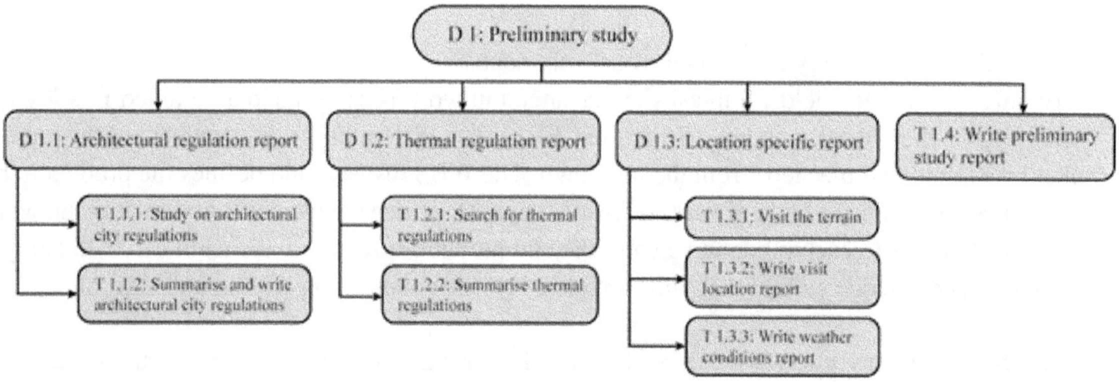

Figure 5 shows the complete D2 branch of the WBS for the same project. In the Figure, 'D' = Deliverable, while 'T' = Task. Eighteen tasks have been identified and defined.

Figure 5. The D2 branch of the WBS shown in Figure 3 (Vandelvelde et al.,2018)

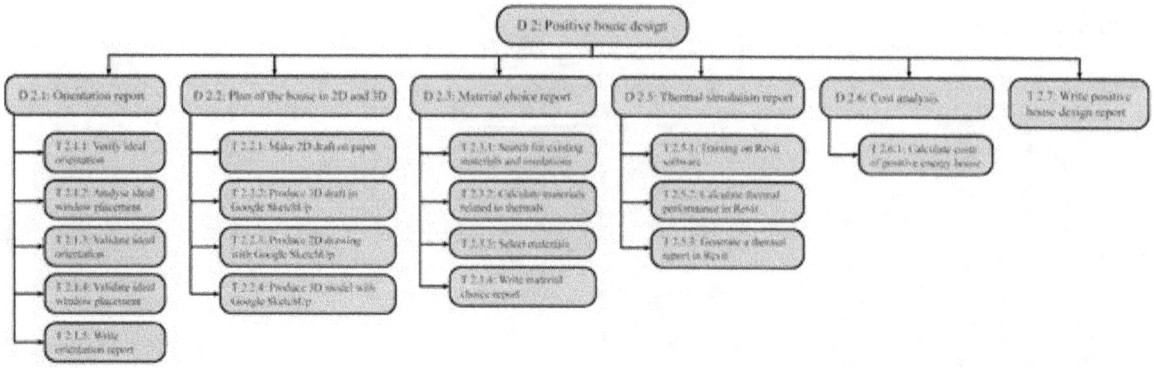

As these figures show, it is helpful to code activities clearly, such as deliverables and tasks. The total number of tasks should remain reasonable to keep a clear global picture of the project. Obviously, project management should be an aid. It should not add unnecessary complexity.

3. Detail activities and sequence. It is necessary to decide in detail what has to be done and in what sequence. Each manageable part can be broken down into more detailed activities. Firstly, the team brainstorms the main task required based on past projects, personal experience, and appropriate information. Each activity is then arranged in order of dependency using the following questions: What do you want to do now? And before that? And next? What can be done simultaneously? More generally, it is essential to organize and plan (and later on monitor) at the right level of granularity (Grenning, 2002).
4. Estimate workloads and the time needed for each task obtained from the WBS. It is important to distinguish between both concepts. Workloads correspond to quantities of work needed to perform tasks, while durations correspond to calendar times or deadlines. For a team or company developing a project for their client, internal tasks are classed as fixed workload, while tasks subcontracted outside have a fixed duration (and cost). Let's consider a non-subcontracted task (performed internally by the team or company for their client). The workload is an input (as well as the availability of the human resources and the calendars). The duration is an output, depending on the workload, resources availability, and calendar. Therefore:
    a. Workloads should be estimated and respected so that the costs estimated for the human resources work are met. Estimating workload is one of the most challenging tasks. It should be performed using experience feedback from former projects.
    b. Durations correspond to calendar times or deadlines. They are important to allow monitoring to make sure all deliverables, and in the end, the whole project, are not running late. The following procedure should be observed for this vital stage: start with fixed dates, assign known duration times, and analyze unknown duration times. The following guidelines are useful to manage uncertainty: consult the responsible person, consider the activities in a random sequence, and beware of optimum performance syndrome (seeing yourself in a very productive light). If an activity is less than one day in duration, look to combine it with another activity and look for an appropriate estimation form. Finally, if difficulty is experienced, ask the following question: is the task poorly defined? If so, clarify the task.
5. Calculate the total project duration. The activity duration estimates can be used to calculate the estimated project duration. They can also be used to associate the relative significance of each activity to the time scale objectives. In practice, the project will either be overrun or under-run, which will affect the planning. Unacceptable results require revisiting the time scales and, possibly, the timing objectives.
6. Match the available resources. Resources must be allocated for each activity, and the finite availability of the resource will be a major issue. Factors to consider are commitments to other projects (also overtime working, recruitment, subcontracting). Capacity charts are a useful aid (mainly in industrial projects where capacity is likely to be an issue). Once the resources have been determined, a review of the costs is necessary to ensure that the cost objectives are being met. The most important resources are qualified people, i.e., efficient, devoted, and skilled at problem-solving. Some possible options to extend the capacity are transferring resources from another project, subcontracting, making better use of time, recruiting more workforce, improving current skills, and

investing in additional equipment. Allocating the human resources to each task obtained from the WBS process allows workloads to be distributed to each human resource. Also, one responsible person should be designated for each task, independently of the quantity of work performed by the responsible person on the corresponding task.

7. Organize the project work. This step is an essential point for the whole project team for the entire project period. The tasks should be identified using simple but straightforward code and name for each in this activity. Detailed knowledge of each individual's technical competence, working speed, accuracy, and special attributes are required. On completion, all team members involved should understand the plan in order to get commitment from everyone. In case of discrepancies, adjustments have to be made.
8. Determine contingency requirements. Contingency plans are alternative strategies that come into effect if the intended plans go wrong. These should be developed ahead of the project execution so that they can be implemented with a minimum of panic and delay. For each activity, managers should ask: What could go wrong? What are the potential consequences? If warranted, can an alternative plan be developed?
9. Choose tools for monitoring and controlling. During the course of a project, it is crucial to monitor whether, for example, deadlines are met, quality is up to standard, workloads are adequate, and costs adhere to those set in the budget. Monitoring these elements helps to keep the client satisfied and the project team efficient.
10. Build the team. As explained in a Teambuilding and Efficient Teamwork chapter of this book, member selection and team-building activities are necessary to create a good team. Indoor construction of Millennium Wheel and Bioethical discussions are two examples of team building activities shown in Figure 6.

Figure 6. Pictures of Indoor construction of Millennium Wheel and Bioethical discussions
(Credit: Pedro Fuentes-Durá)

PROJECT CLOSURE

Wrapping up a project is as important as other processes in Project Management. Releasing all deliverables does not mean a project is complete. Ignoring the project closure process would result in unfinished project management and even unfinished projects. The following activities are involved in the project closing stage: to deliver the products, services, or results; to get a formal sign-off from the customer; to release resources; to assess the project and celebrate.

It is not enough for the team to declare a project done. The customer needs to agree on the project's completion based on a formalized and agreed-upon acceptance procedure.

The team should analyze the product scope as planned – whether it is up to the mark or not. The features of the product scope should meet. Only after clarifying the same, the project should be considered complete. The agreed-upon acceptance procedure must be performed in the presence of the client to assess if the project's objectives are met rigorously.

For example, imagine a worst-case scenario. The customer receives the deliverables and indicates the work is not up to par, and to do some rework is a must — that means additional cost. Another severe issue would be to have to regroup the project team again. Consequently, formal sign-off is essential for closing projects.

After receiving formal sign-off from the customer, the team must release human and material resources so that they can be involved in other projects.

If a third party has been working on the project, those agreements need to be closed. Once they have produced deliverables and delivered them to the end customer, the contracts should be closed; they have no meaning after the deliverables have been accepted.

It is also necessary to compile the project closure documents and convey them to the client (and sometimes to another key stakeholder). The archived files should be useful in the future. Work on documenting the lessons learned from a project can be used as a reference for future projects (What could be improved? What team would do differently if they had to start the project again? What can be reused for future developments?).

Conditions for change requests that were accepted or rejected, cost control and history of the schedule control could be assessed with the lessons learned. The team should involve all the stakeholders during that process. The most commonly made observations are based on the analysis of the differences between real and planned elements (workload, deadlines, cost, etc.). Still, it is also very important to assess the added value for the stakeholders, especially the customer and the team itself (and more specifically for each team member).

In the end, the stage arrives where the team needs to arrange a celebration involving all team members and all stakeholders. They must feel the success they made together as a team. Many project managers usually ignore this stage—and yet it is excellent to improve the morale of the entire team involved in the project.

In an ideal world, teams and organizations should care for milestones and final parties during the course of long-term projects. If at all possible, you have to involve your end customer and third-party vendors to make it a huge success altogether. A celebration indicates the complete, successful closure of a project.

PERSPECTIVES AND METHODS

A PM methodology is the set of conventions (principles, tools, and techniques) that a team agrees to follow, and it should facilitate team collaboration. Rigor, client involvement and satisfaction, team commitment and contentment are PM-based. Concepts standardized and introduced in the literature deal more about what should be done than how it should be done. They do not restrict to particular projects but stand for any project. This means that in practice, a team will develop a specific project using specific practices in the context of their given project.

PMI global standards provide guidelines, rules, and characteristics for project, program, and portfolio management. PMI standards are widely accepted, and when consistently applied, are a valuable aid in achieving professional excellence. A team of over 80 PMI members created the text "A Guide to Project Management Book of Knowledge", the PMBOK® Guide (PMI, 2017), in an effort to standardize project management information and practices. Currently, in its fifth edition, the PMBOK Guide is continually updated by the PMI and shares the fundamental practices used worldwide to achieve the best results. The PMBOK Guide includes a process standard that can be applied to many projects; however, it does recognize that each project is different. It is up to project managers and professionals to apply the techniques and phases covered in the PMBOK Guide to the specific requirements of their projects.

PRINCE2® is a process-based method for effective project management and will give the fundamental skills needed to become a successful project manager (Matos & Lopes, 2013). It is widely used in both the public and private sectors. It stands for PRojects IN Controlled Environments and is used and recognized worldwide. PRINCE2 is entirely scalable, and the most recent update to the framework means it can be more easily tailored to each undertaken project. The main features of this methodology are based on the business focus, an organizational structure directed to the project management team. The planning is done with an orientation toward the final product. There is an emphasis on the division of the project into phases. The roles in the PRINCE2 methodology are clearly defined. The project manager is the person responsible for the organization and controls the project and its processes. The project manager selects the team to execute the project and oversees their work to ensure it is done correctly and on time by following a plan. At least a customer, user, or supplier is also involved in the project. This person (or organization) is paying for the project. They will use the project's results or be impacted somehow by its outcome. On some occasions, the customer is the user, while the supplier is the party who offers expertise to get the project completed successfully. Like the project team, they need organizing to make sure the project meets requirements. Finally, there is a project board, including the user side and the suppliers' point of view. The project manager will regularly report to the project board, letting them know about progress and any problems. In turn, the project board decides how the project will proceed or resolve issues.

Agile is the ability to create and respond to change based on the Agile Manifesto mindset and the 12 Principles behind it. It is a way of dealing with, and ultimately succeeding in, an uncertain and turbulent environment (Poppendieck & Poppendieck, 2003). Agile Alliance® (2022) provides tools, situational guidelines, and an understanding of the various agile approaches available to enable better results. It is beneficial for those project managers accustomed to a more traditional environment to adapt to a more agile approach. It is really about thinking through how the teams can understand what is going on in the environment that they are in today, identify what uncertainty they are facing, and figure out how they can adapt to that as the teams go along.

Scrum (Schwaber & Sutherland, 2020) is probably the most used framework from an agile perspective. Scrum is a process framework used to manage product development and other knowledge work. Scrum is practical in that it provides a means for teams to set up a hypothesis of how they think something works, try it out, reflect on the experience, and make the appropriate adjustments. Scrum is structured to allow teams to incorporate practices from other frameworks where they make sense for the team's context. This kind of framework is necessary when a non-trivial amount of work lends itself to being split into more than one iteration of two to four weeks. Teams following Scrum are expected to learn and adhere to the following values: Commitment, Courage, Focus, Openness, and Respect. The following principles underpin the practical nature of Scrum: Transparency, Inspection, and Adaptation. The Scrum Teams are a cohesive unit of professionals focused on one objective at a time, the Product Goal, and consists of one Scrum Master, one Product Owner, and Developers.

SUPPORTING TECHNOLOGIES

The PM software market is a huge, rapidly evolving area. PM software supply companies provide a set of supporting technologies and management functionalities: virtual teams, virtual channels, document storage and sharing, audio conferencing, online video calling (and screen sharing), online meetings, chats, setting reminders-task notifications, to-dos, project reporting, etc.

For team collaboration, there are a lot of different useful tools. *Evernote©* or *Dropbox©* are well-known note-taking and document organization Apps. Both are powerful for dispersed teams that need online workspaces to discuss ideas, share status updates, and access projects documents. Other tools are more specific.

Slack© is a collaboration hub designed to bring the right people together with the correct information through features like search, shared channels, apps, and integrates pinned items, for example.

Wrike© is a PM tool with online workspaces for faster collaboration, such as a content marketing and design firm that needs to create a workflow approval process to manage project documents. *Wrike* lets the team discuss, proof, and approve design documents, videos, and images.

BaseCamp© is designed for small businesses that need an affordable and "easy to use" project collaboration tool with many task management functionalities.

Google Workspace ©offers a range of tools that help businesses effectively collaborate on projects; for example, *Drive* for managing different document versions, *Meet* for holding online meetings, or *Calendar* to track group task deadlines.

Microsoft Teams© provides an environment to create and manage teams and channels, schedule a meeting, turn on language translations, and share files. It is ready to integrate useful management apps from the company (MS Project) and other sources. It is also prepared to create customized Apps and bots, automate tasks to increase productivity, and use business intelligence.

Asana© is for businesses that want collaboration tools and project management functionalities such as scope planning (identifying risks/determining project deliverables) and project planning (creating task timelines). It is suitable for teams that collaborate on multiple, varied projects. Its functionalities allow users to manage business operations, marketing campaigns, product developments, and more. The tool offers a comprehensive list of third-party integration with file-sharing tools, time track solutions, and more.

Trello© has been widely used for higher education students over the last few years. It organizes projects and day-to-day tasks by creating notecards on a dashboard, where work progress can also be tracked. The team members can sort and organize these cards based on their categories. A quick overview is displayed on the front of the cards. By flipping these cards, the user can dive into more detailed information such as comments, checklists, due dates, attachments, etc. A progress meter adds a checklist to organize a to-do list. Another advantage is that third-party integration is simple.

Airtable© is suitable for businesses that need to manage collaborative databases. It helps teams to plan and execute projects by collaborating via online workspaces. Teams can share files, exchange messages, comment on tasks, make edits to documents, and receive automated notifications of project status.

Monday© is a flexible platform that allows teams to centralize data, leveraging native integrations, third-party apps, and helpful widgets. It enables an efficient follow-up with the relevant team members thanks to an accessible overview of all ongoing processes and the status of essential tasks. It includes an extensive set of templates, ready to be customized and scalable with the workflow.

Jira© is the most popular product management software organizations use to streamline the software development process. It allows everyone to keep aligned during the entire development lifecycle, right from planning and prioritizing the tasks to releasing them. With its at-a-glance view, users can easily understand various projects' state of art. It offers a lot of templates, integration with multiple developer tools, and automation of numerous processes for both webs and Android and iOS smartphones and is free for teams with fewer than ten members.

As standard, all the tools have a free version for a reduced number of members and a limited number of functionalities. Typically, the software is available both on the web as well as the smartphone version.

CONCLUSION

PM has become an indispensable tool in project development. Basically, PM is the structuring of a project in terms of visible and manageable activities, which become the building blocks for planning the lifespan of the project. Project planning must also include the main strategy for a working method within the project team. It should structure the tasks and decide on the tools for monitoring and controlling the progress of the project.

As in any real-life activity, there are three major parameters in every EPS project: task, resources, and time. It is important to know all of these precisely and define them until all three are consistent with each other. However, depending on the nature of the project, environmental (contextual) influences will also significantly impact the project outcomes. A few examples of the kind of questions that might be asked about the project environment are:

- Are there any laws, regulations, or standards the project must conform to?
- What is the state of the market?
- Are there competing products?
- Is there a positive or negative public climate for or against the project?
- Is it possible to define boundaries for the project task and thereby avoid future discussions over controversies regarding responsibilities?

The influence of environmental factors must be known (as far as possible) and included in the planning.

Different methodologies or perspectives (PMBOK, PRINCE, Agile or Scrum) and technologies (Asana, Basecamp, Trello or Slack) have been developed to assist project managers and team members in handling systematically and thoroughly the many and varied activities involved in any project. These should be applied strategically, but thinking strategically requires research, analysis, and forethought in order to create a plan for how to proceed. Designing a solution means drawing on a style of reasoning: endeavoring to formulate inquiries through the apprehension or comprehension of phenomena; that is, questions are posed and answered using information gathered from observation of the various elements involved. The solution does not derive from the problem: it patterns itself around the problem.

It is also important to be critical. Being critical refers to the intellectually well-organized process of actively and skillfully conceptualizing, analyzing, applying, evaluating, and synthesizing information gathered from, or generated by, experience, observation, reflection, reasoning, or communication, as a guide to belief and action (Paul & Elder, 2008). In its exemplary form, being critical is founded on universal intellectual values that transcend subject matter divisions: accuracy, clarity, precision, consistency, relevance, sound evidence, good reasons, depth, breadth, and fairness.

Many decisions will need to be taken throughout the development of a project. Decision-making can be challenging, and almost any decision involves some conflict or generates dissatisfaction. The difficulty is to pick a solution where the positive outcomes can outweigh possible losses.

Finally, there is always room for creativity despite the rules, contexts, restrictions, and constraints. Creative problem solving involves the development of unique and novel responses to problems and opportunities through a complex process of knowledge accumulation, incubation, ideation, evaluation, and implementation. Personal beliefs, fear of criticism, over-management, and stress are the main barriers to creativity. However, there are methodologies (brainstorming, forced analogies, do it, mind mapping, nominal group, etc.) to stimulate the generation of ideas and facilitate other steps in the process. EPS creates a perfect framework in which to train the participants in all the different aspects of PM and in which to develop the many related skills. Most EPS students have little background in practical project management before the start, but the results, performance, and self-assessments confirm a big improvement and consistent learning after this experience.

ACKNOWLEDGMENT

The authors thank David Swetnam, Jack Bradley and David Perry for the inspiration on Project Management, Teambuilding, and Innovation activities.

Besides, the authors would like to thank all the managers who helped and motivated the staff exchange, giving advice and providing the resources needed.

At last but not in least, we would like to thank the European Union Erasmus+ mobility program, which has been a significant source of funding for student and staff exchanges in the EPS network.

REFERENCES

Aghera, A., Emery, M., Bounds, R., Bush, C., Stansfield, R. B., Gillett, B., & Santen, S. A. (2018). A Randomized Trial of SMART Goal Enhanced Debriefing after Simulation to Promote Educational Actions. *The Western Journal of Emergency Medicine*, *19*(1), 112–120. doi:10.5811/westjem.2017.11.36524 PMID:29383065

Agile Alliance. (2022). *What is Agile Alliance and What Does it Do?* https://www.agilealliance.org/the-alliance/

Boyne, G. A., & Chen, A. A. (2007). Performance targets and public service improvement. *Journal of Public Administration: Research and Theory*, *17*(3), 455–477. doi:10.1093/jopart/mul007

Doran, G. (1981). There's a SMART way to write management's goals and objectives. *Management Review*, *70*(11), 35–36.

Eskerod, P., & Riis, E. (2009). Project Management Models as Value Creators. *Project Management Journal*, *40*(1), 4–18. doi:10.1002/pmj.20098

Gantt, H. L. (1919). Organizing for Work. Harcourt, Brace, and Howe.

Gray, C. F., & Larson, E. W. (2021). *Project Management, The Managerial Process* (8th ed.). McGraw – Hill Irwin.

Grenning, J. (2002). Planning Poker or How to avoid Analysis Paralysis while Release Planning. *Hawthorn Woods: Renaissance Software Consulting, 3*, 22-23.

ISO. (1994). *ISO 8402:1994. Quality management and quality assurance – Vocabulary.* https://www.iso.org/standard/20115.html

ISO. (2012). *ISO 21500:2012. Guidance on project management.* https://www.iso.org/standard/50003.html

ISO. (2018). *ISO 21511:2018. Work breakdown structures for project and programme management.* https://www.iso.org/standard/69702.html

Kerzner, H. (2017). *Project Management: A System Approach on Planning, Scheduling, and Controlling* (12th ed.). John Wiley & Sons.

Matos, S., & Lopes, E. (2013). Prince2 or PMBOK – A Question of Choice. *Procedia Technology*, *9*, 787–794. doi:10.1016/j.protcy.2013.12.087

Maylor, H. (2010). *Project Management* (4th ed.). Pearson Education Limited.

Mikkelsen, H., & Riis, J. O. (2017). *Project Management - A Multi-Perspective Leadership Framework.* Emerald Publishing Limited. doi:10.1108/9781787148291

Paul, R., & Elder, L. (2008). *The miniature guide to critical thinking: Concepts and tools.* Foundation for Critical Thinking.

Pinto, J. K. (2007). *Project Management, Achieving Competitive Advantage.* Pearson Education.

PMI. (2017). PMBOK® Guide (Sixth Edition). Project Management Institute.

Poppendieck, M., & Poppendieck, T. (2003). *Lean Software Development: An Agile Toolkit for Software Development Managers.* Addison Wesley.

Schwaber, K., & Sutherland, J. (2020). *The Scrum Guide. The Definitive Guide to Scrum: The Rules of the Game.* https://scrumguides.org/docs/scrumguide/v2020/2020-Scrum-Guide-US.pdf

Shtub, A., Bard, J. F., & Globerson, S. (1994). *Project Management, Engineering, Technology and Implementation*. Prentice-Hall.

Vandelvelde, X., Caesar, N., Suys, M., Kirchhöfer, V., & Suppes, D. (2018). Positive Energy Building Design (EPS final report). École Nationale d'Ingénieurs de Tarbes, France.

Vanhoucke, M. (2018). *The Data-Driven Project Manager. A Statistical Battle Against Project Obstacles*. Springer. doi:10.1007/978-1-4842-3498-3

Zwikael, O., & Chih, Y.-Y. (2014). Project benefit management: formulation and appraisal of target benefits. In *Proceedings of the Project Management Institute Research and Education Conference*. Newtown Square, PA: Project Management Institute.

KEY TERMS AND DEFINITIONS

Client: Someone who receives professional services from an organization.

Deliverable: Something that must be completed or provided as part of a project.

Project Management: The activity of organizing and controlling a project.

Project Planning: The process of organizing a project in a way that it is finished on time and is successful.

Quality: The totality of features and characteristics of a product or service that bear on its ability to satisfy stated or implied needs.

Stakeholders: Persons and organizations such as customers, sponsors, performing organization, and the public that are actively involved in the project or whose interests may be positively or negatively affected by execution or completion of the project.

Team: People working together as a group in order to achieve something.

Work Breakdown Structure: Decomposition of the defined scope of the project or program into progressively lower levels consisting of elements of work.

Chapter 5
Intercultural Communication

David Perry
Universitat Politècnica de València, Spain

ABSTRACT

This chapter explores the notion of effective communication not only in international contexts but in any situation which brings together people with diverse cultures. The chapter begins with a brief look at some intercultural experiences and then explores the notion of 'culture' and how it can shape our behavior and patterns of thought. From there, it moves on to discuss the 'classical view' of human interpersonal communication—and why it is incorrect—and then describes a model of communication which regards this as a 'construction' or 'negotiation of meaning'. The chapter continues by examining the broad concept of intercultural competence and then looks at some of the key skills, attributes and strategies which can contribute to both effective intercultural and intracultural communication. An appendix includes some example activities designed to develop greater cultural awareness.

INTRODUCTION

This chapter offers a general introduction to the topic of intercultural communication, hence, and rather like the chapter in this book on project management, it is not limited solely to the context of the European Project Semester (EPS). The main focus of the chapter is to explore the skills, attributes and attitudes (the competences) which are desirable for effective communication in international contexts, whether these be in educational, social or professional settings and irrespective of the language(s) in which the interactions are carried out. For simplicity we shall call this 'intercultural communication competence' noting that various other terms to describe it or a similar concept ('intercultural effectiveness', 'transcultural communication competence', 'cross-cultural capability' to name just three) exist in the literature.

Many of the competences we describe are generally applicable to any situation which brings together people with diverse cultures, regardless of whether these arise from differences in nationality, region, religion, gender, profession, organization or outlook, or whether they speak the same language or not. In other words, intercultural communication may not only refer to interactions between people from different ethnic or national backgrounds, but also to people from the same country, speaking the same

DOI: 10.4018/978-1-6684-2309-7.ch005

Copyright © 2022, IGI Global. Copying or distributing in print or electronic forms without written permission of IGI Global is prohibited.

language, but who, for reasons such as education, upbringing, occupation, or simply where they live, exhibit differences in how they think, react or behave in a given situation.

Referring specifically to the EPS for a moment, most project teams bring together 3-5 nationalities whose working language together will be English. Depending on the makeup of the team, English may be the first language (L1) of one or more of its members, but more probably will be the second (L2) or even the third (L3) language of most, if not all, of them. Apart from the individual teams there is the EPS peer group as a whole. In Valencia this could consist of around 30-35 students from 8-10 different ethnic backgrounds not just from Europe, but also from Asia, Africa and North, Central and South America. In this situation, an EPS participant may expect to interact sometimes in his or her L1, most of the time in the L2 (i.e., English) and sometimes, either for project work or socially outside the university, in Spanish as an L3. Whatever the case, an EPS group is culturally and linguistically diverse and provides many opportunities to discover and explore appropriate, as well as inappropriate, ways to interact with people from other cultures.

Intercultural communication competence (ICC) has several components; culture is one, communication is another, and the skills and knowledge which make up intercultural competence is a third. However, as we shall see, defining these notions, as well as ICC itself, is not such an easy task. Culture, according to Furstenberg (2010: 329), "...is a highly complex, elusive, multilayered notion that encompasses many different and overlapping areas and that inherently defies easy categorization and classification". Opinions as to what constitutes effective communication and how it takes place differ quite widely, while the rapid transformation of society as a result of globalization and advances in science and technology means that intercultural objectives must continuously evolve in order to reflect the needs of modern citizens and communities (Stewart, 2007). In order to clarify ideas, therefore, and before arriving at our discussion of the competences desirable for intercultural competence, we shall look at these three components.

In Section 3 of this chapter, we explore the notion of 'culture' and how it can shape our behavior and patterns of thought – without forgetting that individual differences and situational factors also influence our reactions, judgements and acts. In particular, we attempt to show why it can be so difficult to communicate in a culture which is not our own; why; that is, agreements may flounder on misunderstandings, why multinational projects may cause high levels of irritation, anxiety, or antagonism; why cultural differences can get in the way of good relationships and even result in increased intolerance and prejudice. Why, in other words, it can be so difficult to cross cultural boundaries. Or why, having done so, it can be so enriching. To do that we need to understand just how much and in what ways we are influenced by our own cultures. Section 4 begins by discussing what might be called 'the classical view' of human interpersonal communication – and why it is incorrect – and then describes a model of communication in which this is seen as a 'construction' or 'negotiation of meaning', while Section 5 explores the broad concept of intercultural competence. We will see that communication management is a major component of intercultural competence, and although specific cultural norms clearly influence certain aspects of communication in those cultures where they apply, attempting to adapt one's communicative style to satisfy some oversimplified, assumed stereotypical behaviors has limitations. Having prepared the ground as it were, in Section 6 we look at some of the key skills, personal attributes and strategies (the competences) which can contribute to both effective intercultural *and* intracultural communication. Putting the emphasis on communication (rather than on culture) seems a more relevant approach to dealing with communication in today's highly complex cultural situations. Following some brief concluding remarks in Section 7, an Appendix provides some sample activities designed to stimulate discussion, to develop greater cultural awareness of one's own cultural identity and groups as well as those of one's peers, and to realize that different communication styles exist which are all equally valid although in certain contexts may not all be equally effective. These activities can be used with groups or individuals. In order to set the scene, however, we begin by exploring what we might call the 'intercultural experience' including a brief look at what students report about their experiences abroad.

INTERCULTURAL EXPERIENCES

In 1580, the French philosopher Michel de Montaigne wrote that:

... everyone gives the title of barbarism to everything that is not in use in his own country. As, indeed, we have no other level of truth and reason than the example and idea of the opinions and customs of the place where we live: there is always the perfect religion, there the perfect government, there the most exact and accomplished usage of all things. (Montaigne, 1580 Ch. XXX)

Of course, it is easy and convenient to reject what we don't know and not take the trouble to understand it. Calling other cultures 'barbaric' justifies exploitation and domination. Three hundred years after Montaigne published his *Essays*, in 1896 to be exact, the British businessman and politician, Cecil Rhodes, declared in a political speech in South Africa:

We have got to treat the Natives where they are, in a state of barbarism. We are to be lords over them. We will continue to treat them as a subject race as long as they continue to be in the state of barbarism. (Cecil Rhodes cited in Mungazi, 1996: 27)

In 1895 Rhodes established the territory which became known as 'Southern Rhodesia' and which was administered by his company (the British South Africa Company) until 1923, when the area became self-governing. The fourth Prime Minister of Southern Rhodesia was, like Rhodes, a firm believer in British colonialism and the superiority of white, Christian Europe, and equally unable to comprehend or admit the richness of a culture other than his own:

We must reject permanently the notion that democracy can consist of mere counting of heads. We must unhesitatingly accept the doctrine that our superiority over the Native rests on the color of our skin, civilization and heredity. We must appreciate that we have a paramount monopoly of these qualities and that the Natives have been denied them by their primitive culture. (Godfrey Huggins in a political speech called 'Partnership in Building a Country' cited in Mungazi, 1991: 167)

Fortunately, such extreme views are not usually found in today's exchange students; nevertheless, the research literature (e.g., Spencer-Oatey, 2011; Ting-Toomey, 1999; Verschueren, 2008; Van Hoof & Verbeeten, 2005) shows that most people beginning some kind of international collaboration, whether this be a work or study period abroad, could benefit from some kind of preparation in, at least, the following areas:

- Communicating across cultures
- Building social or professional relationships across cultures

- Developing personal attributes for living, working or studying across cultures
- Managing across cultures

Since the 1980s, the opportunities in higher education for studying, researching or teaching, as well as for gaining voluntary work experience abroad, have expanded enormously. Greater opportunities for travel are not only found in the field of education: travel is often necessary for business and certainly desirable for holidays or personal development. As a result of all these increased opportunities, the ability to operate and interact across cultures – what we shall call *intercultural effectiveness* – has become an important and valued competence. Indeed, included in the numerous outcomes the Erasmus+ and other exchange programs are meant to produce in students and young people is "enhanced intercultural awareness", while among staff and other workers and professionals in education, mobility activities are meant to result in "greater understanding and responsiveness to social, linguistic and cultural diversity" (European Commission, 2015:31).

Students' Experiences

Most of the findings into students' experiences of their international study period are based on self-reported measures; that is, what the students themselves say about it. As one might expect, students who live with a host family or other students from the host country – as opposed to those who live alone or with other foreign exchange students – report higher levels of integration into the host culture (e.g., Quintrell & Westwood 1994; Abe, Talbot, & Geelhoed, 1998). However, in general terms, whether they live with a host family or not, the vast majority of students report high levels of personal satisfaction with their international study period and believe their ability to interact in a culture different from their own has improved (e.g., King & Ruiz-Gelices 2003; Van Hoof & Verbeeten 2005). That said, we should note that the mere fact of spending time abroad does not necessarily mean that there will be an automatic improvement in a student's (or anyone else's) intercultural effectiveness. As Leask (2015: 64) points out, there is:

... a growing body of evidence that some sort of intervention is required at home and abroad if students are to enter a state of becoming interculturally competent.

Academic performance appears to be largely unaffected by a period abroad, although our own informal surveys of domestic and foreign students (including participants in the EPS) suggest two very positive developments which may have knock-on effects on their *learning* performance. The first is that there is an improvement in soft skills such as language proficiency, communication or adaptability. The second is that most students become more aware of, and more sensitive to, alternative possibilities; that is, they become more open-minded, more creative, more pro-active, better listeners and better problem-solvers. Their transversal skills improve. Both of these outcomes seem to remain with students, even when the study period is over and they have returned to their native countries. This is no more than is claimed by the European Commission (2015) for the Erasmus+ program:

Studying abroad is a central part of Erasmus+ and has been shown to have a positive effect on later job prospects. It is also a chance to improve language skills, gain self-confidence and independence and immerse yourself in a new culture.

As far as students from host universities are concerned, there seems, in general, very little interaction between them and exchange students. When it does occur, domestic students report similar positive benefits to those experienced by foreign students (e.g., Deardorrf, 2006; Jones, 2013; Ward, 2001). On the whole, however, an exchange student's circle of friends and acquaintances is usually other exchange students and only rarely students from the host country. This is a great pity, of course, and an opportunity lost. Although one could argue that it is the exchange student's responsibility to integrate him or herself into campus life, and into the wider society, there are linguistic and cultural factors (on both sides) which make this easier said than done. Moreover, it seems that domestic students are not especially interested in making contact with their international peers (Ward, 2001). In general, one of the biggest barriers to international and domestic students developing more meaningful relationships with each other seems to be the limited opportunities to interact socially (cf. Talbot, Geelhoed & Ninggal, 1999; Smart, Volet & Ang, 2000).

Alred, Byram & Fleming (2003: 6) argue that one of the primary objectives of education is to foster a sense of *interculturality*, and we should note, again, that the term *intercultural* does not mean simply *international*. With this in mind, one would think that if a policy of internationalization as it refers to students and staff is to be meaningful, the skills, knowledge and attitudes making up intercultural competence should be promoted in international *as well as* home members of the university community. In the words of Jones (2013):

A group of students in a contemporary university is likely to include people from differing national, religious, ethnic backgrounds, of different genders, sexual orientation or with physical disabilities. Any of these might offer creative intercultural' opportunities in a domestic curriculum and one route to enhancing intercultural competence, an important objective of curriculum internationalization, maybe on our own doorstep. (Jones, 2013: 7)

Put very simply, universities or other organizations receiving students or staff from abroad could and should take more active measures to promote integration other than the ubiquitous 'International Week', and should not overlook their responsibility to promote greater understanding of cultural diversity in their own staff and students.

WHAT IS 'CULTURE'?

Following his research during the 1970s among mainly male, white collar, IBM employees, the Dutch anthropologist Geert Hofstede (1980; 1991) defined culture as "the collective programming of the mind distinguishing the members of one group or category of people from another".
For our purposes the term 'culture' refers to the way people think, feel and act and we will define it as:

The set of unique characteristics that distinguish the members of one group from those of another group.

Culture in this sense does not refer to individuals. It means rather that we can expect the members of a particular group to behave or react in a certain way. And that they expect us (as members of another group) to behave or react in a certain way. That doesn't mean that we necessarily will. Certain individuals from a group may not act in the way we expect, but culture in the way we are using it refers to a group phenomenon and not to individuals. As Nieto (1999: 48) puts it, culture is:

... the ever-changing values, traditions, social and political relationships, and worldview created, shared, and transformed by a group of people.

National characteristics or attributes are the most common reference framework for describing someone's cultural origins, but in fact, we all belong to many groups. These groups originate from various categories (to use Hofstede's term): nations, ethnicities, religions, occupations, organizations, gender, interest groups, clubs and so on. In fact, the North American psychologist Gordon Allport (1954:29) suggested that as early as the age of 5 a child is capable of understanding that he is a member of various groups, although, Allport added, it is not until the child is 9 or 10 that he will be able to understand what this membership signifies.

'Culture' then, refers to the way we think, feel, and act in given situations. Our individual identities can be viewed as a dynamic synthesis of multiple cultures. We could take the term 'culture' as referring to the unwritten rules or standards controlling how we behave in a given situation within a particular group. It is a learned orientation system for a group of people in a given context (or area of action), and it develops whenever individuals share a context with other people for long enough. The culture grows (as its name suggests!) with them and at the same time shapes their behavior, thought patterns and language. However, even though every individual belongs to several cultural groups – and so has various cultural skills – we tend to assume that everyone thinks, behaves and operates with the same broad cultural orientations as ourselves.

Cultural Orientations

Cultural orientations (or *cultural standards*) are specific ways of perceiving, interpreting, judging and behaving that most members of a given culture consider normal. Anyone who operates in an international context does so according to the background of his or her own cultural values, communication styles and interactive expectations.

Cultural behavior is not something willy-nilly, accidental or whimsical. On the contrary, it is the end product of millennia of collected wisdom, filtered and passed down through hundreds of generations and translated into hardened, undiscussable core beliefs, values, notions and persistent action patterns. As such, a culture cannot be depicted satisfactorily at random or evaluated according to impressions or recent observations. It is a largely finite, predictable and enduring phenomenon – the essential key to survival for a nation or cultural group. (Lewis, 2015)

A number of classifications of cultural orientations exist. Most are broadly based on stereotypes. The Lewis Model (Lewis, 2006: xviii), for example, "classifies the world's cultures in three rough categories": *Linear-Active* where tasks tend to be followed sequentially. *Multi-Active* which are centered on relationships and where people often pursue multiple goals simultaneously, and *Re-Active* which follow harmonizing, solidarity-based strategies. According to this Model (Lewis 2006: xviii-xix), members of Linear-active cultures "plan, schedule, organize, pursue action chains, [and] do one thing at a time". Multi-actives do many things at once, "planning their priorities not according to a time schedule, but according to the relative thrill or importance that each appointment brings with it". While Re-actives "prioritize courtesy and respect, listening quietly and calmly to their interlocutors and reacting carefully to the other side's proposals". Table 1 shows more details of the three categories according to the Lewis Model.

Table 1. The Lewis Model of Cultural Types (Lubin, 2013)

Linear-Active	Multi-active	Re-active
Talks half the time	Talks most of the time	Listens most of the time
Does one thing at a time	Does several things at once	Reacts to partner's action
Plans ahead step by step	Plans grand outline only	Looks at general principles
Polite but direct	Emotional	Polite, indirect
Partly conceals feelings	Displays feelings	Conceals feelings
Confronts with logic	Confronts emotionally	Never confronts
Dislikes losing face	Has good excuses	Must not lose face
Rarely interrupts	Often interrupts	Doesn't interrupt
Job-oriented	People-oriented	Very people-oriented
Uses mainly facts	Feelings before facts	Statements are promises
Truth before diplomacy	Flexible truth	Diplomacy over truth
Sometimes impatient	Impatient	Patient
Limited body language	Unlimited body language	Subtle body language
Respects officialdom	Seeks out key person	Uses connections
Separates social and professional	Interweaves social and professional	Connects social and professional

Another influential classification was made by Hofstede (1984). His most recent model of national culture consists of six dimensions which represent national (not individual) preferences for one state of affairs over another and which distinguish countries from each other. Other classifications emphasize whether a culture can be regarded as, for example, more hierarchical (formal) or more egalitarian (informal), or whether time is seen as rigid (monochronic) or fluid (polychronic). Gesteland (2005) offers descriptions of the behavior to expect when doing business in more than 40 countries. While interesting, these classifications are not our main concern here. For one thing, they can lead us to over-generalize and make the kinds of statements below, both found in Chaney & Martin (2011: 102, 103) and both cited by Spencer-Oatey (2011):

Japanese: They converse without responding to what the other person says. Emphasis is on nonverbal communication, so they do not listen.

German: Germans usually do not use first names unless they are close friends (of which they have few).

Secondly, culture is not the only parameter influencing behavior. Let us look at two simple examples, although both occur in intercultural contexts very frequently.

First example

Imagine that a meeting has been scheduled for 10 am and Person X arrives 20 minutes late. There could be several reasons for this; it may be because s/he personally doesn't care about being on time. This would be an individual characteristic. It may be because s/he was stuck in a traffic jam, in which case there is a situational factor to be considered, or it may be because s/he comes from a culture that does not refer to chronological time when defining the start of a meeting. This would be a cultural factor. Whatever the cause, the people waiting for Person X may well perceive, interpret and judge his or her behavior in particular ways, and how each one interprets and judges the behavior – and how they then react – may be influenced by their own cultures. They may, for example, see Person X's absence as causing them to waste valuable time, and react by telephoning them to find out where they are. On the other hand, they might simply think Oh, not everyone is here yet, I'll chat to a colleague while we're waiting.

Second example

The meeting finally begins. Most of the participants express their opinions except Person Y. This could be because:

(i) Y is shy;
(ii) Y is not confident that s/he is able to express his/her opinions in the working language of the meeting;
(iii) Y's boss is also attending the meeting and it would be inappropriate for Y to appear more knowledgeable than his/her superior.

Thus, although causes of disagreement or conflict in multinational situations *may* lie in differences of culture, conflict may just as well be caused by individual differences, particular circumstances, or situational constraints such as insufficient resources, conflicting or unclear objectives, lack of time, and so on. Thus, culture is only one of three interacting parameters, the other two being individual characteristics and attributes of the situation. Each parameter, either separately or more usually in combination, may explain the behavior of someone in any given situation.

Focusing on supposed cultural differences, therefore, is not of much help at an operational level; it emphasizes differences in *nationality* and may lead people to make generalized predictions about how this or that person will behave based on the stereotypical ways associated with a particular country or culture. Nor does it take into account the possible impact of situational factors or of individual personalities. People, we should remember, are already "… culturally complex beings with repertoires encompassing national, regional, organizational, professional and gender influences" (Friedman & Berthoin Antal, 2005: 82). As mentioned above, any group of people will tend to develop its own culture; in fact, any group of people from different national or organizational backgrounds often find it necessary to establish a team culture (i.e., an agreed common understanding) of what the team will try to accomplish and how it will go about this. Furthermore, our main focus is on the competences and strategies which will not only contribute to effective *communication*, but also, we hope, facilitate understanding and enrich the participants. It is to communication that we now turn our attention.

COMMUNICATION

The first point to clarify is that in the context of human interpersonal communication, communication is not the simple 'exchange of meaning' as Gibson (2000: 9) in a textbook for students of intercultural business communication, defined it. Even more recently, Nordquist (2015) in the Education section of the website About.com asserts that communication is:

The process of sending and receiving messages through verbal or nonverbal means. (Nordquist, 2015)

Underlying both these definitions is a prevalent view of communication as the *transmission of meaning* in which a sender transmits a message to a receiver (cf. Shannon & Weaver, 1949).

Communication is successful in this model when senders and receivers encode and decode language signals in exactly the same way. Any discrepancy between the message sent and the message received is explained in terms of either different degrees of familiarity with the language code or some kind of interference in the encoding/decoding process.

A similar view was described by Reddy (1979) in what he called *the conduit metaphor*, in which:

(i) language functions like a conduit, transferring thoughts bodily from one person to another;
(ii) in writing and speaking, people insert their thoughts and feelings in the words;
(iii) words accomplish the transfer by containing the thoughts or feelings and conveying them to others;
(iv) in listening or reading, people extract the thoughts or feelings once again from the words. (Reddy, 1979: 290)

According to Reddy (1979: 285), "English has a preferred framework for conceptualizing communication" which, through the semantic structures of the language, can bias our thought processes into believing that communication is like a conduit, and he gives many examples including (p. 286):

(i) Try to get your thoughts across better.
(ii) None of Mary's feelings came through to me with any clarity.
(iii) You still haven't given me any idea of what you mean. (Reddy, 1979: 286)

Thus, in the framework of the conduit metaphor as well as in the transmission of meaning model of communication, the receiver (a reader or a listener) has the task simply of 'extracting' the meaning contained in the words transmitted by the sender; taking it out of them so that it gets "into his or her head" (Reddy, 1979: 288).

However, as Reddy (1979: 286) also pointed out, "… no one *receives* anyone else's thoughts directly in their minds when they are using language" (emphasis in the original), and probably most linguists nowadays (e.g. Akmajian et al., 2001; Knapp, 2004; Zegarac, 2008) would define communication as a process of meaning construction and negotiation. As an example of what may be involved in comprehending a *written* text, consider this short article which appeared some years ago in the British newspaper *The Guardian.*

If we want a new king and queen of England – a Michael Howard or Theresa May – then find us a suitable Tudor or Stuart (say Alex Tudor and Alec Stewart, to open the bowling and batting). Find us a stray Bourbon to take the biscuit or a Battenburg on free transfer, with cake. Do not any longer assume that the lot we currently endure can be somehow transformed into objects of reverence or patriotic desire. They can't. The Queen may or may not be beyond reproach, but the family she leaves us with and the courtiers who surround her are just one damned reproach after another. They are the Conservative dilemma writ large across the coming century. (The Guardian, October 27, 2003: 11)

In order to understand this text, in order to establish some kind of coherent representation of it in memory, the reader must not only connect the different ideas or concepts (text elements) appearing in different parts of the passage, but also he or she must make connections between these text elements and previous background knowledge. Without entering into a complete analysis, it should be clear that it will mean very little to any reader who has no prior knowledge of the very culturally-specific themes, or for whom the idiomatic phrases 'to take the biscuit', 'on a free transfer', 'to have your cake and eat it' (to take just three examples), mean nothing. Thus, a reader without the necessary linguistic and/or background knowledge will have nothing to guide his or her comprehension. They may find the text almost incomprehensible, and will find it difficult to place it in its sociocultural and historical context. Clearly, then, a reader has an active role in processing and interpreting written texts. As Rosenblatt (1994: 1063) put it in his Transactional Model of reading:

Instead of two fixed entities acting on one another, the reader and the text are two aspects of a total dynamic situation. The 'meaning' does not reside ready-made 'in' the text or 'in' the reader but happens or comes into being during the transaction between reader and text. (Rosenblatt, 1994: 1063)

Understanding the short text above is difficult enough, but at least a reader can go back and re-read the text. In real-time face-to-face speaking, body language, intonation and other paralinguistic features are used, intentionally or not, to express subtle nuances of meaning.

We speak with our vocal organs, but we converse with our entire bodies ... Paralinguistic phenomena occur alongside spoken language, interact with it, and produce together with it a total system of communication ... the conversational use of spoken language cannot be properly understood unless paralinguistic elements are taken into account. (Abercrombie, 1968)

The more technical discussion of what is loosely described as tone of voice involves the recognition of a whole set of variations in the features of voice dynamics: loudness, tempo, pitch fluctuation, continuity, etc. ... Among the most obvious non-vocal phenomena classifiable as paralinguistic, and having a modulating, as well as punctuating, function is the nodding of the head (in certain cultures) with or without an accompanying utterance indicative of assent or agreement ... One general point that has been continually stressed in the literature is that both the vocal and non-vocal phenomena are to a considerable extent learned rather than instinctive and differ from language to language (or, perhaps one should say, from culture to culture). (Lyons, 1977)

We have seen, then, that effective communication involves not only language knowledge but also draws on general world knowledge, as well as on specific subject-matter knowledge. Effective comprehension entails huge amounts of cognitive processing involving evaluating, inferring, integrating information both within and across written or spoken discourse, monitoring the comprehension process, being aware of inconsistencies and anomalies, applying appropriate strategies when comprehension breaks down, and modifying one's existing knowledge structures in response to new information that promote the construction of meaning.

In intercultural contexts, such 'meaning construction' can be especially problematic: people may focus on different clues when inferring meanings, or they may arrive at different meanings from the same clues. As a result, mismatches may occur in the messages that people think have been communicated. This is why agreeing on the meanings of terms and concepts is necessary; especially at the beginning of a project if problems are to be avoided later on. It is equally important to check and monitor understanding. This does not only apply to people's use of terms and concepts, it also applies to the general discourse – during meetings for example. Four simple strategies for checking understanding are:

- asking for repetition;
- asking for clarification;
- asking for confirmation;
- summarizing – for everyone.

In sum, although human interpersonal communication uses language codes (English, Spanish, Japanese, Swahili, Yoruba, and so on), languages by themselves do not transfer exactly the thoughts, emotions or feelings of the writer or speaker. Many meanings are not contained explicitly in the language code and a reader or listener needs to work out the message through inferencing, elaboration and association. Communication is "not a success without effort" system. It is an "energy must be expended system" (Reddy, 1979: 308), or as George Bernard Shaw suggested: "The single biggest problem in communication is the illusion that it has taken place."

INTERCULTURAL COMPETENCE

We should now be in a better position to explore how we might conceptualize intercultural competence. The first thing to note is that there are a number of terms which are used to refer to it. They include:

- cross-cultural capability
- cultural fluency
- intercultural competence
- intercultural communication competence
- intercultural effectiveness
- intercultural sensitivity
- transcultural communication competence

There seems to be no consistent distinction between these terms. Indeed, they seem to be largely interchangeable as the following definitions show.

We conceive of intercultural communication competence as 'the ability to effectively and appropriately execute communication behaviors to elicit a desired response in a specific environment.' This definition shows that competent persons must not only know how to interact effectively and appropriately with people and the environment, but also know how to fulfil their own communication goals using this ability. (Chen & Starosta,1998: 241–2)

Transcultural communication competence (TCC) refers to an integrative theory-practice approach enabling us to mindfully apply the intercultural knowledge we have learned in a sensitive manner. Specifically, it refers to a transformation process connecting intercultural knowledge with competent practice. (Ting-Toomey, 1999: 261)

A wealth of interaction skills that permit individuals to cross cultural boundaries flexibly and adaptively. (Ting-Toomey, 1999: 261)

Intercultural competence is the ability to communicate effectively in cross-cultural situations and to relate appropriately in a variety of cultural contexts. (Bennett & Bennett, 2004: 149)

The ability to communicate effectively and appropriately in intercultural situations based on one's intercultural knowledge, skills and attitudes. (Deardorff 2006: 247)

A dynamic ongoing, interactive self-reflective learning process that transforms attitudes, skills and knowledge for effective communication across cultures and contexts. (Freeman et al. 2009: 13)

Intercultural competence is ... a state of becoming rather than a destination. (Leask, 2015: 63)

What emerges from these definitions is that intercultural competence is a highly complex concept, and although they do provide some answers as to what it might be, they also raise many questions. For example:

a) What is 'effective communication' and how do we measure it?
b) What does it mean 'to be appropriate'?
c) If I am an effective communicator in one culture (say Finland), will I also be an effective communicator in another (for example, Spain)?
d) Is intercultural competence a cluster of skills or a combination of attitudes and attributes? How much of each?
e) If I learn a set of culture-specific skills and knowledge and am competent in that culture, will it make me culturally incompetent in another?
f) Do I need to be proficient in the language of the country to be interculturally competent? What does it mean 'to be proficient in a language'?
g) How does intercultural competence transform attitudes, skills or knowledge?
h) How could I know when and if I ever become interculturally competent?

We can say, however, that effective intercultural competence:

- is not about knowledge of a specific culture or even several cultures but refers to operating effectively across cultures;
- is an ongoing, mindful, transformational process which leads us to challenge our assumptions, and values;
- consists of skills and knowledge as well as personal attributes and attitudes;
- has communication as a major component.

INTERCULTURAL COMMUNICATION COMMUNICATIVE COMPETENCE

There are numerous examples and models in the literature which attempt to map out the competences needed for effective intercultural interaction (e.g. Chen & Starosta, 2005; Glaser et al., 2007; Prechtl & Davidson-Lund, 2007; Ting-Toomey, 1999), all of which identify communication as a major component of intercultural competence. Other studies (e.g., Diamond et al. 2011; Fitch & Desai, 2012; Joshi & Lazarova, 2005) into what skills and knowledge employers value highly, have shown that the ability to communicate well in both spoken and written contexts, as well as an awareness of cultural nuances and protocols and an ability to embrace multiple perspectives and challenge thinking or the capacity to develop new skills and behaviors, are rated by employers higher than foreign language proficiency. Many of these studies were carried out in English-speaking contexts, which might reduce the importance given to knowledge of foreign languages. On this point, the choice of a working language for a multinational team should be an important one as it carries with it a number of implications. In the words of Bournois & Chevalier (1998: 207):

Choosing the working language can create winners and losers. Language is clearly associated with power, influence and emotional issues. Let us put it bluntly: the ability to master or not master English can create an unequal playing field. Sometimes language differences can be interpreted as personality problems and you can be treated as a deviant or simply ignored just because you do not seize all the linguistic subtleties (Berry, 1990). Beyond discouraging participation, you tend to assimilate language fluency with scientific competencies. If you want to be the leader of the project, it is better if you have English as your mother tongue. Conversely, it is not always the scientific stars that represent a country or an institution but those who are relatively more apt in English. (Bournois & Chevalier, 1998: 207)

Project managers would do well to bear in mind the proficiency level in the working language of the team members. However, not all communication problems in international contexts are a result of insufficient command of the working language. As we have discussed above, differences in communicative conventions across cultures can cause people to interpret messages in very different ways from what the speaker or writer intended. And nor are disagreements on the meanings of fundamental terms and concepts restricted to international contexts. Even between academic staff who are native speakers of English yet work at different universities (and, hence, from organizations with different cultures) there is often confusion between the terms *module*, and *unit*, as well as *chapter* and *section* or *online learning*, *distance learning* and *e-learning*. We should remember, here, that *intercultural communication* does not refer solely to situations between people from different countries, and the need for negotiating agreement on the meanings of terms and concepts is often necessary if problems are to be avoided later on.

Effective communication management, then, requires more than merely choosing a common working language. It needs an approach which tries to maximize understanding. Table 2 identifies a number of linguistic, cultural and social skills, competences, attitudes and strategies which might contribute to more effective communication in intercultural contexts. These are derived from several sources, although this form of visualization is based on a framework given by Spencer-Oatey (2011). The left-hand column presents the intercultural competences, while the right-hand column shows their associated strategies for more effective communication in intercultural contexts. The list is not exhaustive and nor are the sections mutually exclusive.

Table 2. Intercultural communication competences

Intercultural competence	Intercultural communication strategy
Communication management	• Uses the communication channel most appropriate for the purpose (face-to-face; e-mail, telephone, etc.). • Attends to who should receive what information and how (e.g. in some cultures it is not appropriate to send an email to one's superior or to include a high-ranking official in a mass email. An individual oral or written report given directly to the person concerned would be more appropriate in these circumstances. • Recognizes when communication breakdowns or misunderstandings occur and takes action to resolve them. • Sensitivity to working practices; e.g., organizational decision-making processes and hierarchy, interaction style, access to information, etc., and takes these aspects into account.
Active listening	• Listens with attention and shows this through nods, agreement, questions, repetition of words. • Checks and clarifies the meaning of important terms, concepts and ideas • Notices potential misunderstandings and seeks clarification or negotiates meaning until a common understanding is reached. • Draws attention to important ideas and messages to ensure that all participants attach the same meaning to them.
Attuning	• Attends to indirect signals of meaning, such as intonation, hesitation, eye contact and body language, • Actively tries to deepen their knowledge of the meaning of such indirect signals. • Learns to interpret indirect signals appropriately in different cultural and communicative contexts
Use of language (language adjustment)	• Adapts use of language to the proficiency level of the recipient(s) so as to maximize comprehensibility • Attends to, and adapts where necessary, aspects such as: – vocabulary, including idioms. – length and grammatical complexity of sentences – speed of delivery – pronunciation and/or accents. • Structures and highlights information by using 'scaffolding' (such as discourse markers), sequencing, pausing to allow assimilation, and appropriate visual or written aids. • Makes salient the most important aspects. • Uses the most appropriate style and channel of communication for the context and audience (e.g. formal/informal, telephone/email).
Building relationships	• Attends to hierarchy and power relations, and how they may influence behavior in different contexts. • Sensitive to the partner's cultural values and practices, to role relations and role obligations in the partner's culture. • Pays attention to people's personal sensitivities and avoids making them 'lose face'. • Builds networks and connections on a personal as well as a professional basis. • Shows care and concern for other people's welfare. • Willing to try out words and expressions in partners' languages.
Personal attributes and attitudes	• Open-minded with a broad outlook. • Willing to learn. • Patient. • Observant – without being judgmental. • Adapts behavior to suit the circumstances. • Tolerant of others - accepts ideas and behavior which is different from his/her own. • Self-confident and with a sense of identity. • Respects local values without abandoning his/her own. • Aware that his/her own behavior may seem strange or 'incorrect' to others. • Prepared to make mistakes – and to learn from them. • Recognizes negative emotions and takes steps to manage them. • Sensitive to communication needs and to potential breakdowns or misunderstandings.

CONCLUSION

Intercultural communication competence (ICC) refers to the ability to handle multicultural encounters and operate effectively in intercultural contexts. It is becoming increasingly recognized as an important ability which can:

- increase understanding and acceptance of diversity (as opposed to mere tolerance)
- increase effective communication
- help good relationships to develop
- help to create and maintain international partnerships
- allow creativity to flourish
- improve group and individual effectiveness

ICC itself appears to consist of a complex combination of linguistic, cultural and social skills, attitudes and strategies. Perhaps the most important aspect, though, is that as ICC develops, changes in behavior or knowledge or attitudes reflect a change in the individual's worldview, from regarding without question his or her own primary beliefs and behaviors as the only reality, the only way to do things, to accepting his or her primary culture as just one possible reality among many, equally valid, possibilities (Bennett, 2004). And Bennet goes on to say that a person's worldview must shift from *avoiding* cultural difference to *seeking* cultural difference. This transformation itself generates greater sensitivity and potential for ICC to develop further.

EPS students may or may not become interculturally competent on their own, but they will go on to live in a global society, one in which they will need to function effectively in multi-cultural situations. They will need to accept and appreciate the cultural differences and be aware of the historical experiences that divide people, as well as show the common values and the understanding that brings people together. EPS projects and coursework create the conditions for the development of this appreciation.

REFERENCES

Abe, J., Talbot, D. M., & Geelhoed, R. J. (1998). Effects of a Peer Program on International Student Adjustment. *Journal of College Student Development, 39*(6), 539–547.

Abercrombie, D. (1968). *Elements of General Phonetics*. Edinburgh University Press.

Akmajian, A., Demers, R. A., Farmer, A. K., & Harnish, R. M. (1990). *Linguistics. An Introduction to Language and Communication*. MIT Press.

Allport, G. W. (1954). *The Nature of Prejudice*. Perseus Books.

Alred, G., Byram, M., & Fleming, M. (Eds.). (2002). *Intercultural Experience and Education*. Multilingual Matters. doi:10.21832/9781853596087

Bennett, J. M., & Bennett, M. J. (2004). Developing intercultural sensitivity: An integrative approach to global and domestic diversity. In D. Landis, J. Bennett, & M. Bennett (Eds.), *Handbook of Intercultural Training* (pp. 147–165). Sage. doi:10.4135/9781452231129.n6

Bennett, M. J. (2004). Becoming Interculturally Competent. In J. Wurzel (Ed.), *Toward multiculturalism: A reader in multicultural education* (pp. 62–77). Intercultural Resource Corporation.

Berry, J. W. (1990). Psychology of acculturation. In *Nebraska Symposium on Motivation: Cross-cultural perspectives. Current theory and research in motivation* (vol. 37, pp. 201–234). Lincoln, NE: University of Nebraska Press.

Bournois, F., & Chevalier, F. (1998). Doing research with foreign colleagues: A project-life cycle approach. *Journal of Managerial Psychology, 13*(3/4), 206–212. doi:10.1108/02683949810215011

Chaney, L. H., & Martin, J. S. (2011). *Intercultural Business Communication* (5th ed.). Prentice Hall.

Chen, G.-m., & Starosta, W. J. (2005). *Foundations of Intercultural Communication* (2nd ed.). Allyn & Bacon.

de Montaigne, M. (1580). *Essays of Michel de Montaigne* (C. Cotton, Trans. & W. C. Hazlitt, Ed.). https://www.gutenberg.org/files/3600/3600-h/3600-h.htm

Deardorff, D. K. (2006). Identification and assessment of intercultural competence as a student outcome of internationalization. *Journal of Studies in Intercultural Education, 10*(3), 241–266. doi:10.1177/1028315306287002

Diamond, A., Walkley, L., & Scott-Davies, S. (2011). *Global graduates into global leaders*. CIHE.

European Commission. (2015). *Erasmus+ Programme Guide* (English Version 3). https://ec.europa.eu/programmes/erasmus-plus/documents/erasmus-plus-programme-guide_en.pdf

Fielden, J. (2007). *Gobal Horizons for UK Universities*. CIHE.

Freeman, M., Treleaven, L., Ramburuth, P., Leask, B., Caulfield, N., Ridings, S., & Sykes, C. (2009). *Embedding the Development of Intercultural Competence in Business Education*. Final Report to the Australian Learning and Teaching Council.

Friedman, V. J., & Berthoin Antal, A. (2005). Negotiating Reality. A Theory of Action Approach to Intercultural Competence. *Management Learning, 36*(1), 69–86. doi:10.1177/1350507605049904

Furstenberg, G. (2010). Making culture the core of the language class: Can it be done? *Modern Language Journal, 94*(2), 329–332. doi:10.1111/j.1540-4781.2010.01027.x

Gesteland, R. (2005). *Cross-Cultural Business Behavior*. Copenhagen Business School Press.

Gibson, R. (2000). *Intercultural Business Communication*. OUP.

Glaser, E., Guilherme, M., Garcia, M. C. M., & Mughan, T. (2007). *Intercultural Competence for Professional Mobility*. Council of Europe.

Hofstede, G. (1980). *Culture's Consequences: International Differences in Work-Related Values*. Sage Publications.

Hofstede, G. (1984). *Culture's Consequences: International Differences in Work-Related Values* (2nd ed.). SAGE Publications.

Hofstede, G. (1991). *Cultures and Organizations: Software of the Mind: Intercultural Cooperation and its Importance for Survival*. McGraw Hill.

Jameson, D. (1993, March). Using a Simulation to Teach Intercultural Communication in Business Communication Courses. *The Bulletin*, 3–11.

Jones, E. (2013). *Internationalization and employability: the role of intercultural experiences in the development of transferable skills. Public Money and Management*. March Issue.

Joshi, A., & Lazarova, M. (2005). Do "Global" Teams Need "Global" Leaders? Identifying Leadership Competencies in Multinational Teams. In Managing Multinational Teams: Global Perspectives (Advances in International Management) (vol. 18, pp. 281 – 302). Emerald Group Publishing Limited. doi:10.1016/S0747-7929(05)18011-1

King, R., & Ruiz-Gelices, E. (2003). International student migration and the European 'Year Abroad': Effects on European identity and subsequent migration behaviour. *International Journal of Population Geography*, *9*(3), 229–252. doi:10.1002/ijpg.280

Leask, B. (2015). *Internationalizing the Curriculum*. Routledge. doi:10.4324/9781315716954

Lewis, R. D. (2006). *When cultures collide: leading across cultures*. Nicholas Brealey Publishing.

Lewis, R. D. (2015). *Introduction to Cross-Culture*. Richard Lewis Communications. https://www.crossculture.com/services/cross-culture/introduction-to-cross-culture/

Lubin, G. (2013). The Lewis Model Explains Every Culture in The World. *Business Insider*. https://www.businessinsider.com/the-lewis-model-2013-9

Lyons, J. (1977). *Semantics* (Vol. 2). Cambridge University Press.

Mungazi, D. A. (1991). *Colonial Policy and Conflict in Zimbabwe: A study of cultures in collision 1890-1979*. Taylor & Francis.

Mungazi, D. A. (1996). *The Mind of Black Africa*. Wetsport, Greenwood Publishing Group.

Nieto, S. (1999). *The Light in Their Eyes: Creating Multicultural Learning Communities. Multicultural Education Series*. Teachers College Press.

Nordquist, R. (2015). *Glossary of Grammatical and Rhetorical Terms*. About.com. http://grammar.about.com/od/c/g/communicaterm.htm

Prechtl, E., & Davidson-Lund, A. (2007). Intercultural competence and assessment: perspectives from the INCA project. In H. Kotthoff & H. Spencer-Oatey (Eds.), *Handbook of Intercultural Communication* (pp. 467–490). Mouton de Gruyter. doi:10.1515/9783110198584.5.467

Quintrell, N., & Westwood, M. (1994). The influence of a peer-pairing programme on international students' first year experience and the use of student services. *Higher Education Research & Development*, *13*(1), 49–57. doi:10.1080/0729436940130105

Reddy, M. J. (1979). The Conduit Metaphor: A Case of Frame Conflict in our Language about Language. In A. Ortony (Ed.), *Metaphor and Thought*. Cambridge University Press.

Rosenblatt, L. (1994). The transactional theory of reading and writing. In R. B. Ruddell, M. R. Ruddell, & H. Singer (Eds.), *Theoretical models and processes of reading* (4th ed., pp. 1057–1092). International Reading Association.

Shannon, C. E., & Weaver, W. (1949). *A Mathematical Model of Communication*. University of Illinois Press.

Smart, D., Volet, S., & Ang, G. (2000). *Fostering social cohesion in universities: Bridging the cultural divide*. Australian Education International Department of Education, Training and Youth Affairs.

Spencer-Oatey, H. (2011). *Achieving mutual understanding for effective intercultural management*. Working Paper, Centre for Applied Linguistics, University of Warwick. https://www2.warwick.ac.uk/fac/cross_fac/globalpeople/resourcebank/researchpapers

Stewart, V. (2007). Becoming citizens of the world. *Educational Leadership*, *64*(7), 8–14.

Talbot, D., Geelhoed, R., & Ninggal, M. (1999). A qualitative study of Asian international students' attitudes toward African Americans. *NASPA Journal*, *36*(3), 210–221. doi:10.2202/1949-6605.1081

Ting-Toomey, S. (1999). *Communicating across Cultures*. The Guilford Press.

Van Hoof, H. B., & Verbeeten, M. J. (2005). Wine Is for Drinking, Water Is for Washing: Student Opinions About International Exchange Programs. *Journal of Studies in International Education*, *9*(1), 42–61. doi:10.1177/1028315304271480

Verschueren, J. (2008). Intercultural Communication and the Challenges of Migration. *Language and Intercultural Communication*, *8*(1), 21–35. doi:10.2167/laic298.0

Ward, C. (2001). *The impact of international students on domestic students and host institutions. Export Education Policy Project*. New Zealand Ministry of Education.

Weigand, E. (1999). Misunderstanding: The Standard Case. *Journal of Pragmatics*, *31*(6), 763–785. doi:10.1016/S0378-2166(98)00068-X

Widdig, B., & Breslow, L. (2005). *Communicating Across Cultures 21G.019*. Spring 2005. Massachusetts Institute of Technology. https://ocw.mit.edu

Zegarac, V. (2008). Culture and communication. In H. Spencer-Oatey (Ed.), *Culturally Speaking. Culture, Communication and Politeness Theory* (pp. 48–70). Continuum.

ADDITIONAL READING

Aslan, E. (2017). The impact of face systems on the pragmalinguistic features of academic e-mail requests. *Pragmatics and Society*, *8*(1), 61–84. doi:10.1075/ps.8.1.04asl

Council of Europe. (2001). *Common European framework of reference for languages*. University Press.

de Guzman, M. R., Durden, T. R., Taylor, S. A., Guzman, J. M., & Potthoff, K. L. (2016). *Cultural Competence: An Important Skill Set for the 21st Century*. https://extensionpublications.unl.edu/assets/html/g1375/build/g1375.htm

Holliday, A., Kullman, J., & Hyde, M. (2021). *Intercultural communication: an advanced resource book for students* (4th ed.). Routledge. doi:10.4324/9780367482480

Hua, Z. (2019). *Exploring intercultural communication: language in action*. Routledge.

Moloney, R., & Harbon, L. (2010). Making intercultural language learning visible and assessable. In *Proceedings of Intercultural Competence Conference* (*vol. 1*, pp. 281–303). Academic Press.

Scollon, R., Scollon, S. W., & Jones, R. H. (2011). *Intercultural Communication: A Discourse Approach* (3rd ed.). Wiley-Blackwell.

Sinicrope, C., Norris, J. M., & Watanabe, Y. (2007). Understanding and assessing intercultural competence: A summary of theory, research, and practice. *Second Language Studies*, *26*(1), 1–58.

Vescio, T. K., Gervais, S. J., Heiphetz, L., & Bloodhart, B. (2014). The stereotypic behaviors of the powerful and their effect on the relatively powerless. In T. D. Nelson (Ed.), *Handbook of prejudice, stereotyping and discrimination* (pp. 247–266). Psychology Press.

Villalobos-Buehner, M. (2020). Fostering Cultural Competence Awareness and Dispositions to Reach "Thirdness" or "Decentering" by Engaging in Intercultural Dialogue. *The EUROCALL Review*, *28*(1), 26–39. doi:10.4995/eurocall.2020.12206

Wilson, J., Ward, C., & Fischer, R. (2013). Beyond culture learning theory: What can personality tell us about cultural competence? *Journal of Cross-Cultural Psychology*, *44*(6), 900–927. doi:10.1177/0022022113492889

KEY TERMS AND DEFINITIONS

Communication: A transaction between two or more interlocutors requiring effort and the joint construction of meaning. There is no guarantee that what is understood will be the same as what is meant.

Communication Style: The verbal and non-verbal behavior that is used by a person in a given situation and context to express their ideas. A person´s communication style may influence how other participants in the situation perceive that person.

Cultural Profile: The significant features or characteristics of a group.

Culture: The collective values, norms, attitudes, traditions, beliefs, ways of doing things, shared by a group of people.

Intercultural: Inclusive of people from different backgrounds, whether these be national, regional, educational, social, religious or ethnic, as well as of different genders, sexual orientation and physical abilities. Not the same as 'international'.

Intercultural Communication Competence: The ability to handle multicultural encounters and operate effectively in intercultural contexts.

Worldview: The set of presuppositions, beliefs and values held by an individual that are used to perceive and interpret experiences and form opinions, attitudes, and behavior.

APPENDIX 1. EXAMPLE ACTIVITIES FOR RAISING AWARENESS OF CULTURAL DIFFERENCES

Fielden (2007) suggests that if all graduates developed intercultural awareness and sensitivity they would be better able to live and work in global society. The following activities provide some intercultural sensitivity training. They are not specifically communication activities, nor are they language learning activities. They range through topics such as history, geography, food, etiquette, social institutions, time, the family, religion, the role of destiny, treatment of the elderly and many others. They serve to raise awareness of one's own and what might be other worldviews and some of the factors which have formed them.

Written Assignment

Cultural profile

To 'profile' someone or something is to describe the significant features of that person or thing. In this assignment you need to write a cultural profile of your country and of yourself, and relate this to your communication style.

In other words, you need to do two things: (*i*) to reflect on your own cultural identity and to describe yourself as a cultural being, and (*ii*) to show how your cultural background may influence your way of communicating (your communication style).

This is an individual assignment. You need to write about 3 pages (typed not hand written). The questions below in Table 3 are there to help you and give you ideas. You don't need to answer every question, and you may include other ideas or topics of your own choosing, but you can use these as a starting point to set you thinking.

Table 3. Cultural profiling (Some questions have been taken or adapted from Widdig & Breslow, 2005)

Some aspects of your own cultural identity you could comment on are:
Are most people in your country descendants of 'natives' of the country? Or do the people have a mixed ancestry? Are there any regional and/or ethnic groups in your country? What are relations like between the different groups? What historical legacies (events, people or periods from the past) still have a strong influence in your country? Are there any special physical features in your country (e.g., forests, rivers, mountains, plains, etc.) which have special significance in your history? Are there any traditional beliefs or ceremonial practices which people still perform? How does the climate of your country influence the behavior/personality of the people? Does living in a different climate affect you now? Many countries use an animal or bird as a symbol; does yours? Do you identify with it? Are the boundaries of your country natural (that is, a geographical feature such as a river, sea or mountain range) or political? When/how were the boundaries of your country established? How would you describe a typical person from your country? In some countries (even those with democratic forms of government), the birth of a boy is celebrated more than the birth of a girl. Boys are valued more. What does your gender contribute to your definition of yourself? What does your economic background or social class contribute to your definition of yourself? Are there particular myths or stories you remember from childhood that have helped to define you? Are there particular rituals (for example family or public celebrations) that have helped to define you? What are the relationships between your country and the countries around it? How do people from your country see themselves (e.g., helpful, reserved, with a sense of humor, etc.)? What do the inhabitants of the countries surrounding your own think of your country? Are they right? Within the family, who is expected to do what? How have cultural habits (e.g., music, food, religious practices, economic ambitions, social organization, etc.) in your country changed in the last 50 years? What would you say are the core values of your society? What aspect of your culture do you think would be most difficult for an outsider to assimilate/understand?
Some aspects of communication style you could comment on are:
Topics you tend to avoid talking about or which you talk about frequently (e.g., the English are well-known for talking about the weather). How you use non-verbal communication such as hand gestures, facial expressions, body language and personal space. Whether it is considered rude to interrupt. How loud or how quietly people speak in public (the only people you can hear speaking on London buses are the Spanish tourists!) Whether people allow a person who is speaking to finish what s/he says before speaking themselves. Whether men and women speak in the same way – or whether there are some words or expressions only used by men and others only used by women. How you speak to different people in different ways, e.g., to family members, to friends, to older people, to people of your own age, etc., to professors. How your use of language differs from e.g., your parents, grandparents, other social groups?

Class Activities

Although I have called these class activities 'interviews' they should be more properly regarded as conversations or discussions with the participants, exchanging information prompted by the questions asked. They can be carried out in pairs or better still in groups of up to five students, preferably from different nationalities and genders. It is not necessary for the participants to work through all the questions, nor to follow the order given in Tables 4 and 5. Some time should be given at the end of the session for each group to share what they have learned with the whole class.

First interview

The aim of this task is for you to get to know something about your classmates' cultural identities – and maybe something about your own you hadn't realized. Form into groups of three (three to five) and find out from each other where you're from. Then interview one another using the questions below. Try to get as much detail as possible about the culture(s) of your group members. You'll be asked to report back to the class what you have learned.

Table 4. Suggested questions for the first interview (Some questions have been taken or adapted from Widdig & Breslow, 2005)

Nº	Questions
1	What are some of the typical foods or drinks in the culture?
2	How or when are they traditionally served (for example, at any time or only on special occasions such as weddings or holidays)?
3	Why are they typical? Have they always been produced in the country or were they introduced from somewhere else? For example, the British drink a lot of tea but tea isn't grown in Britain. There are other, historical, reasons why tea is popular.
4	Are there any typical styles of dress?
5	Is there anything (or any color) people don't generally wear?
6	How do people typically spend public holidays or weekends?
7	Do buildings have identifiable styles or features? In other words, what makes buildings in your own culture different or recognizable from the buildings in this or other countries?
8	How do people greet one another? Is it the same for men and women?
9	How close together should people stand when they are having a conversation?
10	Is it usual/unusual to invite people into your home?
11	If you invited someone for dinner, would you expect them to bring a gift; for example, flowers, chocolates, something to drink?
12	How would a visitor be welcomed to someone's home?
13	How are weddings/births typically celebrated?
14	What about deaths? What kind of ceremonies or rituals take place?
15	Are men and women regarded as having the same roles; for example caring for children, doing housework, shopping, work and jobs?
16	How is time valued (e.g., how late can you be when meeting friends, or keeping a business appointment)?
17	Are family relationships as important now as they were, say, 50 years ago?
18	Can you give a cultural 'fun fact' about your country; something that a visitor might need to know in order to avoid misunderstandings or embarrassment? For example, in some countries, it is considered rude to place your hands on your lap during a meal. Both hands should be kept on the table at all times.

Second interview

After the first interview, the aim of this task is for you to get to know something more about one another's cultural identities. Using the questions below, try to get as much detail as possible about the culture(s) of your group members. You'll be asked to report back to the class what you have learned.

Table 5. Suggested topics and questions for the second interview (Some questions have been taken or adapted from Widdig & Breslow, 2005)

	Climate and geography
1	How does the climate of your country influence the behavior/personality of the people?
2	Are there any special physical features in your country (e.g., forests, rivers, mountains, plains, etc.) which have special significance in your history?
3	What about plants or animals? Many countries use an animal or bird as a symbol; does yours? Do you identify with it?
4	Are the boundaries of your country natural (that is, a geographical feature such as a river, sea or mountain range) or political?
5	When were the boundaries of your country established?
6	Social relations, government and education
7	Do animals have rights in your country and are they well-cared for?
8	How is the country governed?
9	What is people's opinion of the police force in your country?
10	Are family relationships as important now as they were, say, 50 years ago?
11	How important is the individual in the culture? How important is the group?
12	In schools, universities and in society in general, are individuals encouraged to ask questions and put forward new ideas or to accept the already existing state of affairs?
13	How important is it to have good connections; for example, to find a job or get promotion, deal with administrative affairs or bureaucracy, have a medical operation and so on?
14	How important is hierarchy; for example, in the family, in business, in social meetings?
15	How are elderly people treated? Are they asked for their opinion on family issues and listened to with respect or are they simply tolerated? On public transport, do younger people give up their seat to an elderly person?
Economics and trade	
16	What are the primary and secondary industries? Have these always been important economic activities or are they new industries? How have they affected people's way of life?
17	Have you traditionally imported food products from other countries? If so, what influence has this had on your country's cuisine?
18	Apart from the product itself, what benefits has commerce (or contact) with other countries brought to yours? Has it brought any drawbacks?
History	
19	What historical legacies (events, people or periods from the past) still have a strong influence in your country?
20	Many countries use an animal or bird as a symbol; does yours? Do you identify with it?
21	Are there any especially famous artists (e.g., actors/writers/painters/musicians/etc.) who have influenced the culture of your country? For example, Cervantes in Spain, Goethe in Germany and Shakespeare in Britain).
22	What about scientists, engineers or inventors? (Scotland, for example, has produced many famous engineers). Have these had any influence?
Religion and beliefs	
23	How important is religion in the culture of your country?
24	How is divine power viewed in relation to human effort? What is the role of destiny in people's lives? What about luck?

Simulation

This simulation is loosely based on an activity described by Jameson (1993). Basically, the participants are put in groups of three and each one given a card on which are written the characteristics of the culture to which they belong for the purpose of this exercise: either blue (see Table 6), yellow (see Table 7) or green (see Table 8). The colors themselves are not important, almost any three colors could be chosen. Participants are then asked to discuss together a project while maintaining the cultural role they have been assigned (see the instructions below). As with the interviews, above, time should be given at the end for students to comment on the experience and share their thoughts.

The instructions for this activity are the following:

1. Read the description of your assigned culture. Think about and practice how you will talk and behave. Be sure to practice conversation distance, greeting rituals, and nonverbal behavior. Practice until you are reasonably familiar with your cultural orientation.
2. Then you will be put into a 'multi-cultural' group where you have to agree on what to do in a project. During this meeting you should maintain the cultural role you have been assigned while you interact with the other people in your multi-cultural group.
3. Please do not discuss the characteristics of your culture with anyone. Just behave in the way you have practiced and discuss the topic you are given. Notice how other people react to you, and how you react to them.

Be ready to discuss what you have observed about your own behavior and reactions as well as the behavior and reactions of others.

The task assumes that:

1. You are a student on an exchange in Belgium [or whichever country you choose] and have been placed in a group with other students to do a short project which involves researching and then reporting on 3 aspects of Belgian culture. The information you gather will be based on your own readings and observations and, in addition, you will need to prepare questionnaires in order to survey Belgian people about their views on the topics.
2. Together with the other students in your multi-cultural group decide on the 3 topics you will research from the list below – or you may include any others you would like to do. Please maintain the cultural role you have been assigned while you interact with your group and do not discuss the characteristics of your culture with anyone. Just try to reach agreement through consensus on the 3 topics.

The set of possible topics to address is vast, but could include: family; education in Belgium; eating & food; women at work; gender equality; national holidays; academic expectations; language; the Belgian beer culture; religion; sexual relationships in modern Belgium; death and funerals; the three regions of Belgium: Brussels, Flanders & Wallonia, etc.

Table 6. Characteristics of Blue culture

Beliefs, values and attitudes that underlie your culture's communication	You believe that fate and luck control most things. You believe in feelings more than reasoning. In your culture, an authoritarian leader makes the final decisions.
Nonverbal traits of your culture	You treat time as something that is not important. Your conversation distance is close (about 35 cm to 40 cm, face-to-face). You touch others on the arm occasionally when in conversation.
Verbal traits of your culture	You frequently express gratitude and other emotions to others. You interrupt others who are speaking and are not offended if they interrupt you.
Social & business customs in your culture	You treat women as superior to men. You avoid conflict.
Greetings and courtesies	When meeting someone or leaving them, you hug warmly but gently. Your most common method of greeting is to say, "Hi there!" You call everyone by their first names, and you use these frequently in conversation (e.g., "As I was telling you, George . . .")
Food	You enjoy eating and drinking. In your culture, eating and drinking are considered some of the greatest pleasures in life. You think it is a sign of friendship and sharing to take food from another person's plate to eat yourself.

Table 7. Characteristics of Yellow culture

Beliefs, values and attitudes that underlie your culture's communication	You believe that people determine their own destinies. You make decisions in a democratic manner. You believe in reason over feelings.
Nonverbal traits of your culture	Your conversation distance is far (about 90 cm, face-to-face). You believe that time is a valuable commodity, and you don't want to waste any. You openly express emotions (e.g., anger, dissatisfaction, happiness).
Verbal traits of your culture	You are direct in expressing your ideas and opinions. You ask many questions, and expect clear answers.
Social & business customs in your culture	You welcome conflict as a way of reaching a better decision. You treat women and men as equals.
Greetings and courtesies	When meeting someone or leaving them, you shake hands firmly. Your most common greeting phrase is, "Good morning" or "Good afternoon." You call people (except for children) by their title (Mr., Mrs., Ms.) plus their family name.
Food	It is considered uncivilized and uncouth — and thus is taboo — to eat any raw food such as uncooked fruits and vegetables. It is also considered uncivilized to let your hands touch your food — all food is eaten either with a fork or spoon or by holding it with a napkin.

Table 8. Characteristics of Green culture

Beliefs, values and attitudes that underlie your culture's communication	Members of your culture are strongly religious and believe that a higher power directly and frequently intervenes in human affairs. Your culture makes decisions by consensus of the group and only after thorough discussion.
Nonverbal traits of your culture	You avoid eye contact with strangers and people you don't know well. Your conversation distance is medium (about 65 cm, face-to-face).
Verbal traits of your culture	You never contradict others directly in conversation. Silence is considered a positive and affirmative signal.
Social & business customs in your culture	You treat men as superior to women. In business settings, men and women often segregate themselves.
Greetings and courtesies	You call people by the courtesy title, 'Brother' or 'Sister', followed by their first names. Your most common greeting phrase is 'Peace'. You bow slightly when greeting someone.
Food	In your culture, people are very conscious of cleanliness, so you always tuck a napkin in your collar before eating or drinking anything. In a group, you always make sure that everyone else has enough to eat before you eat yourself.

Chapter 6
Technical and Professional Communication in the European Project Semester

Joseph Barr
Universitat Politècnica de Catalunya, Spain

Elisabet Arnó-Macià
Universitat Politècnica de Catalunya, Spain

Ignasi Perat
Universitat Politècnica de Catalunya, Spain

ABSTRACT

In EPS, communication is key. Project teams plan and develop projects involving stakeholders in internationalized contexts. Outcomes are defined in a written report and an oral presentation, both assessed for content and communicative value. The chapter situates technical and professional communication (TPC) in engineering education and identifies communication practices as part of the professional practices of engineers. Main approaches to TPC in engineering are presented, and how TPC is approached from a problem-solving perspective is discussed. The rationale and organization of the TPC course is presented with examples based on a practical methodology that promotes familiarization, practice, reflection, and dialogue. A proposal for integrating TPC in EPS and similar programs is made. From an interdisciplinary view, the chapter reflects on the importance of TPC in engineering education, the needs of globalized professional settings, and the ubiquity of digitalization to discuss how engineering students can effectively engage in international engineering communication.

INTRODUCTION

The European Project Semester (EPS) at UPC Vilanova was launched in the academic year 2007-08 and has been offered since then once a year, in the spring semester. Throughout the years, the importance of Technical and Professional Communication (TPC) in the student learning process has become evident.

DOI: 10.4018/978-1-6684-2309-7.ch006

Copyright © 2022, IGI Global. Copying or distributing in print or electronic forms without written permission of IGI Global is prohibited.

TPC plays a cornerstone role in the training of future engineering professionals since engineers need to clearly understand (existing problems) and be understood (proposed solutions). As such, the acquisition of TPC skills within EPS is crucial for project development and effective interaction with numerous stakeholders, as teams go through the milestones of their projects, as well as to prepare student transition from the academic to the professional world. The goal of this chapter is to describe the TPC module offered at UPC Vilanova and analyze how it contributes to improve student TPC skills.

The remainder of this chapter will discuss the approach to TPC in EPS at UPC. After drawing on theoretical backgrounds that inform appropriate pedagogies TPC in engineering education, special emphasis will be placed on the interdisciplinary nature of TPC, drawing on the perspectives of both communication and engineering faculty. The rationale for the design of the TPC course will be discussed, with a detailed presentation of its contents and methodology. Taking into account the position of the TPC course in the EPS program, the last section of this chapter will present the impact of TPC on the program over the years, the challenges encountered in the evolution of EPS at UPC as well as lessons learned and implications for the integration of TPC in EPS and other capstone programs.

THE ROLE OF TPC IN EPS AT UPC-VILANOVA

In EPS, which replicates a real-life engineering project and places emphasis on transversal (or soft) skills, communication is key. It is through continuous communication that project teams can plan and develop their projects, involving a number of stakeholders, such as collaborators, supervisors, companies and clients, in an internationalized context. Furthermore, project outcomes are defined in the form of communication artifacts, typically a written report and an oral presentation, both of which are assessed for contents and for their communicative value. As engineering students experiencing a real-world project for the first time, they also become aware of the importance of TPC, as engineering practice requires multiliterate engineers, who can grapple with the complexities of international professional communication. The role of TPC in engineering has attracted the attention of education and researchers, with a growing body of literature devoted to it over the years (e.g., Ford & Paretti, 2013; Cleary et al., 2017; Eggleston & Rabb, 2018; Yong & Ashman, 2019). Parallel to this interest among academics, TPC has also attracted the attention of corporate and professional sectors, where there is growing awareness of the importance of TPC in professional practice. Examples of this greater focus on TPC are found in the resources for learning technical writing created by Google (2021), preceded by the explicit motto "Every engineer is also a writer" or the resources provided by IEEE (2021), which stress the importance of technical communication in English for engineers on a global scale.

Adopting a problem-based learning approach, focusing on the development of a real project, the EPS curriculum includes some "key elements" (European Project Semester, 2022, under 'Concept'), common to all EPS programs, which basically involve project-supportive courses that train students in a variety of transversal skills for professional practice. Among them, language courses are included (specifically, an English course and a course in the local language) as well as project management and teambuilding. Given the focus of EPS on project development, the English course is, in fact, a course in technical and professional communication, which has a direct impact on the outcomes, both in terms of communication with partners and supervisors during the development of the project and of presentation and dissemination of project results (project report and presentation as well as other dissemination materials). Because of the pivotal role of TPC in engineering projects, in the EPS program offered by the Universitat Politècnica de Catalunya (UPC) at Vilanova i la Geltrú, the communication course is, together with project management (PM), one of the main project supportive courses. Looking at the configuration of other EPS courses, one can find that, unsurprisingly, they are approached from the perspective of project-based learning and focused on practice and improvement for the project. The course listings that appear on the websites of EPS providers vary in their degree of explicitness, titles such as English and communication skills and explicit descriptions of some of the courses, it can be seen that they focus on academic and report writing as well as oral presentation skills so as to prepare students for their successful reporting and presentation of the project, with continuous assessment of students' progress and attention to midterm as well as final reports and presentations[1].

The EPS at UPC started in 2008 (Segalàs, 2012) and since then, it has been offered once a year (in the spring semester), featuring different types of real projects (in collaboration with companies, research groups or public institutions) that involve multidisciplinary work. The program is offered at the campus of Vilanova i la Geltrú (a coastal town located 45 km south of Barcelona) and is integrated in the school of engineering's offer to final-year undergraduate students. As a full-time capstone program that runs over one semester, it is worth 30 ECTS credits, 20 of which are devoted to the project and 10 to the courses. Following the founding principles of the EPS (e.g., Andersen, 2004), the focus is not only on project outcomes, but especially on the process. That is to say, students are encouraged to reflect on their learning and to collaborate with others. Therefore, assessment takes into account progress, interim reports and also self- and peer-assessment, reflecting a view of engineering education, based on cooperative teams, that enforces positive interdependence as well as individual and group accountability (Smith, 1995).

One of the characteristics of the School of Engineering of Vilanova i la Geltrú is that it offers a variety of bachelor's degrees in the field of general engineering (mechanical engineering, industrial design and product development, electrical engineering, industrial electronics and automation) and Information Technology (Computer Science). This variety of degrees leads lecturers and students to embark on multidisciplinary projects, which resonates with the nature of EPS projects.

Although the focus of EPS projects at UPC is on engineering, most of them (reflecting real-life engineering) involve social, business and ethical issues, adopting sustainable and eco-design approaches (Duarte et al. 2020). Following general EPS guidelines, the teams comprise four to six students from at least three different nationalities and, in addition, one of the team members is a local student, as EPS is offered to local students who want to engage in an Internationalization at Home (IaH) experience (Crowther et al. 2000; Beelen & Jones, 2015). The EPS student cohort at UPC usually consists of 40 students, who are divided into international, multidisciplinary teams and are assigned a real project, proposed by a company (or research group) and co-supervised by two tutors, one assigned by the company and the other by the university. Such projects are presented as real-life professional challenges so that students have to collaborate with other team members, take on different roles and solve real problems. Effective communication, thus, becomes essential to solving these problems. On the one hand, project development requires students to communicate with each other and with their supervisors. On the other hand, they will have to report on their progress and results, so they will need to communicate with a variety of audiences, similar to what happens in authentic professional communication. The texts that students submit have to conform to the requirements of the types of texts (or genres) that are common in professional engineering practice. The technical reports that accompany the projects have to include the technical, economic, administrative and environmental aspects that are expected in professional engineering projects, which students will have to deal with once they have graduated. The TPC course is crucial towards these objectives and the challenge faced by the TPC instructors is to guide students towards producing documents that draw heavily on engineering knowledge and practices and that have to be accepted in professional contexts. Complementary to the TPC course, lecturers from the same department teach the intercultural communication course, offered as an intensive seminar, and which is complemented by the "study buddy" program in which local students are paired up with EPS students for cultural and linguistic exchange.

The organization of the EPS program at UPC Vilanova is as follows. The first four weeks of the program are devoted to intensive short seminars on a range of subjects on project-oriented transversal skills, such as teambuilding, innovation, intercultural communication and eco-design among others, offered by a combination of local and guest lecturers, which gives the program a true international atmosphere. Two semester-long courses, technical and professional communication (TPC) and project management (PM) are offered in an integrated manner to accompany students in the different stages of the project, especially in relation to the two project milestones, the midterm report and oral presentation, in which students present their interim results and receive constructive feedback from an international, multidisciplinary panel of lecturers, and the final report and presentation, at the end of the term. TPC training is designed to be conducive to help students succeed at such crucial stages. The TPC course is approached from a meaningful and situated perspective, not only in relation to students' projects, but also interacting with the subjects of other seminars, such as intercultural communication, sustainability, teambuilding, similarly to what could be expected in a real engineering project.

BACKGROUND

Engineering curricula increasingly include a broad range of competences that go beyond specialized technical skills and knowledge, reflecting a view of engineering that incorporates social and ethical dimensions. Since the turn of the century, guidelines for designing engineering curricula, such as the 'Dublin' descriptors (Joint Quality Initiative Informal Group, 2004), in the European university reform, and the ABET Engineering Accreditation Commission (ABET, 2021), in the U.S., have established that current curricula should develop awareness of the social impact of engineering, communication skills, critical thinking, ethical commitment, and skills for lifelong learning. In engineering education, Technical and Professional Communication (TPC) plays a key role, considering that engineers spend much of their work time communicating (Gopsill et al., 2015). Even if engineering students do not think that learning communication is part of their job (Maylath et al., 2013), they need to become aware of the role of communication in engineering and to develop skills for presenting reports and other types of written and oral information to different audiences, which will be assessed by different stakeholders. This section will situate TPC in engineering education, by identifying communication practices as part of the professional engineering practices, and will lay out the theoretical frameworks that support the proposed approach to TPC in capstone programs in general and the EPS program in particular.

TPC in Engineering Education

The fact that communication forms part of professional engineering practices may not be so apparent to engineering students, who, when starting college are generally attracted to mathematics and physics and, as they progress through their studies, they become more interested in technological matters and may prefer doing calculations or manipulating objects rather than writing reports. In addition, and unlike engineering curricula in other universities, in which TPC is a compulsory component, at UPC Vilanova, TPC courses are elective, which means that not all students take them. In such situations in which TPC courses may be perceived as an add-on, it is not uncommon to find the paradox that while students are interested in learning disciplinary communication in English, they sometimes hold stereotypical views of TPC (in our context, known as English for Specific Purposes) as "traditional" English language courses (Arnó-Macià & Mancho-Barés, 2015). However, despite this lack of knowledge of what TPC in engineering education involves, engineering students who have taken such courses generally perceive the value of TPC training at university, as a form of induction in specialized communication in English and as a gateway to international academic and professional practice (Arnó et al., 2020). One of the proposals made in this chapter is that TPC should be a compulsory course in all engineering curricula, which would help students develop the different types of written and spoken texts that they are required to produce throughout their studies, especially their final thesis. In fact, it is in their final thesis (a requisite for graduation) that students have to incorporate what they have learnt during their studies, with an oral presentation and a long report that are assessed by a panel of examiners, usually composed of engineering lecturers, that assess not only the technical contents, but also the communication skills that students demonstrate in the report and the presentation.

While students give great importance to designing, calculating, programming or building a device, they tend to neglect the importance of explaining how it works or how it was developed. For example, the design of a prototype must be accompanied by a user manual that covers a number of aspects ranging from its composition to its operation and must be understandable to multiple audiences (not only engineers) that need to know about, use or operate the machine or system.

As a result, one of the main challenges faced by TPC lecturers (usually with a background in linguistics) is not only to train students in language and communication, but also to make engineering students aware of the importance of TPC for their professional future. A key issue that emerges in TPC pedagogy in engineering is interdisciplinary collaboration, so that communication and engineering faculty (like the authors of this chapter) find common ground to improve students' communication skills. In the well-established fields of Writing Across the Curriculum (WAC), Writing/Communication in the Disciplines (WID/CID) and Integrating Content and Language (ICL), a considerable body of research has been developed to align communication instruction with disciplinary contents, not only to improve students' much needed communication skills, but also to "support engineering learning through assignments that prompt students to select information in ways that are consistent with disciplinary values and stakeholders' needs, and make and justify decisions about approaches and solutions in ways that demonstrate sound engineering judgment" (Paretti et al., 2019: 28). In other words, TPC instruction in engineering not only helps students become better communicators, but also helps them become better engineers.

A key issue in effective TPC pedagogy is student engagement, that is, attracting engineering students to TPC and making them aware of its importance in engineering practice. Apart from interdisciplinary collaboration, it is also useful to approach TPC from an engineering perspective, designing TPC programs that draw on engineering contents and design thinking (Shalamova, 2016; Lane, 2021). Such approaches can facilitate dialogue in interdisciplinary faculty collaboration as well as help students draw on their own skill sets and knowledge, integrating engineering knowledge and communication so as to produce written and spoken texts that can fulfil the expectations from both engineering and communication lecturers (apart from other audiences to which they may be addressed). In addition, as part of this interdisciplinary faculty dialogue, having shared engineering frameworks to approach technical communication (e.g. design thinking, usability and user experience, problem-based approaches, etc.) can help both engineering and TPC faculty find common ground to integrate TPC in engineering education.

This integration of TPC and engineering education becomes even more apparent in engineering capstone courses (Paretti et al., 2007), like EPS, which take a holistic, project-based approach to training students by familiarizing them with the professional skills needed in international contexts. In the EPS, based on international project-based collaboration, TPC is a crucial component. Accordingly, the EPS program at UPC features the TPC course as an essential component to assist students in the production of the main deliverable and evaluated artifacts of the program, namely the Written Report and the Final Oral Defense presentation. These deliverables are assessed both in terms of their professional content and communication value by multidisciplinary expert audiences. As a capstone course that involves the collaboration between university and companies, students are guided through the stages of developing a real project, the TPC in EPS can help bridge the gap between university and workplace situations, as capstone courses involving real companies can contribute to creating situations and contexts that university TPC classes cannot fully simulate (Ford et al. 2021).

Theoretical Background

The main approaches to teaching TPC in engineering draw on well-established disciplines for which there is a considerable body of literature. On the one hand, there is a long tradition in the field of English for Specific Purposes (ESP), which started in the second half of the 20th century with an increased interest in the language of science and technology, partly due to the growing economic and technological power of the US after the Second World War. Thus, the first work on ESP was in fact English for Science and Technology (EST) (Swales, 1985). Key distinguishing features of ESP are that it approaches the teaching of language and communication from the perspective of the needs of the students to whom it is addressed and that its pedagogical practices draw on the activities of the disciplines (Dudley-Evans & St. John, 1998). On the other hand, in the United States, a tradition that has gained momentum has been termed technical communication, professional communication or technical and professional communication (TPC), with such established journals and associations as the IEEE Transactions on Professional Communication or the Society for Technical Communication (STC). Precisely, the definition of technical communication posited by the Society for Technical Communication (2021) stresses that "The value that technical communicators deliver is twofold: They make information more useable and accessible to those who need that information, and in doing so, they advance the goals of the companies or organizations that employ them."

Both frameworks stress the importance of situated learning—gearing the teaching of technical communication towards disciplinary contents— promoting those skills and literacies that are part of the academic and professional practices that future professionals will engage in (Zhang & Chang, 2017; Wingate, 2018), and using the types of written and spoken texts (i.e. genres) of those disciplines (Kedrowicz & Taylor, 2013). Following this trend, Lockwood (2019) proposes a framework for designing workplace English courses, which can be placed on a continuum of situatedness, from more general (educational settings) to more specific (workplace settings), and suggests that the design of such courses should incorporate the perspectives of professionals while taking into account real workplace practices, especially as more and more communicative situations are based on the use of ICT (information and communication technology). Thus, the teaching of TPC goes beyond linguistic skills to embrace professional activities and the development of broader competencies, related to the ethical and social dimensions of the technology professions. Along these lines, TPC is defined by Lannon & Gurak (2021: 23) as "a humanistic endeavor, not just a set of job-related transcription tasks, with broad societal implications." This trend towards approaching communication from a situated, disciplinary-specific perspective has developed, over the past years, as opposed to an approach to communication based on teaching general communication skills and types of texts common to different disciplines (Hyland, 2016).

In addition, as engineering communication contexts become increasingly globalized, greater emphasis is placed on intercultural aspects--which become essential in the EPS program, with projects developed by international teams. Batova (2018) argues for the integration of intercultural communication in technical communication from an experiential approach that incorporates students' experience and perspectives. A global perspective on TPC goes hand in hand with the adoption of an English as a *Lingua Franca* (ELF) model, accommodating both native and non-native speakers of English (Cogo & Jenkins, 2010).

APPROACH TO TPC IN EPS

Based on the principles of TPC presented above, what follows is the approach to TPC in the EPS at UPC, based on the integration of TPC in engineering and the adoption of a problem-solving approach to technical communication, an approach that suits students' engineering mindset (Woolever, 2005). This section presents the rationale and organization of the TPC course in the EPS at UPC. At this point, it should be mentioned that the TPC course has been designed and is being taught by lecturers of English for Specific Purposes (ESP), who, in addition to the TPC course in the EPS, also teach ESP courses that form part of the engineering curricula in the different bachelor's degrees offered, as part of a set of elective courses focusing on technical writing, academic and professional speaking skills and project communication, each of which is worth six European Credit Transfer System (ECTS) credits, which amount to four weekly hours. The design of the TPC course for the EPS is similar to the regular project communication course in the sense that both share the same goals, prepare students to write a report on their projects and present them orally. The main difference between both courses is that the TPC course in the EPS is attached to a real project and helps students develop the communication artifacts that will be evaluated by an external audience, following the conventions set for project texts, while in the regular project communication course students can choose the project they want to communicate about and the texts that they want to develop (targeted to their chosen audience and purpose). The EPS course is shorter in weight, 2 ECTS, and currently it only has a weekly session of 1.5 hours (in past editions it used to be a longer course, with the opportunity to have small group coaching sessions). In spite of these differences, the TPC course in the EPS is based on the same rationale and approach as other technical communication courses taught by the same department.

Specifically, the TPC course in the EPS at UPC has been designed with the aim of raising students' awareness of the importance of technical communication in an engineering project, which, in the case of EPS, has a dual focus: (*i*) professional, as the EPS involves developing a real project in collaboration with industry, (*ii*) and academic, as students have to submit written assignments and, especially, work on their final thesis, which will be assessed by a panel of examiners. The course has been designed to give students exposure to and practice in the aspects required to communicate their projects professionally and effectively, which in the context of EPS involves helping students transfer their technical communication skills from university to the workplace. Below is a detailed account of the organization of the course in terms of the contents, the methodology and the types of activities used, followed by the challenges encountered and the evolution of the EPS over the years.

The TPC Course

Before describing the course objectives, contents and methodology, it is worth highlighting a couple of issues in relation to the EPS students' profiles, their initial expectations of what a technical professional communication course in English is, and why they are required to take it over the whole semester.

EPS Student Profiles and Expectations

As detailed earlier in the chapter, participants on the EPS are final year undergraduate students from a wide range of cultures, nationalities, and academic backgrounds. 'Native' English speakers from the USA, UK and Ireland are regular participants on the course alongside those whose first language is not English. Students who wish to take part in the EPS should already possess advanced skills in spoken and written communication. The TPC course is defined in the EPS coursebook as a course *in English* rather than an 'English' course.

Despite the description of the TPC course in the EPS coursebook, it is perhaps inevitable that some students' initial expectations for the TPC course may be of a language course focusing on linguistic competence rather than the TPC driven communicative competences the course attempts to foster. This can cause students, for whom English is not their first language, to initially feel they may be at a disadvantage compared to those whose first language is English. However, perceptions change quickly and over the duration of the course. Students begin to realize that those 'native' English-speaking engineering students, like themselves, have generally had little training in TPC skills and what is more, unlike themselves, lack training in using English as a *lingua franca*. On the EPS, the language playing field is, in fact, more level than some students may initially believe.

TPC Course – Aims and Organization

The milestones of the EPS mean that students are continuously shaping, reshaping, adding, adapting and ultimately finalizing the evaluated deliverables; i.e. the midterm and final written reports and the midterm and final defense presentations. Along with the course content, the TPC course offers not only a regular time and space for EPS students to meet with their teams at the school, but also an established feedback loop that allows support and guidance from the TPC tutor from project beginning to end.

As stated previously in the chapter, engineering communication contexts have become increasingly globalized. The nature of the EPS in replicating such a context – a collaborative international team project - makes it clear to students from the beginning that communication, with teammates, project supervisors, tutors and other stakeholders in their projects, is vital. Students attend seminars in teambuilding and intercultural communication, among others, during the first four weeks of the semester to incubate and encourage said communication.

The overarching objective of the TPC course therefore is to further cultivate and indeed champion the need for successful communication in the development and delivery of the projects. Nevertheless, the communicative skills needed to explain the particular technical details of a project in writing for an expert audience are not the same as those needed to engage a tribunal in an oral defense situation. The

TPC course attempts to encourage students to reflect on how these and other types of communicative situations are different, which communication strategies need to be adopted in these situations and to develop these strategies to be able to manage the situations successfully.

Overall, and through a practical methodology that involves familiarization, practice, dialogue with and assessment from the course instructor and external audiences, the TPC course intends to provide students with support in order to achieve the following aims:

- to develop strategies for effective technical and professional communication in English, both in speech and writing;
- to carry out a collaborative project;
- to develop critical thinking skills;
- to integrate TPC into engineering content/projects.

TPC course contents

The TPC course consists of six modules that cover the principles, planning, process and product of both the written and oral texts, as well as an overall post-project reflection and self-evaluation. Recognition of and practice in written and oral communication with different audiences and for different purposes is emphasized from the beginning of the course and constantly returned to as students make progress in the confection of the written report and oral defense. Following a process-genre approach, the course combines the formal models of genre with the communicative process carried out through "dialogical modeling" (Dressen-Hammouda, 2008), that is, dialogically constructed by the teacher and the students through interaction and discussion. A process-genre approach is also that espoused in the textbook that serves as the main reference (Bombardó et al., 2007). On the one hand, the notion of genre (Swales, 1990) is essential, as students need to become familiar with the conventions and structure of the typical written and spoken texts used in their field. On the other hand, process, as an essential part of the EPS, also becomes essential in the TPC course, both referring to the stages of producing the document and to the reflective and interactive processes that facilitate effective communication and project development. In turn, reflective and process-oriented skills can help students become aware of the strategies and resources they can use to succeed in future professional scenarios. For this reason, in the module contents presented below, *process* and *reflection* appear as key words.

The course not only focuses on the formal deliverables that are communicated to external audiences, but also focuses on team communication and project development, so that students can develop effective communication skills such as *participating in meetings, time and conflict management* and *cross-cultural communication.*

Although the contents of the modules (detailed below) are covered during the course, depending on the immediate needs of the students, the order in which themes are introduced is flexible. For example, the midterm defense comes around quickly and students are given support in delivering a team presentation and responding to questions and comments from the midterm tribunal despite this content appearing in module 5.

A detailed explanation of the contents of the course modules is given below.

Module 1: Principles and Characteristics of Technical Communication

During an initial introductory session in which the course is presented, students are asked to reflect on their own experience of different types of written and spoken texts, both in English and their own languages. They reflect on the wide variety of genres they already have experience with, from message chats on online gaming platforms to formal academic writing. This is done to raise awareness of the skills they already possess in recognizing the different requirements needed for such different types of communication and to reflect on how these skills can be transferred to the requirements of the type of technical communication they will need to produce during the EPS. The first module then covers the following themes:

a. What is technical communication? Characteristics and a problem-solving approach;
b. Audience;
c. Purpose;
d. Tone and style;
e. Organizing a team and participating in meetings.

Class activities seek to relate the concepts involved in the themes of the module to their own projects. They are encouraged to adopt a problem-solving approach to their own projects not only in relation to the intricacies of the project itself, but also as they consider questions of defining the scope of their project (*Why are they doing it? Where? Who is the project for? Who will be affected and how? What limitations do they have? etc.*), deciding which team member(s) is/are responsible for which tasks, keeping a record of progress, decisions made, next steps etc.

The importance of audience and purpose and their influence on communicative choices is central to the module. A variety of sample texts are analyzed for intended audience, including analysis of the use of language as well as tone and style. Students also practice producing short texts for different audiences and for different purposes.

The final component of the module focuses on organizing the team and participating in meetings. In terms of organization of the team, concepts from the teambuilding seminars are reinforced. Regarding meetings, students are encouraged to hold regular team meetings both in person and virtually and to prepare agendas for them, take minutes and keep a record of when they were held and who attended. It should be taken into account that virtual meetings and online collaboration became more of a fixture of our lives during the pandemic and are a permanent feature of the professional world. Students on two EPS courses affected by the COVID-19 pandemic have needed to adapt to collaborating virtually and the TPC course content has attempted to reflect this reality.

Class time is given over to meeting role plays with each team member taking turns to be responsible for chairing a meeting (or at least one of the topics on the agenda for the meeting), preparing the agenda and taking minutes. This often leads to the intended outcome of disagreements arising between the participants in the meeting. Reflection on strategies for dealing with disagreements, resolving conflicts, cultivating cooperation and positive working relationships results from these practice meetings with the intention of them being put into practice in the project team meetings.

Module 2: Planning and Organizing a Collaborative Project

The second module continues with reflection on the differences between written and oral communication and scenarios are presented in which students decide how best to react – via written or oral communication – and again gain practice in producing written and oral texts for different communicative situations.

Again in this module, as the teams begin to make some progress with their projects, it is valuable for them to reflect on not only what they are being asked to do in their projects, but also to consider how they are going to explain what they are doing, why, and for whom. They look at sample situations that generate communicative tasks which allow them to consider technical communication as solving interrelated problems. Thus, following the problem-solving approach proposed by Lannon (2000), students are encouraged to consider which information needs to be communicated, how this information can be presented to persuade the reader to take the action needed, how potentially conflicting stakeholder interests also need to be considered as well as deciding how the team collaborates to make sure the information is communicated. In other words, students gain insight into applying their engineering problem-solving skills to technical communication. This module addresses:

a. Oral and written communication;
b. Definition of a communicative task: analyzing the communicative situation.

Module 3: The Process – Researching and Gathering Information, Citation, Outlining and Drafting.

This module focuses on the process of confection of both the report and the oral defense. Students have access to a template for the written report which details compulsory components such as a title page, an abstract, team members, bibliography, appendices, etc. However, the number and focus of the sections of the report vary depending on the project and teams need to make decisions about what to include and what to leave out.

As can be seen from the main components of this module, the focus is on gathering and organizing information from a variety of sources and how to use them appropriately in both the written report and oral defense. Emphasis is placed on the importance of avoiding plagiarism and the referencing system used - all reports are required to use the IEEE referencing format.

Importantly, explicit focus is given to patterns of organization for the project texts. Choosing the most appropriate pattern of organization is an important factor that contributes to clarity and readability. Patterns such as problem-solution or cause-effect seem appropriate for the content of many of the project reports and students are encouraged to consider alternative patterns for different sections of their reports. Thus, the module works on:

a. Main stages in the communication process;
b. Gathering information from written and oral sources;
c. Taking notes and expressing one's ideas (avoiding plagiarism);
d. Organizing ideas and developing an outline.

Module 4: Written Communication: Preparing and Reviewing a Written Project.

It is worth noting that the written report makes up 50% of the students' final evaluation. It is evaluated both for its professional content and its communication value. It is therefore essential to devote time to its confection. The elements covered in module 4 give students this support when reviewing and editing their written reports, both at the midterm and final delivery stage. Again, the concepts of audience and purpose are returned to as students consider the primary and secondary readers of their reports. Among other aspects, students need to think about the appropriateness of their reports' format, structure, content, visual information, and layout. Consequently, the module focuses on:

a. Academic documents in engineering;
b. Drafting and revising a document (content, register, appropriateness);
c. Elements of language and style in academic writing.

Module 5: Oral Communication. Preparing and Delivering a Team Oral Presentation

As mentioned previously, although the material that focuses on oral presentations appears in module 5, focus on oral presentations comes before and after both the midterm and final defenses. Along with the components outlined below, explicit focus is given to engaging an audience, dealing with questions 'tactfully' as well as advice on aspects of the delivery of the presentations such as intonation, pausing, signposting, handovers and body language. As a result, the list of topics addressed are:

a. The process of preparing and delivering oral presentations in English;
b. Strategies and techniques for preparing an oral presentation;
c. Structure of an oral presentation;
d. Elements of language (introducing the presentation, signposting, dealing with questions, levels of formality, etc.);
e. Team presentations.

Module 6: Evaluating and Reflecting on the Project, Both the Written Document and the Oral Presentation and Developing Tools for Lifelong Learning.

Another component of the EPS assessment is self and peer evaluation. It is therefore useful for students to have some time to think about how they should evaluate themselves and their teammates. In a perfect world equal credit would be given to each member of the team, nevertheless, in practice this is not always the case. This module therefore promotes concepts of critical reflection. Specifically, it explores:

a. Analysis of the written document and oral presentation;
b. Critical reflection on one's production, the overall project and other projects;
c. Peer review;
d. Strategies and resources for learner autonomy, including internet resources (language skills, technical communication, teamwork, etc.

Methodology and Types of Activities

In this section, the aims, presented above, together with examples of tasks and an account of challenges encountered will be presented and discussed. In addition, and as a reflection of the growing digitalization of society in general, and workplaces in particular, a reflection will be included on the more informal modes of communication that have also been incorporated into EPS in the past years (promotional texts, graphics and videos), alongside videoconferencing and chat communication which are currently becoming part of newer generations' repertoires (Janssen & Carradini, 2021).

The TPC course adopts a highly student-centered approach in attempting to achieve its aims. It is task-based with a constructive feedback loop that includes tutors, project supervisors as well as the students themselves. The clear milestones of delivering the midterm and final oral defenses and written reports provide students with deadlines which promote the use of the tools, strategies and skills offered to the students in the seminars taken in the early weeks of the course.

The midterm milestone provides students with the opportunity not only to explain what the scope and objectives of their project are, but also to explain how they have managed its organization up to that point, what they plan to accomplish by the final milestone and how. Details of their project management skills and strategies are presented at midterm; their use of tools such as Gantt charts for timeline monitoring, as well as other PM apps for idea management, milestone tracking and collaboration are explained to the midterm panel. The TPC course devotes class time to practicing for the midterm defense in collaboration with the PM course. The TPC and PM tutors offer feedback as do the other teams who are encouraged to learn from, comment on and offer constructive feedback to their peers. The midterm defense itself is a valuable opportunity for students to again put the skills and strategies for oral presentations focused on in the TPC classes into practice. We are hugely fortunate to be able to count on panels of international and local experts to hear the midterm presentations and read the midterm reports. The comments and questions from the midterm panel never fail to provide valuable feedback and focus for the students and are immensely helpful.

For the written reports, especially the final report, a well written introduction is vital. Apart from the abstract, the introductory section will usually be the part of the report that is read by the most people. A series of tasks in the TPC classes challenge students to consider what a 'good' introduction is, what it should include and exclude and in what order. We are fortunate to have a bank of previous EPS project reports and, having obtained permission from the authors of those reports, their contents are sometimes exploited in the TPC class. In considering how to write the introduction for the written report, first, students read several introduction samples from previous years' projects, discuss their content and identify any common functions within the organization of the content and questions that the content answers. They then perform a 'reverse outlining' task to extract a plan for the introductory samples to further focus on organization and function. Students then agree on a list of functions that an EPS written report introduction should contain and the questions that should be answered in the introductory section. The following list is that agreed upon by students on the EPS in the Spring of 2021.
The introduction of the written report should include:

1. General context of the problem that the project hopes to solve. *Why is the project necessary? What are we doing? What is the gap in existing / previous projects that this project fulfils?*

2. Background of the company / research group / other and context of the project, i.e. previous EPS projects with the same company, etc. *Who wants the project done? Where is it being carried out? Who is doing it? How is it being done?*
3. Identification of objectives of this project. *What are we doing? Why are we doing it? Who will benefit? How? Who will be affected? How?*
4. A road map for the rest of the report. *How is the project being carried out / has it been carried out? How is the rest of the report organized?*

Following the identification of the functions and questions, students are then asked to write an introduction for their own project. They do this in their teams, collaboratively. They then read other teams' introductions, provide some comments and feedback and then tweak their own introductions to produce a version that is submitted for tutor evaluation. By focusing on the process, patterns of organization and thinking of questions that need to be answered, students gain insight into the writing process and develop writing strategies in the context of their own project that they can then apply to other sections of the written report.

Evolution and Challenges

Although the general structure of the EPS program has been fairly stable since it started to be offered at UPC Vilanova around 15 years ago, the projects developed have evolved over the years, reflecting recent societal changes and new concerns. EPS program coordinators are sensitive to these changes and select projects that involve social responsibility, sustainability and that foster ethical, collaborative and interpersonal values, in accordance with the principles of the EPS creators, of familiarizing students with "integrated engineering contexts" (e.g., Andersen, 2004; Andersen, 2012).

Proposing meaningful and socially responsible projects is essential to engage and motivate students. The words expressed by a former EPS student at the end of the team's final project presentation reflect not only the values that we try to instill in our students, but also the very origin of the EPS program, to foster collaboration and mutual understanding:

Our job as engineers is not only to help companies earn money, but especially to improve people's lives.

Therefore, projects are rooted in current societal concerns and seek solutions that are interesting to companies, respect the environment and facilitate people's lives. We have witnessed an evolution from projects focusing on technical and economic aspects to projects that focus on the UN's Sustainable Development Goals (SDG). There is greater awareness of the need to find solutions for cooperation and the eradication of poverty, to protect our planet and guarantee peace and prosperity. The focus on sustainability, inclusive design and accessibility have characterized our EPS projects over the years. However, an evolution has been observed in terms of technological changes, like the internet of things, smart cities, wearable and embedded technologies as well an eco-design. Other changes that we have witnessed have to do with the digitalization of teaching, accelerated as a result of the COVID-19 situation, but which has had a profound impact on the way we teach, communicate and collaborate and which will stay in the future. For example, the visiting lecturers who taught some of the seminars or participated in examination panels had to switch to videoconferencing during the pandemic, thus leading to hybrid project presentations (students presenting onsite and the tribunal attending online). Instant messaging or collaborative writing tools, which have been useful for teams for some years, will be used even more in the future. Hybrid meetings or classes will probably become a possibility for those visiting lecturers for whom it is not feasible to travel physically, and videoconferencing will probably be kept as a convenient option for more frequent and focused meetings or to seek the occasional collaboration of external experts. The possibilities are endless, but important lessons have been learnt about flexibility and adaptability, applicable to globalized professional contexts.

INTEGRATING TPC IN EPS AND OTHER CAPSTONE COURSES

Throughout this chapter, it has been made apparent that as an integral part of engineering practice, TPC should also be stressed in engineering education programs. Specifically, in the case of the EPS and other project-based capstone programs that immerse students in real professional practices, TPC needs to be integrated in such a way that students can become familiar with those situations that are common in the professional practice of engineering. With its dual focus on academic and professional communication, the TPC course in the EPS aims to help students cope with the multiple communication demands posed in the program (communication within teams, problem-solving, producing technical documentation and manuals to accompany the design of the prototype, and producing reports and presentations to be defended before a panel). In addition, another key objective of this course is to go beyond the specific EPS demands and prepare students for effective and efficient professional communication as part of their future professional lives. This is probably the most pressing challenge, making the TPC course project-related enough so that students see its immediate applicability and at the same time, making it open enough so that learning outcomes can be transferred to future professional situations. With this dual view in mind, this section makes a specific proposal for integrating TPC in EPS and other capstone programs, especially in relation to other program components, namely related courses such as project management and intercultural communication as well as project-based work.

Based on what has been discussed in this chapter, the first point that should be mentioned is that TPC should be one of the core courses or modules in EPS and presented in a clearly identifiable way. Taking into account the golden rules of EPS that establish that an English course has to be offered alongside a local language course, our first recommendation would be not to label this course as "English", which could lead students to think it is a remedial language course, or even those students who are native speakers of English or have a high level of proficiency in the language may feel that this course is not necessary or relevant to their needs. Quite on the contrary, we follow Cogo & Jenkins (2010) in that English as a *Lingua Franca* is also necessary for native speakers of English to interact in globalized settings. Even more so, both native and non-native speakers alike need to develop their technical and professional communication skills and produce documents that are effectively received by different audiences.

Another recommendation that derives from the approach to TPC presented above is that it should be configured as a semester-long course, accompanying students throughout the project, as they state, develop, and refine their ideas, which build up throughout the different project development phases. Similarly, regarding the dialogical modelling type of instruction, it is advisable to organize classes in teams (with reduced numbers of students), so that class sessions resemble a seminar or workshop rather than a lecture type of class.

Perhaps one of the main challenges that the TPC course designer and teacher has to face is how to combine the specificity of the course, helping students verbalize and construct their project deliverables, and its transferability, so that the TPC instruction received as part of the EPS can help students improve their communication skills and be able to adapt to other projects and audiences, especially in the present time, with an increasing number of international student networks, contests and competitions (e.g. Formula Student), which require effective international communication.

Related to the point above, two interrelated principles on which the TPC course should be based are modularity and reflectiveness. Modularity refers to the structure of the course, in the form of modules that form a coherent structure but that can also be approached independently. This way, the course can be made more flexible depending on student needs or project development. The fact that each module deals with one aspect of TPC can help students focus on it (e.g. planning the text, reviewing or citing sources, for example) when needed. Reflectiveness, related to raising awareness of TPC processes, helps develop students' ability to critically evaluate one's own work as well as that produced by others. Such an approach allows students to gradually develop a set of skills and strategies that they can apply in future projects.

An important part of the TPC course is the integration of digital and hybrid forms of communication as well as the use of a variety of ICT tools to assist in the communication, teaching and collaboration processes (e.g. language tools, collaborative writing platforms or learning management systems). These are found not only during project development, but especially in the international workplace. For some time now, and accentuated after the COVID-19 pandemic, videoconference and hybrid meetings as well as all types of online communication and the creation of digital artifacts (videos, posters and other graphic material) are acquiring more and more importance in the presentation and display of projects, both in the EPS and in a wide range of academic and professional contexts.

From the lessons learned through our experience in designing and implementing the TPC course in EPS at UPC Vilanova, we can also derive lessons for designing or reappraising the EPS program, from the perspective of communication. Because of its central role in project development, the TPC course should be coordinated with the project management (PM) course. A twofold benefit would derive from this approach. First, project management and communication are strongly linked in international project management (e.g., Mikhieieva & Waidmann, (2017) so that good communication contributes to better project management and more satisfactory results. Apart from better results, the second benefit is that the alignment of tools and methods used in both TPC and PM provide students with a disciplined and integrative vision of how projects can be effectively managed and developed, providing appropriate documentation for different milestones, a learning outcome directly applicable to the workplace. In the EPS at UPC, both TPC and PM are the only semester-long courses that accompany students throughout the semester.

The TPC course should also be integrated with other short seminars given in the first weeks of the EPS. The most apparent connection is with intercultural communication, taught by the same lecturer, and devoted specifically to raising students' awareness of the notion of culture, cultural diversity and about those skills and attitudes that can facilitate effective interactions on a global scale. Teambuilding is also connected to TPC. Similar to intercultural communication, the intensive seminar provides students with immersion in concepts and principles related to forming teams and effective teamwork, while they apply and develop teamwork skills during the process of project development, with the TPC class helping them towards more effective meetings, the presentation of shared ideas, and conflict resolution. Innovation and sustainability are also related to TPC and to the project, as students should be able to identify and present, in a convincing way, both the innovations that their projects bring and the sustainability issues that any engineering project must take into account nowadays. These issues are discussed in the TPC class, as students work on how they should present them in their reports.

Whether it is the EPS or another capstone program, the integration of different skills, methodologies and concepts that need to be applied globally for effective project development is paramount in preparing students for globalized engineering environments. The challenge remains, as presented in the literature review, as to what extent we can replicate the challenges of authentic professional environments at university. There will be an inevitable transition from study to work contexts in which students will have to demonstrate their ability to transfer their communication, collaboration and professional skills. However, an EPS program rooted in real projects proposed by industry can help students experience some of the scenarios, types of communication and challenges that they are likely to encounter in international engineering practice.

CONCLUSION

This chapter has adopted a broad interdisciplinary perspective on the importance of TPC in engineering education and, especially, in capstone programs like the EPS, which can help students transition between university and the workplace. Designed to help students cooperate and communicate to solve a problem posed by industry, the EPS constitutes a real scenario in which students can apply their engineering knowledge while they put into practice the transversal skills needed in international settings. In this context, TPC skills are crucial for project development and effective interaction with numerous stakeholders, as teams go through the milestones of their projects. This chapter proposes that a specific TPC course is included in the EPS, which draws on engineering skills and practices (for example, using a problem-solving approach). The principles of EPS program design establish that EPS programs should include an English course and work on communication skills, so it can be assumed that the kind of English course that is appropriate in EPS is, in fact, a TPC course in English (also known as English for Specific Purposes). Using this nomenclature is important for two main reasons, for students, to avoid misinterpretations about the nature of the course (not an ordinary English course), and for course designers, to have access to a long tradition of research and practice in a well-established field in applied linguistics, known as English for Specific Purposes or Technical and Professional Communication (the latter term commonly found in the United States). The international atmosphere of the EPS, with cohorts of students and lecturers from different countries, makes it an ideal setting for a focus on English as a Lingua Franca, which is appropriate for both native and non-native speakers to become effective participants in international professional communication. As an important part of engineering education, a focus on TPC can help students in their studies (to produce academic assignments) and in the workplace. The combination of both an academic and a professional orientation makes the EPS an ideal scenario for helping students transfer skills between university and the workplace.

In a central position because of its relationship with project-work and other transversal courses, the TPC course can help students integrate different skills (project management, teambuilding, etc.), progress smoothly throughout the project and produce appropriate texts. To this end, the course presented in this chapter, TPC in the EPS at UPC, has been designed with different principles in mind: based on a problem-solving approach, promoting collaboration, structured in a modular sequence, taking a process-genre approach to communication (so that students can produce appropriate types of texts), encouraging reflection, and integrating digital tools for communication and learning. An appropriate TPC course designed to prepare students for effective international communication with the multiple stakeholders addressed in the EPS (in the university, company, and internationally) can help towards the global EPS goals envisioned by its founder: "It is important that engineering students acquire an international dimension and it is important to strengthen links with the world outside the university i.e. the industry and the international society" (Andersen, 2012: 27).

REFERENCES

ABET. (2021). *About ABET.* https://www.abet.org/about-abet/

Andersen, A. (2004). Preparing engineering students to work in a global environment to co-operate, to communicate and to compete. *European Journal of Engineering Education, 29*(4), 549–558. doi:10.1080/03043790410001711243

Andersen, A. (2012). The European Project Semester: A Useful Teaching Method in Engineering Education. In L. C. de Campos, E. A. T. Dirani, A. L. Manrique, & N. van Hattum-Janssen (Eds.), *Project Approaches to Learning in Engineering Education: The Practice of Teamwork* (pp. 15–28). Sense Publishers. doi:10.1007/978-94-6091-958-9_3

Arnó, E., Aguilar, M., & Tatzl, D. (2020). Engineering students' perceptions of the role of ESP courses in internationalized universities. *English for Specific Purposes, 58*, 58–74. doi:10.1016/j.esp.2019.12.001

Arnó-Macià, E., & Mancho-Barés, G. (2015). The role of content and language in content and language integrated learning (CLIL) at university: Challenges and implications for ESP. *English for Specific Purposes, 37*, 63–73. doi:10.1016/j.esp.2014.06.007

Batova, T. (2018). Global Technical Communication in 7.5 Weeks Online: Combining Industry and Academic Perspectives. *IEEE Transactions on Professional Communication, 61*(3), 311–329. doi:10.1109/TPC.2018.2823598

Beleen, J., & Jones, E. (2015). Redefining Internationalization at Home. In A. Curaj, L. Matei, R. Pricopie, J. Salvi, & P. Scott (Eds.), *The European Higher Education Area* (pp. 59–72). Springer. doi:10.1007/978-3-319-20877-0_5

Bombardó, C., Aguilar, M., & Barahona, C. (2007). *Technical Writing: A Guide for Effective Communication*. Edicions UPC.

Cleary, Y., Karreman, J., Closs, S., Drazek, Z., Engberg, J., Ghenghea, V., Meex, B., Minacori, P., Müller, J., & Straub, D. (2017). TecCOMFrame: A competence framework for technical communication. In *Proceedings of the 2017 IEEE International Professional Communication Conference (ProComm)* (pp. 1-5). 10.1109/IPCC.2017.8013937

Cogo, A., & Jenkins, J. (2010). English as a lingua franca in Europe: A mismatch between policy and practice. *European Journal of Language Policy, 2*(2), 271–294. doi:10.3828/ejlp.2010.16

Crowther, P., Joris, M., Otten, M., Nilsson, B., Teekens, H., & Wächter, B. (2000). *Internationalisation at Home. A position paper*. European Association for International Education.

Dressen-Hammouda, D. (2008). Aligning EAP writing pedagogies across European universities: A case study from France. In I. Fortanet-Gómez & C. Räisänen (Eds.), *ESP in European Higher Education. Integrating Language and Content* (pp. 75–96). John Benjamins Publishing. doi:10.1075/aals.4.06dre

Duarte, A. J., Malheiro, B., Arnó, E., Perat, I., Silva, M. F., Fuentes-Durá, P., & Ferreira, P. (2020). Engineering education for sustainable development: The European Project Semester approach. *IEEE Transactions on Education*, *63*(2), 108–117. doi:10.1109/TE.2019.2926944

Dudley-Evans, T., & St John, M. J. (1998). *Developments in English for specific purposes: A multidisciplinary approach.* Cambridge University Purposes.

Eggleston, A. G., & Rabb, R. J. (2018) Technical Communication for Engineers: Improving Professional and Technical Skills. *Proceedings of the 2018 ASEE Annual Conference and Exposition.*

European Project Semester. (2022). http://europeanprojectsemester.eu/

Ford, J. D., & Paretti, M. (2013). Introduction to special issue on international engineering communication. *Connexions: International Professional Communication Journal*, *2*(1), 3–8.

Ford, J. D., Paretti, M., Kotys-Schwartz, D., Howe, S., & Ott, R. (2021). New engineers' transfer of communication activities from school to work. *IEEE Transactions on Professional Communication*, *64*(2), 105–120. doi:10.1109/TPC.2021.3065854

Google. (2021). *Technical Writing.* https://developers.google.com/tech-writing

Gopsill, A., Hamish, C., McAlpine, B., & Hicks, B. J. (2015). Supporting engineering design communication using a custom-built social media tool – PartBook. *Advanced Engineering Informatics*, *29*(3), 523–548. doi:10.1016/j.aei.2015.04.008

Hyland, K. (2016). General and specific EAP. In K. Hyland & P. Shaw (Eds.), *The Routledge Handbook of English for Academic Purposes* (pp. 17–29). Routledge. doi:10.4324/9781315657455

IEEE. (2021). *IEEE Learning Network.* https://iln.ieee.org/public/trainingcatalog.aspx

Janssen, D., & Carradini, S. (2021). Generation Z Workplace Communication Habits and Expectations. *IEEE Transactions on Professional Communication*, *64*(2), 137–153. doi:10.1109/TPC.2021.3069288

Joint Quality Initiative Informal Group. (2004). *Shared 'Dublin' descriptors for Short Cycle, First Cycle, Second Cycle and Third Cycle Awards.* https://www.aqu.cat/doc/doc_24496811_1.pdf

Kedrowicz, A. A., & Taylor, J. L. (2013). Engineering communication and the global workplace: Preparing professionals and global citizens. *Connexions: International Professional Communication Journal*, *1*(2), 81–105.

Lane, L. (2021). Interstitial design processes: how design thinking and social design processes bridge theory and practice in TPC pedagogy. In M. J. Klein (Ed.), *Effective Teaching of Technical Communication Theory, Practice, and Application* (pp. 29–43). WAC Clearinghouse.

Lannon, J. (2000). *Technical Communication* (8th ed.). Longman.

Lannon, J., & Gurak, L. (2021). *Technical Communication, EBook* (Global Edition). Pearson Education.

Lockwood, J. (2019). What do we mean by 'workplace English'? A multilayered syllabus framework for course design and assessment. In K. Hyland & L. L. C. Wong (Eds.), *Specialised English. New Directions in ESP and EAP Research and Practice* (pp. 22–35). Routledge. doi:10.4324/9780429492082-3

Maylath, B., King, T., & Arnó, E. (2013). Linking engineering students in Spain and technical writing students in the U.S. as co-authors: The challenge and outcomes of subject-matter experts and language specialists collaborating internationally. *Connexions: International Professional Communication Journal*, *2*(1), 150–185.

Mikhieieva, O., & Waidmann, M. (2017). Communication Management Tools for Managing Projects in an Intercultural Environment. *PM World Journal*, *6*(8), 1–15.

Paltridge, B. (2014). Genre and second-language academic writing. *Language Teaching*, *47*(3), 303–318. doi:10.1017/S0261444814000068

Paretti, M. C., Eriksson, A., & Gustafsson, M. (2019). Faculty and Student Perceptions of the Impacts of Communication in the Disciplines (CID) on Students' Development as Engineers. *IEEE Transactions on Professional Communication*, *62*(1), 27–42. doi:10.1109/TPC.2019.2893393

Paretti, M. C., McNair, L. D., & Holloway-Attaway, L. (2007). Teaching technical communication in an era of distributed work: A case study of collaboration between U.S. and Swedish students. *Technical Communication Quarterly*, *16*(3), 327–352. doi:10.1080/10572250701291087

Segalàs, J. (2012). The EPS experience at UPC-BARCELONA TECH. In L. C. de Campos, E. A. T. Dirani, A. L. Manrique, & N. van Hattum-Janssen (Eds.), *Project Approaches to Learning in Engineering Education: The Practice of Teamwork* (pp. 109–124). Sense Publishers. doi:10.1007/978-94-6091-958-9_8

Shalamova, N. (2016). Blending engineering content with design thinking and UX to maximize student engagement in a technical communication class. In *Proceedings of the IEEE International Professional Communication Conference (IPCC)* (pp. 1-5). 10.1109/IPCC.2016.7740493

Smith, K. A. (1995). Cooperative Learning: Effective Teamwork for Engineering Classrooms. *IEEE Education Society/ASEE Electrical Engineering Division Newsletter*, 1-6.

Society for Technical Communication. (2021). *Defining Technical Communication*. https://www.stc.org/about-stc/defining-technical-communication/

Swales, J. (1985). *Episodes in ESP*. Pergamon Press.

Swales, J. (1990). *Genre Analysis. English in Academic and Research Settings*. Cambridge University Press.

Wingate, U. (2018). Academic literacy across the curriculum: Towards a collaborative instructional approach. *Language Teaching*, *51*(3), 349–364. doi:10.1017/S0261444816000264

Woolever, K. R. (2005). *Writing for the Technical Professions* (3rd ed.). Pearson Longman.

Yong, E., & Ashman, P. J. (2019). Integration of the Structured Development of Communication Skills within a Chemical Engineering Curriculum at the University of Adelaide. *Education for Chemical Engineers*, *27*, 20–27. doi:10.1016/j.ece.2018.12.002

Zhang, Z., & Chang, E. (2017). Editorial: Current research and practice in teaching disciplinary literacies. *ESP Today*, *5*(2), 132–147. doi:10.18485/esptoday.2017.5.2.1

KEY TERMS AND DEFINITIONS

Communication Skills: The presentation and exchange of information between individuals considering the context and the set of accepted symbols, rules and conventions.

Engineering Education: Training of future engineers, considering both hard and soft skills, to be able to solve future challenges affecting individuals, society and planet.

English for Specific Purposes: The teaching and learning of English based on the language, activities and texts of the students' disciplines or contexts, which can be academic or professional.

European Project Semester: A semester-long capstone learning program for engineering undergraduates based on project-based learning and integrating international and multidisciplinary teamwork.

Integrating Content and Language (ICL): An approach to teaching an academic subject (usually in higher education) in a foreign language, which includes both content and language learning goals (e.g., an engineering course taught in English).

Writing/Communication in the Disciplines (WID/CID): An approach that promotes the learning of disciplinary writing/communication by requiring students to produce writing/communication that reflects disciplinary practices and conventions, with the aim to help students become familiar and fluent with specific genres and formats typical of their discipline.

Writing Across the Curriculum (WAC): An approach that promotes and caters for students' academic writing and in every course offered in a program of studies.

ENDNOTE

1 See individual descriptions of EPS course listings by the different providers at the EPS website (European Project Semester, 2022). Under 'Providers' one can access the different EPS websites, some of which include more detailed course descriptions.

Chapter 7
Design Thinking at EPS:
An Interdisciplinary and International Experience

Claudia Alquezar Facca
https://orcid.org/0000-0003-0113-5516
Instituto Mauá de Tecnologia, Brazil

ABSTRACT

The purpose of this chapter is to present the experience of applying the design thinking methodology in the development of projects during the European Project Semester. Through an interdisciplinary approach, projects of diverse natures were developed by teams composed of students from different areas of education (mainly engineering and design) and from various nationalities, supported by a team of coaches from distinct disciplines, which demonstrated in practice how the integration between coursework and disciplines is essential and works well in project-based learning. The methodology used, its steps, techniques, and methods applied in the development of projects will be presented, as well as the achieved results.

INTRODUCTION

This chapter reviews an interdisciplinary and international experience of applying a Design Thinking approach to product development projects carried out in the European Project Semester (EPS) at the *Instituto Superior de Engenharia do Porto* (ISEP), the school of engineering of the *Instituto Politécnico do Porto* (IPP), in Portugal, in the period from 2019 to 2021. During this time, the author was undertaking a "sandwich" doctorate in Portugal. The experience resulted from an international partnership between two higher education institutions: ISEP, in Portugal, and *Centro Universitário do Instituto Mauá de Tecnologia* (CEUN-IMT), in Brazil.

Although the agreement previously established between the institutions initially encompassed opportunities for academic mobility among the respective students, in this case, a new form of partnership was opened when, after an institutional visit to ISEP carried out by the author, the possibility of more

DOI: 10.4018/978-1-6684-2309-7.ch007

Copyright © 2022, IGI Global. Copying or distributing in print or electronic forms without written permission of IGI Global is prohibited.

active participation was identified. An invitation by the EPS coordinator (Prof. Dr. Benedita Malheiro) for a workshop on design thinking to be performed by the author was agreed upon.

Through an interdisciplinary approach, integrating different areas of knowledge and guided by specialist professors, projects of diverse natures were developed by teams composed of students from different education disciplines (engineering, design, etc.) and various countries. The intention was to demonstrate the importance of integration between programs and disciplines in project-based learning, in finding innovative solutions to real project problems.

BACKGROUND

The intense changes that the world has witnessed in recent decades, towards cultural, social and economic globalization in the 21st century, are reflected in education as an object of study, concern, focus, strategy and discussion in all nations of the globe (Facca, 2020). The study carried out by the Organization for Economic Co-operation and Development (OECD) highlights that the concept of competences in education goes beyond the acquisition of knowledge, also involving the mobilization of this knowledge with skills, attitudes and values to meet complex demands. Future students will need broad and specialized knowledge plus a wide range of cognitive and meta-cognitive skills (critical thinking, creative thinking, learning to learn and self-regulation), social and emotional (empathy, self-efficacy, and collaboration), as well as practical and physical skills (using new information and communication technology devices) (OECD, 2018). Now and in the years to come, it is fundamental to ensure that motivated young women and men concerned about problems in the developing world continue to enter the field in enough numbers (Facca, Alves, & Barbosa, 2019).

Current challenges faced by engineering education are multiple and hard to address. Institutions and students are experiencing constant changes, including financial constraints, focus shifting and the need to provide and acquire new competences (Malheiro, Silva, Ribeiro, Guedes, & Ferreira, 2014).

The performance of new professionals is closely associated with the improvement of products and processes, optimization and innovation of production management models and Research & Development (R&D) efforts of the companies. By necessity, therefore, higher education requires a central place in the discussions involving the formulation and implementation of strategies for industrial development. They are part of the range of challenges that countries must face to sustain productivity gains and to strengthen their competitive positions in the dynamic knowledge-based economy of the world. The economic growth should be accompanied by a qualitative and quantitative improvement of employment and greater social cohesion. This demands a change in higher education systems, promoting interdisciplinarity, stimulating entrepreneurship, emphasizing the cultural and social involvement of universities, considering the education field as an incubator for the technology-based entrepreneurial talent that will drive economic growth (CNI, 2018).

The world is getting smaller and more complex, but there is more technology, there is more diversity and more intercultural exchange. To address this, today's young people need to have more intercultural competences. They need to be sensitive to different countries' organizational and business cultures. They also need to 'think out of the box' (EPS, 2021).

EUROPEAN PROJECT SEMESTER

In this context, in order to meet the demand resulting from globalization, there is the European Project Semester (EPS), a program designed and created by Prof. A. Andersen in 1995 in Denmark (Malheiro, Silva, Ribeiro, Guedes, & Ferreira, 2014) where the students can gain international experience in a multidisciplinary and multicultural project for one semester in another university (EPS, 2021).

EPS is a mixture of "project-related courses" and project-organized/problem-based learning. It is crafted to address the design requirements of the degree and prepare engineering students with all the necessary skills to face the challenges of today's world economy. Projects are completed in cooperation with commercial businesses and industries or with research centers. There are minor differences between the administration of EPS-projects and the various EPS-providers, but all providers have agreed about the following key elements. EPS is a program offered by several European universities to students who have completed at least two years of appropriate study and is created with engineering students in mind, but students from other disciplines can participate and often enhance the group dynamic (EPS, 2021).

The EPS program is based on the following main features (EPS, 2021): project-centered and accompanying subjects that must support the project subject; offered to 3rd and 4th year engineering, business, and product design students; teambuilding and project management; multinational and multi disciplinarily teams; project supervision focused on the process as well as the product.

The program is offered by a select group of European higher education institutions and was launched by the *Instituto Superior de Engenharia do Porto* (ISEP), at the *Politécnico do Porto* (P. PORTO), in 2011 regarded as an additional tool to apply the best engineering education practices, increase the employability of its graduates and promote internationalization (Malheiro, Silva, Ribeiro, Guedes, & Ferreira, 2014).

The European Project Semester at ISEP (EPS@ISEP) challenges students from multiple educational backgrounds and nationalities to combine competences and embrace ethically- and sustainability-driven design for solving multidisciplinary real-life problems in close collaboration with industrial partners and research institutes. This approach stimulates students to contribute and apply their specific knowledge and develop transversal skills; namely social and intercultural communication, during the different stages of team collaboration (ISEP, 2021)

The curriculum of the EPS@ISEP program consists of the following curricular units (ISEP, 2021): Project Management and Team Work; Marketing and Communication; Foreign Language; Energy and Sustainable Development; Ethics and Deontology. The goal is to offer a program not only focused on developing scientific and technical competences, but also focused on soft skills, which are in high market demand (Malheiro, Silva, Ribeiro, Guedes, & Ferreira, 2014).

PROJECT-BASED LEARNING

Another important issue, as a theoretical reference in this chapter, concerns active learning methodologies. These propose different ways of working in the classroom, whose choice will depend on the objectives and the target audience, among other issues. Since EPS is project-centered, it is essential to understand the principles and values involved in Project-Based Learning (PjBL) to ensure students are offered the best didactic and pedagogical approach during the project development.

Project-Based Learning (PjBL) is similar to Problem-Based Learning (PBL). Both are considered learning approaches that pursue shared goals: they emphasize student independence, encourage research and collaboration, authentically apply content, focus on open-ended questions, develop the skills needed for the 21st century and seem adequate to improve their interdisciplinary skills (Brassler & Dettmers, 2017).

PjBL is defined as a pedagogy that involves two components: an issue or problem that serves to organize and direct (project) activities and a resulting series of artifacts or products that addresses the main issue (Brassler & Dettmers, 2017, p. 3).

PjBL is a student-centered educational model that organizes learning around projects and allows students to learn by doing, applying ideas, and engaging in real-world activities. Adding the concept of interdisciplinary learning to PjBL, iPjBL arises, that is, Interdisciplinary Project-Based Learning, whose focus also involves the application of different information, data, techniques, tools, and perspectives in the search for an innovative and effective product (Brassler & Dettmers, 2017).

So, it can be said that the implementation of active learning strategies in curriculum redesign, including especially iPjBL, enhances learning and the development of interdisciplinary competences (Facca, 2020). Project-based learning (PjBL) and problem-based learning approaches are particularly suited to engineering education, but they no longer provide all competences required in future engineers. Engineers are expected to be able to function in the global market, work in multicultural multidisciplinary teams, hold expertise in a specific engineering field and be proactive lifelong learners (Malheiro, Silva, Ribeiro, Guedes, & Ferreira, 2014).

The curriculum offered at EPS@ISEP is comprehensive. It is supported by PjBL, focusing on important issues such as technology, sustainability, marketing, management, and the development of transversal skills like ethics and teamwork. However, there is no specific discipline focused on product design or design thinking. Hence, there is a latent need or good opportunity to insert the design thinking-based approach in order to offer new creative paths in solving real, complex and human-centered problems.

KNOWLEDGE AND INTERDISCIPLINARITY

The current times are characterized by a great "cognitive inflation", allied to a profusion of disciplines never seen before (Domingues, 2005). No time in history has produced as much knowledge as the 20th century – a real disciplinary big bang. The increase in the number of disciplines, fields of knowledge, scientists, publications, and bibliographic collections has grown dramatically and is responsible for both the evolution of science and technique and for the appearance of true epistemological obstacles to the expansion of knowledge (Facca, 2020).

The great challenge posed to thought and education at the beginning of this century and millennium is the contradiction between the increasingly global, interdependent, and planetary problems on the one hand and the persistence of a mode of knowledge that still privileges the fragmented and compartmentalized knowledge of another hand. The ultra-specialization of the individual and the hyper-fragmentation of knowledge have become responsible for the fact that today no one is able to dominate their field of knowledge or specialty. The consequence of this is the arrival of the end of the specialist, as occurred at the beginning of modern times when the generalist disappeared, precisely because of the emergence of the disciplinary specialty (Domingues, 2005).

The need for change in higher education and the promotion of interdisciplinarity stimulates entrepreneurship. It emphasizes the cultural and social involvement of universities in building a higher education system, driven by strategic government investment in engineering education as an incubator for technology-based business talent (CNI, 2018; Graham, 2018). Over the past two hundred years, Engineering has faced many challenges ranging from production efficiency, cost reduction, quality improvement, pollution control, and safety concerns to automation, computerization, miniaturization, integration of complex systems and resource constraints, among others (Ertas, 2018). Such challenges are contemplated in an evolving and expanding disciplinary structure that has demanded a performance influencing virtually every aspect of modern life. Although the multidisciplinary and interdisciplinary approaches initially developed have produced interesting results, the focus has always been on the artifact as a complex system and not as part of a complex adaptive system. With the current changes resulting from globalization, more engineers are being demanded, and new forms of collaboration and integration directly address the complex design of adaptive systems. "Engineering education must produce a new type of engineer, one who breaks the barriers of disciplinary thinking. Transdisciplinarity must become an integral part of the future of this venerable profession" (Ertas, Gatchel, Rainey, & Tanik, 2007, p. 6).

Traditional disciplinary approaches have often proved insufficient, and researchers and educators are exploring complementary strategies. In this system, products have become integrated, and design and production requirements cross these boundaries, which requires the contribution of various disciplines in Engineering, as well as other disciplines such as business, social sciences, and biological sciences. With the acceleration of the development of new technical systems, the demand has shifted from multi or interdisciplinary project teams to transnational and transorganizational work (Ertas, Gatchel, Rainey, & Tanik, 2007).

In this context, interdisciplinarity can be considered as the approach of different disciplinary fields of knowledge for the solution of specific problems through the sharing of methodology (Domingues, 2005). It is one of the pillars of the EPS@ISEP program.

DESIGN THINKING'S INTERDISCIPLINARY APPROACH

In the scenario of global competition, companies began to realize that it was no longer enough to offer only technological superiority or excellence in performance as a market advantage (Brown, 2009; Vianna, Vianna, Adler, Lucena, & Russo, 2012). Seeking new paths for innovation has coined what is now known as Design Thinking (DT): "an approach focused on the human being that sees in multidisciplinarity, collaboration and tangibility of thoughts and processes, paths that lead to innovative solutions for business" (Vianna, Vianna, Adler, Lucena, & Russo, 2012, p. 12). DT is a user-centered approach to creative problem solving and innovation (Brown, 2009; Vianna, Vianna, Adler, Lucena, & Russo, 2012; Melo & Abelheira, 2015).

According to Melo and Abelheira (2015), DT can be considered a methodology that provides Design tools to solve complex problems by balancing associative reasoning, which leverages innovation and analytical thinking, reducing risks and placing people at the center of the process. By enabling its principles to be adopted by professionals working in different areas, DT ends up decentralizing the practice of Design from the hands of specialized professionals and enables new complementary behaviors.

The integration between disciplines and multidisciplinary teamwork has become increasingly necessary given the complexity of the present-day challenges. Thus, by bringing Design closer to Engineering and Business, both the breadth of interdisciplinary competences and the depth of disciplinary specialization can be considered. Understanding people's problems, designing solutions, implementing the best option, and prototyping to test and validate the best path are steps adopted in DT in an iterative and recursive process (Guzzo & Facca, 2018). Thoughts such as divergence and convergence, analysis and synthesis, deduction, induction and abduction, materialization and experimentation, individuality and collaboration can represent alternative ways to produce new ideas and innovative design solutions (Brown, 2009). The primary benefit of using Design as a way of thinking is its holistic approach to developing projects and finding innovative solutions.

Although there is no single definition, Design Thinking is an idea, a strategy, a method, and a way of seeing the world; a way to solve problems through creativity. According to Tim Brown (2009), CEO of IDEO, "Design Thinking is a human-centered approach to innovation, which starts with the designer's tools kit to integrate people's needs, the possibilities of technology and the requirements for business success". This approach combines what is desirable from a human point of view (desirability) with what is technologically feasible (feasibility) and economically viable (viability) (IDEO, 2020).

Design Thinking begins with "the skills that designers have learned over several decades in their quest to meet human needs with available technical resources, within the practical constraints of business" (Brown, 2009, p. 10). Putting its tools in the hands of people who may never have considered themselves designers and applying them to a wider range of problems, Design Thinking takes the next step, breaking paradigms, approaching innovation effectively and affordably, and integrating all aspects of society (IDEO, 2020).

DESIGN THINKING AT EPS@ISEP

Although the EPS program addresses issues related to the humanities, technologies and project management, there was the need to include the design thinking approach as a way to understand people's needs, define problems, encourage user-centered research, substantiate the conceptualization and provide new tools to develop creative and innovative solutions since the beginning of the project. So, based on the DT's principles and values approach, a workshop was inserted to allow students to develop such activities and, consequently, improve their skills.

The design thinking workshop utilized in the EPS program at ISEP has been expanded and consolidated since its inception. Initially, in 2019, experimentally, it was delivered in 4 hours over one morning but now has a more significant contribution to the program (16 hours over four days). According to the feedback given by the program coordination, the insertion of the DT was very well accepted by the students, contributing positively to the development of the projects, mainly regarding the definition of the users' problems and needs to be attended to.

At the beginning of the semester, the Design Thinking Workshop is included in the program to introduce students to a new user-centered approach to problem-solving, based on the principles, values, processes, techniques, and tools from Design, as a way of changing their mindset. Instead of starting out thinking strictly about the product or the technological solution, students were encouraged to initially think about the needs and desires of people who are not yet being considered. Usually, in an engineering-driven scenario, teams of researchers will discover a new way of doing something, and only afterward will they think about how the technology might fit into an existing business system and create value. By focusing their attention on short-term viability, they may be trading innovation for increment. Design teams will cycle back through all three considerations - desirability, feasibility and viability - throughout the life of a project, but the emphasis on fundamental human needs is what drives design thinking to depart from the status quo (Brown, 2009).

If IDEO is most famous for business-oriented Design Thinking, the d.school at the Hasso Plattner Design Institute at Stanford University is the main reference in the education sector (Björklund & Keipi, 2019). The d.school does not seek to educate traditional designers and, in fact, does not offer "Design" courses. Instead, it serves as a unique environment where students from Medicine, Business, Law, or Engineering can come together to work on collaborative design projects of public interest. d.school encourages human-centered research, brainstorming, and prototyping in all projects. Still, it also applies these fundamental principles of Design Thinking to itself: spaces are dynamic, academic rankings are irrelevant, the curriculum is in permanent flux – it is, in short, a continuous prototype of the educational process itself (d.school, 2020).

Before beginning the practical activities, the definitions, concepts, principles, and values related to design thinking are presented to the students to align the knowledge and introduce the theme and the designer's way of thinking. Values like empathy, collaboration, and experimentation, besides the principles of divergent and convergent thoughts, are fundamental to clarify ideas and perceive other points of view.

The practical activity starts from a question: how to improve the user experience related to a product or service, related to one of the project themes presented by the program.

The workshops held in the EPS@ISEP program are based on the stages of the d.school design process, which is divided into the five modules shown in Figure 1: (*i*) Empathize, to get to know the target audience; (*ii*) Define the key issues; (*iii*) Ideate and create solutions; (*iv*) Prototype, to build and represent ideas; and (*v*) Test ideas and possible feedback from users (d.school, 2020).

Figure 1. d.school design process
(d.school, 2020)

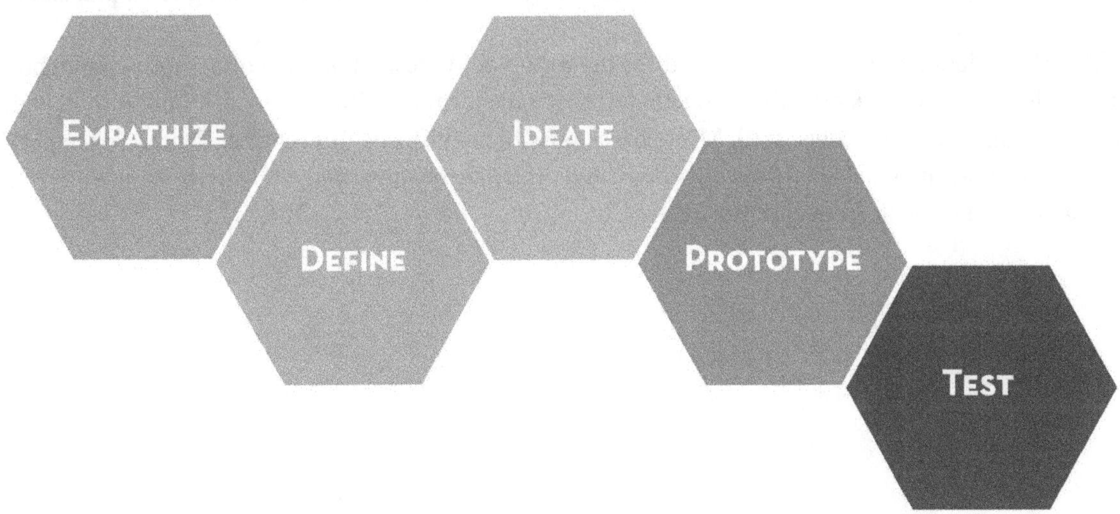

We teach the phases of design thinking as linear steps, but in practice, the process is not always linear – the phases combine to form an iterative approach that you can try out and adapt to suit your specific challenge (IDEO, 2020).

At each stage of the process, students, gathered in multidisciplinary teams, are challenged to use some tools belonging to the design thinking development method, always considering the systemic view of the project, including the human, technical and business aspects, in addition, of course, to sustainability issues.

For example, in the first stage (Empathize Mode), each team has to observe and view users and their behavior in the context of their lives; engage and interact with and interview users through both scheduled and short 'intercept' encounters and immerse and experience what their user experiences. Some techniques applied were: Empathy Map and Journey Map (d.school, 2020).

The second stage (Define Mode) is when each team will unpack and synthesize its empathy findings into compelling needs and insights and scope a specific and meaningful challenge. The define mode is critical to the design process because it explicitly expresses the problem being strived to address through the efforts (d.school, 2020). The techniques used in this step are based on the analysis and synthesis of the information collected, generating insights to be developed in future ideas, such as: Persona and User Profile, Target Audience, etc.

The Ideate Mode, the third stage, is the mode of the design process in which it is possible to generate radical design alternatives, exploring a wide solution space – both a large number of ideas and a diversity among those ideas (d.school, 2020). The main method used here is brainstorming, a mental and collective exercise that could generate as many ideas as possible, so that they are subsequently evaluated and selected, until a viable and feasible path to be prototyped is defined.

The last two stages (Prototype and Test Mode) are about getting ideas and explorations out of our heads and into the physical world. In early explorations, it is possible to keep the prototypes rough and rapid to quickly learn and investigate many different possibilities when people (the design team, the user, and others) can experience and interact with them. Those interactions can help drive deeper empathy, as well as shape successful solutions (d.school, 2020). Due to the pandemic and the online classes, mediated by technology, the prototyping phase was also carried out digitally through the realization of digital prototypes. Students had to create videos and animations with the project storytelling to be presented to everyone at the end of the process.

The Miro Platform (Figure 2) was used for the group activities, an online collaborative whiteboard that enables teams to work effectively together, from brainstorming with digital sticky notes to planning and managing agile workflows. With Miro, it was possible to utilize a full set of collaboration capabilities, make cross-functional teamwork effortless, and organize meetings and workshops using video chat, presentation, sharing, and many other features.

Figure 2. Miro Platform
(MIRO, 2021)

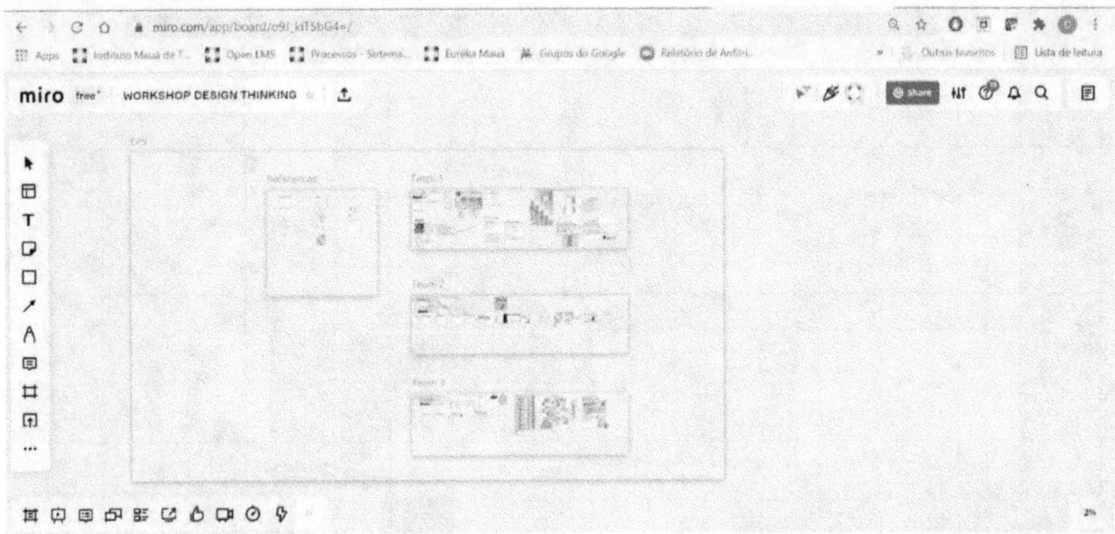

On the Miro Platform, a panel was shared with all students where each team could develop the tasks in a dedicated space, sharing their ideas and observing the development of both their team and other workgroups (Figure 3).

Figure 3. Example of a board developed by one of the teams on the Miro Platform
(MIRO, 2021)

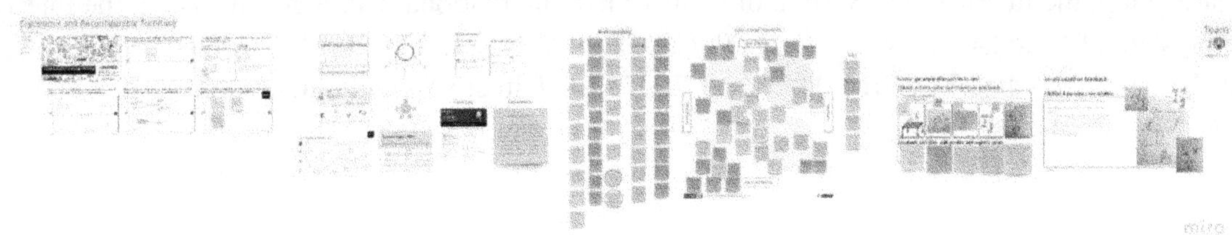

Due to the Covid-19 pandemic, classes in 2021 were taught remotely and totally online through the Zoom Platform (Figure 4). Zoom is a cloud-based video conferencing service that can be used to meet with others virtually – either by video or audio-only or both, all while conducting live chats – and it lets record those sessions to view later.

In the 2021 edition, three working groups were formed, with five members of different nationalities and complementary specialties. As always, the program coordination then presented a list of project proposals for each group to choose what they would like to work on. The themes chosen were: Crowd Orchestration, Smartification of Everyday Objects and Ergonomic and Reconfigurable Furniture.

Figure 4. A view of all students active in 2021on the Zoom platform
(Personal Collection, 2021)

The objectives of the projects were: design, build and test a solution or a system, following sustainable and ethical design practices and the purpose, target user segment and the full set of device requirements were to be defined by the team based on the state of the art, marketing, sustainability and ethical analyses. The requirements were defined as: use or reuse low-cost hardware solutions; use open source software; adopt the International System of Units (NIST International Guide for the use of the International System of Units); comply with the 2006/42/CE 2006-05-17, 2004/108/EC 2004-12-15, 2014/35/EU 2016-04-20, 2014/53/EU 2014-04-16 and ROHS EU Directives; and a maximum budget of 100 € (EPS@ISEP, 2021).

CASE STUDY: FREE DESK, A RECONFIGURABLE AND ERGONOMIC SMART DESK

This chapter introduces, as a case study, one of the projects developed in Spring 2021 entitled "FREE Desk, a reconfigurable and ergonomic smart desk" (Cărăşel, Vandoorne-Feys, Nicoară, & Michał, 2021), created by the students Ioana-Silvia Cărăşel (Information and Communication Technologies, RO), Anastasia Vandoorne-Feys (Product Development, BE), George-Gabriel Nicoară (Industrial Engineering, RO), Marcel Michał Karpiak (Mechanical Engineering and Applied Computer Science, PL) and Nikola Kocheski (Information and Communication Systems for Business and Management, MK). This work was chosen because, among the developed projects, this was the one that kept the same concept presented in the design thinking workshop until the end of the process, being used in the final product proposal.

"In this project, the team designed a sit-to-stand desk that enables the user to define which position allows him/her to focus the best and be comfortable at the same time. The FREE product starts off as a bench, providing usability for other daily activities that may occur before or after working. It can be transformed into a sitting desk, and eventually into a standing desk. The objective is to provide a simple but functional and ergonomic design" (Cărășel, Vandoorne-Feys, Nicoară, & Michał, 2021).

In the first stage of the process – Empathize and Definition – the initial exploratory research was carried out in which each group member reported their own experiences related to the theme to start the discussion on the subject and raise the necessary parameters for the in-depth research that would follow (Figure 5).

Figure 5. Initial exploratory research
(MIRO, 2021)

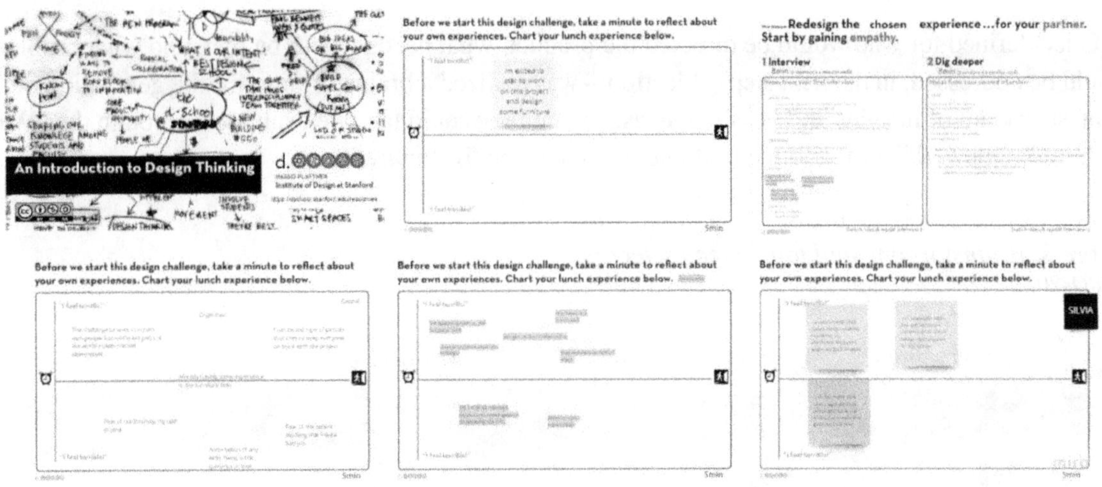

Next, the research was conducted regarding potential users, investigating what their needs would be, what problems were involved, what would be the situations of use. The empathy map and the customer journey map contribute to analyzing the tasks performed until the definition of the persona and the target audience (Figure 6).

Figure 6. User Journey, Empathy Map, Persona, and Target Audience
(MIRO, 2021)

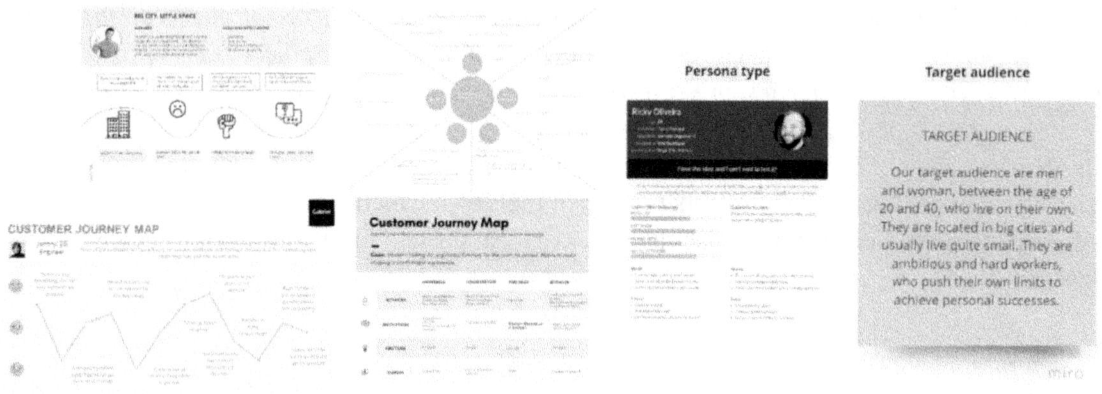

Once defined for who would be directed the product, what needs would be met, and what problems should be addressed, in the next step – Ideation – was realized a brainstorming for the generation of the greatest amount of insights and possible ideas; so that they could later be evaluated (through the "Ways to Grow Framework" tool) and the best one, selected (top 5) (Figure 7).

Figure 7. Brainstorming and top 5 ideas selection
(MIRO, 2021)

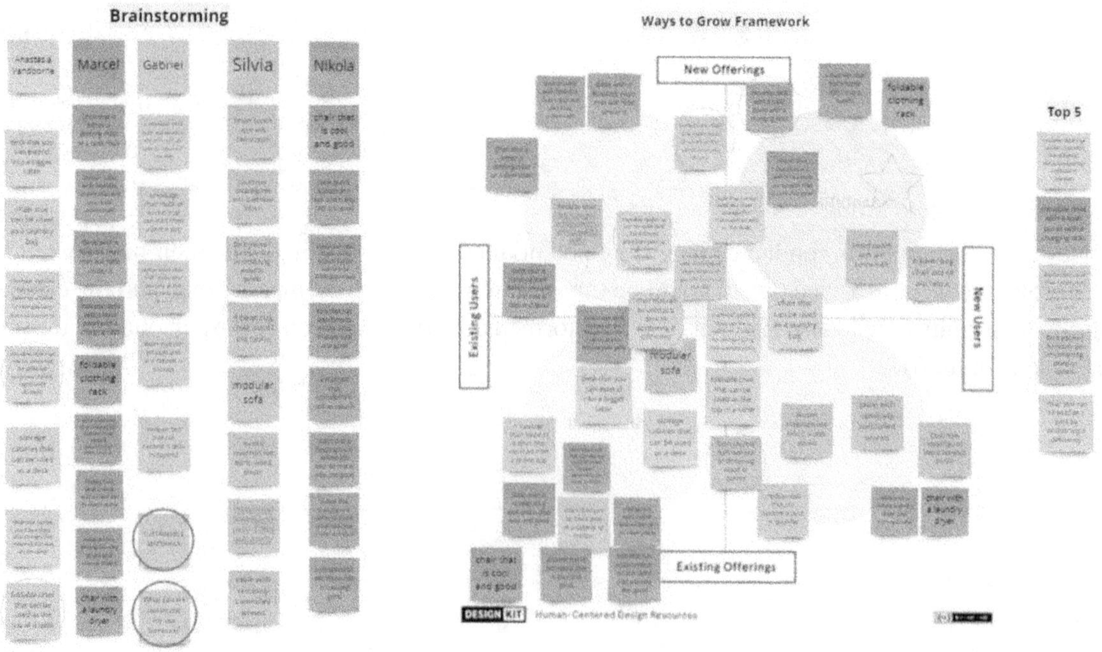

At the end of the ideation phase, the group should illustrate the five best ideas through drawings and graphic representations and, after an evaluation of the positive and negative points, combine the best solutions and propose the final concept of the project (Figure 8).

Figure 8. Ideas generated and the final proposal
(MIRO, 2021)

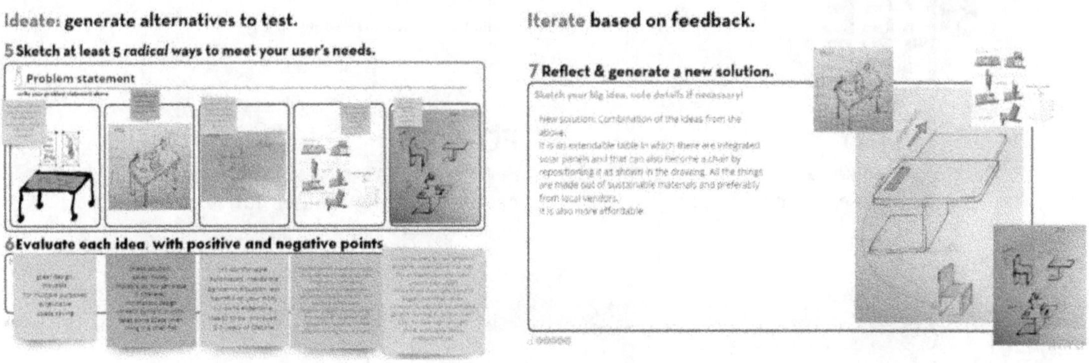

The last two stages – Prototyping and Testing – due to the pandemic were carried out digitally, as physical presence and laboratory activities were suspended during this period. Thus, the prototype was made by creating a video where the group could present their idea and even test the reaction of their audience. The final video is even available on the official project page on the EPS Program Wiki Platform (EPS@ISEP, 2021).

At the end of the process, it is possible to observe the connection of the initial concept – developed during the design thinking workshop – with the final proposal presented (Figure 9), helping to direct the track, defining the problem to be solved, defining the target audience and designing the central idea of the project. It is natural that the project, being dynamic, would continue to be developed, evolving, undergoing several changes and adjustments throughout the entire process. The issue to be highlighted here is precisely related to the contribution of design tools, before the product's technical development, considering people and their needs, before the engineering solution.

Figure 9. Free Desk Final Project
(Cărăşel, Vandoorne-Feys, Nicoară, & Michał, 2021)

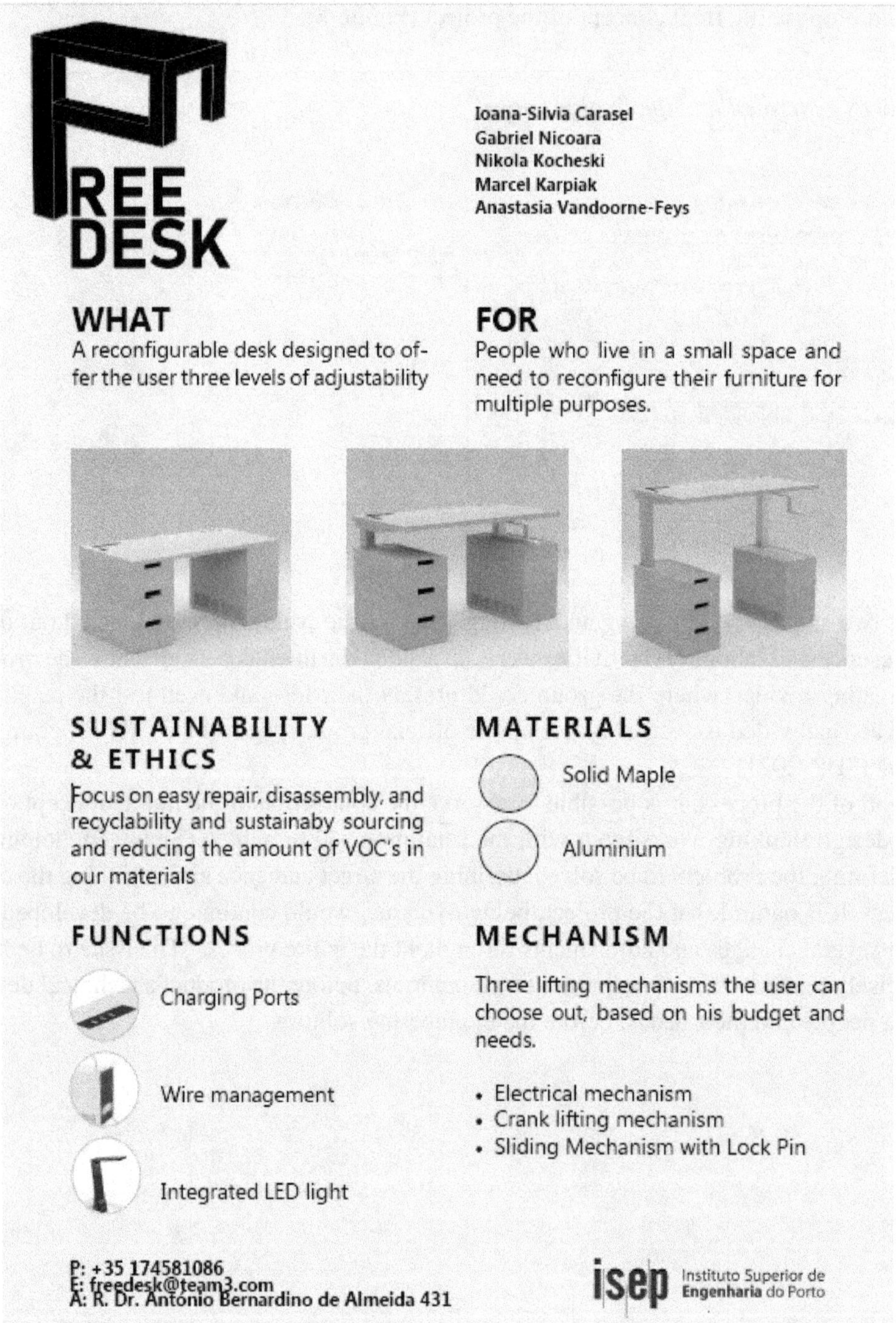

FUTURE RESEARCH DIRECTIONS

The EPS@ISEP program is visionary in the sense that it integrates different areas of knowledge, based on interdisciplinarity; creates opportunities for connection between students of different nationalities, fosters excellent mobility and consequent internationalization; it is centered on project-based learning, developing solutions to humanity's significant problems and improving people's lives.

Design Thinking could be permanently incorporated into the program as a subject, sharing its approach, methodology and tools, participating from the beginning to the end of the project and being part of the entire development process.

As a future perspective, the program, while a model of project development and management, can be replicated by other countries, communities and even continents, spreading its intrinsic innovation culture worldwide. And, of course, taking, exploring and integrating design thinking.

CONCLUSION

According to the experience reported, it is possible to observe that there are favorable conditions for the insertion of the design issues in the EPS@ISEP program, complementing the content being administered, incorporating the subjects most related to people and users, their needs and desirability, allied to the feasibility and viability, which are already considered previously and are essential do the success of the project development.

Through methodologies such as Design Thinking, design can significantly collaborate in the training of engineers, enriching the process of developing project solutions in the search for new forms of innovation both from an educational and design point of view, forming a base of inter or even transdisciplinary knowledge.

It is hoped that this research has contributed to studies related to the understanding and application of Design Thinking approach, to the evolution of higher education, to more complete training of engineers, to the integration between different areas of knowledge, in favor of a better society.

ACKNOWLEDGMENT

The author would like to thank specially to Prof. Dr. Benedita Malheiro for the invitation and the possibility of being part of the EPS@ISEP Program.

This research was supported by Instituto Mauá de Tecnologia (IMT 2017) – institutional and financial support – and by *Coordenação de Aperfeiçoamento de Pessoal de Nível Superior* (CAPES), Brazil, (Finance Code 001) – partial financing.

REFERENCES

Björklund, T., & Keipi, T. (Eds.). (2019). *Design +: Organizational renewal and innovation through design.* Aalto University - Aalto Design Factory. Retrieved from https://aaltodoc.aalto.fi/handle/123456789/41115

Brassler, M., & Dettmers, J. (2017). How to Enhance Interdisciplinary Competence—Interdisciplinary Problem-Based Learning versus Interdisciplinary Project-Based Learning. *The Interdisciplinary Journal of Problem-Based Learning*, *11*(2). Advance online publication. doi:10.7771/1541-5015.1686

Brown, T. (2009). Change by design. How Design Thinking transforms organizations and inspires innovation. New York, NY: HarperCollins ebooks.

Cărăşel, I.-S., Vandoorne-Feys, A., Nicoară, G.-G., & Michał, M. (2021). *FREE Desk, a reconfigurable and ergonomic smart desk.* Polytechnic of Porto, *Instituto Superior de Engenharia do Porto*, Porto. Retrieved from https://www.eps2021-wiki3.dee.isep.ipp.pt/doku.php?id=docs

CNI. (2018). *Ensino de engenharia: fortalecimento e modernização.* Propostas da indústria eleições 2018, Confederação Nacional da Indústria (CNI), Brasília. Retrieved from https://bucket-gw-cni-static-cms-si.s3.amazonaws.com/media/filer_public/5e/ec/5eec09ca-2b12-4880-8a0b-804411795ea7/ensino_de_engenharia_web.pdf

Domingues, I. (Ed.). (2005). *Conhecimento e Transdisciplinaridade II: Aspectos metodológicos.* Humanitas.

EPS. (2021). *EPS - European Project Semester.* Retrieved from European Project Semester: http://europeanprojectsemester.eu/

EPS@ISEP. (2021). *Ergonomic Reconfigurable Home-Desk.* Retrieved from EPS@ISEP | The European Project Semester (EPS) at ISEP: https://www.eps2021-wiki3.dee.isep.ipp.pt/doku.php

Ertas, A. (2018). *Transdisciplinary engineering design process.* John Wiley & Sons.

Ertas, A., Gatchel, S., Rainey, V., & Tanik, M. M. (2007). Networked Approach to Transdisciplinary Research & Education. *The Academy of Transdisciplinary Learning & Advanced Studies, 3*(2), 1–12.

Facca, C. A. (2020). A contribuição do pensamento do design na formação do engenheiro: o espaço do Fab Lab como experiência transversal. Doctoral Thesis, Universidade Anhembi Morumbi, PPG Design, São Paulo.

Facca, C. A., Alves, J. L., & Barbosa, A. M. (2019). Overview of Design Teaching on Engineering Courses: A Comparative Study between Brazil and Portugal. In *International Conference on Education and New Developments* (pp. 129-133). Porto: World Institute for Advanced Research and Science. doi:10.36315/2019v2end027

Graham, D. R. (2018). *The global state of the art in engineering education.* Massachusetts Institute of Technology, School of Engineering. Cambridge, MA: New Engineering Education Transformation. Retrieved from https://rhgraham.org/resources/Global-state-of-the-art-in-engineering-education---March-2018.pdf

Guzzo, F., & Facca, C. A. (2018). *O Design Thinking como Metodologia de Projeto Aplicada na Disciplina de Introdução à Engenharia. In Anais do 18º Congresso Nacional de Iniciação Científica - CONIC-SEMESP. 6.* Universidade Paulista. Retrieved from http://conic-semesp.org.br/anais/files/2018/trabalho-1000002444.pdf

IDEO. (2020). *Design Thinking.* Retrieved from IDEO Design Thinking: https://designthinking.ideo.com/

ISEP. (2021). *European Project Semester at ISEP (EPS@ISEP).* Retrieved from ISEP - Instituto Superior de Engenharia do Porto: https://www.isep.ipp.pt/Course/Course/44

Malheiro, B., Silva, M., Ribeiro, M. C., Guedes, P., & Ferreira, P. (2014). The European Project Semester at ISEP: The challenge of educating global engineers. *European Journal of Engineering Education, 40*(3). Advance online publication. doi:10.1080/03043797.2014.960509

Melo, A., & Abelheira, R. (2015). *Design Thinking & Thinking Design*. Novatec.

MIRO. (2021). *Miro Help Center*. Retrieved from MIRO: https://help.miro.com/hc/en-us/articles/360017730533-What-is-Miro-

OECD. (2018). *The Future of Education and Skills - Education 2030.* Organisation for Economic Co-operation and Development (OECD). Paris, France: OECD Publishing. Retrieved from https://www.oecd.org/education/2030/E2030%20Position%20Paper%20(05.04.2018).pdf

d.school. (2020). *Design Thinking Bootleg*. Retrieved from d.school - Stanford Design School: https://dschool.stanford.edu/resources/design-thinking-bootleg

Vianna, M., Vianna, Y., Adler, I., Lucena, B., & Russo, B. (2012). *Design Thinking: inovação em negócios* (2nd ed.). MJV Press.

ADDITIONAL READING

Buchanan, R. (1992). Wicked Problems in Design Thinking. *Design Issues, 8*(2), 5–21. Retrieved from https://www.jstor.org/stable/1511637. doi:10.2307/1511637

Cross, N. (1982). Designerly ways of knowing. *Design Studies, 3*(4), 221–227. doi:10.1016/0142-694X(82)90040-0

Cross, N. (2005). *Engineering Design Methods: Strategies for Product Design* (3rd ed.). John Wiley & Sons, Ltd.

Cross, N. (2011). *Design Thinking: Understanding How Designers Think and Work.* Berg Publishers. doi:10.5040/9781474293884

KEY TERMS AND DEFINITIONS

Design Thinking: The designer's way of thinking and acting.

Desirability: People's needs and desires, which makes sense to users.

Empathy: Putting yourself in the other's shoes.

Feasibility: What is possible to be manufactured or produced in the near future, with the technology we know today.

Interdisciplinarity: Integration between two or more areas of knowledge, where one modifies the other.

Problem-Based Learning (PBL): Learning method intrinsically based on developing solutions to complex problems.

Project-Based Learning (PjBL): Learning method intrinsically based on developing solutions to complex projects.

Viability: What can be part of an economically and financially sustainable business.

Chapter 8
Active Learning Strategies for Sustainable Engineering:
The Case of the European Project Semester

Abel José Duarte
Instituto Superior de Engenharia do Porto, Instituto Politécnico do Porto, Portugal

Benedita Malheiro
https://orcid.org/0000-0001-9083-4292
Instituto Superior de Engenharia do Porto, Instituto Politécnico do Porto, Portugal

Manuel Fernando Silva
https://orcid.org/0000-0002-0593-2865
Instituto Superior de Engenharia do Porto, Instituto Politécnico do Porto, Portugal

Paulo Duarte Ferreira
https://orcid.org/0000-0001-6432-6366
Instituto Superior de Engenharia do Porto, Instituto Politécnico do Porto, Portugal

Pedro Barbosa Guedes
https://orcid.org/0000-0001-9254-7692
Instituto Superior de Engenharia do Porto, Instituto Politécnico do Porto, Portugal

ABSTRACT

Engineering aims to improve the quality of daily life on a planet with limited resources. This chapter describes how the European Project Semester offered by the School of Engineering of the Porto Polytechnic contributes to make engineering undergraduates conscious of their environmental and societal impact. The program is a project-based active learning framework where problem-solving is supported by intercultural communication and creativity, ethical and sustainable reasoning, and global contextual analysis. This analysis shows that it fosters scientific, technical, and interpersonal competencies, emphasizing ethical and sustainable design as recommended by several reference institutions through the testimonials left on the project reports and the evolution perceived by the participants during the semester. These findings are supported by a sample of 45 projects conducted by 228 students from 23 countries and 65 degrees. All projects aimed to have a lasting impact on the future engineering practice of participants and, somehow, transform the world positively.

DOI: 10.4018/978-1-6684-2309-7.ch008

Copyright © 2022, IGI Global. Copying or distributing in print or electronic forms without written permission of IGI Global is prohibited.

INTRODUCTION

Life on Earth is an endless quest to perpetuate species in time through reproduction. However, reproduction does not provide a default mechanism to pass on knowledge to the next generations. From rudimentary chemical information to the demonstration of experiences and procedures by trial and error, nature finds in humanity the most complex and elaborate way of transmitting information to offspring (Besson, 2014).

Nowadays, the challenge is not only to convey knowledge, but to create an educational framework where the student can be actively committed to his/her own education and, at the same time, contribute to the preservation of life and the planet.

This chapter reports how the European Project Semester (EPS) offered by Instituto Superior de Engenharia do Porto (ISEP) combines active learning and sustainability-driven engineering education. On the one hand, active learning, which is founded in constructivist epistemology, states that learning requires the active participation of the student and comes about through transactions and dialogue among students and between faculty and students, in a social setting (Panitz, 1999). On the other hand, educating engineers for sustainable development involves fostering critical and ethical thinking and instilling a lifelong search for equity, solidarity and preservation of natural resources, cultural and genetic diversity (Duarte et al., 2020). This combination is particularly well-suited since student motivation and engagement tend to increase when challenged to design ethics and sustainability-driven solutions to real problems.

EPS@ISEP relies on project-based learning and teamwork as a means to promote the development of technical-scientific and soft skills in undergraduates, while designing sustainable and ethical solutions. This learning framework educates engineer undergraduates to become agents of change, contributing to a joint, peaceful and sustainable future on Earth.

This document is organized into five sections. Section 2 provides the background. Section 3 describes the European Project Semester and its implementation at ISEP. Section 4 presents EPS@ISEP facts and testimonies. Section 5 draws the conclusions.

BACKGROUND

Active Learning in Engineering

According to Michael (2006), well-known active learning pedagogies include cooperative learning (Johnson et al., 1991), collaborative learning (Leigh & MacGregor, 1992), experiential learning (Kolb, 2014), problem-based learning (PBL) and project-based learning (PjBL) (Bell, 2010). In the case of engineering education, the adoption of PjBL pedagogies is essential to develop the required skills in young graduates. However, to motivate engineering undergraduates to drive their learning process it is necessary to adopt attractive methodologies and inspiring problems. The desire to find, together with peers, solutions for real-world problems triggers in students the will to search for knowledge, learn complex subjects and share findings. These joint search, design, development and test activities are intended to provide students with a fun, motivating and educational experience, increasing student self-confidence, interest in learning and, ultimately, in engineering (Elger et al., 2000). The Committee on the Engineer of 2020, Committee on Engineering Education, and National Academy of Engineering (2005) stated that engineering education must be realigned to promote technical excellence as well as an understanding of work strategies, team, communication, ethical reasoning, societal and global contextual analysis skills in practicing engineers. This implies that humanities, economics, political science, language, and interdisciplinary technical subjects must be part of the undergraduate education to prepare professionals to communicate with the public, engage in a global engineering marketplace and become lifelong learners. Thus, the essential professional competencies of an engineer are critical thinking and problem solving, effective communication, collaboration and team building, and creativity and innovation – the four C – as well as socio-professional ethics and sustainable development – the two S (Malheiro et al., 2019). These abilities are often required by the employers of the alumni and are therefore included in the curricula to educate competent practitioners (Vos & de Graff, 2004; de Graff & Christiensen, 2004). Moreover, the CDIO Initiative, which is an open, innovative educational framework for engineering graduation degrees set in the context of Conceiving – Designing – Implementing – Operating real-world systems and products, identifies two main types of PjBL: (*i*) lecture-based, including such methods as a partner and small-group discussions, demonstrations, debates, concept questions and feedback from students about what they are learning; and (*ii*) experiential, when students take on roles that simulate professional engineering practice, for example, design, build and test projects, simulations and case studies (Crawley et al., 2014). PjBL corresponds, in the context of this paper, to the promotion and adoption of an autonomous self-learning attitude among staff and students. For future engineers, it supports the essential lifelong learning posture and for staff the motivation for continuous research regarding innovative student learning methodologies.

Sustainability Driven Engineering

Sustainability-driven engineering plays two equally important roles: it drives sustainable development and motivates students to become actors of change as well as active learners. As of the Second World War, population, knowledge, and quality of life have experienced exponential growth. At the same time, society was confronted with the fact that human decisions cause irreversible consequences on Earth, namely climate change, pollution, depletion of resources and extinction of species. Engineering undergraduates are usually drawn by the will to tackle real-world problems directly or indirectly related to the preservation of the planet if design decisions are based on ethical and sustainability concerns. According to several international, engineering education and professional engineering organizations, engineering is a key profession for achieving sustainable development. To solve problems considering the economic, social, human, and cultural perspectives, the engineer of the future needs to be creative, mobile, and master communication as well as its field of expertise. This set of professional skills has been identified by international institutions like the Organisation for Economic Cooperation and Development (2014, 2018) and the United Nations Educational, Scientific and Cultural Organization (2010), engineering education societies as the European Society for Engineering Education (2016) and the American Society of Engineering Education (2010), engineering education accreditation associations, namely, the European Network for Accreditation of Engineering Education (2015) and the Accreditation Board for Engineering and Technology (2000), and national engineering academies such as the Royal Academy of Engineering (Lucas et al., 2014) and the National Academy of Engineering (2012). In this chapter, sustainability encompasses the environmental impact, social implications, and economic viability (Duarte et al., 2020). Sustainability-driven engineering refers to the design of solutions considering the involved ethical and sustainability dimensions.

European Project Semester

EPS is a student-centered integrated learning framework for undertaking engineering capstone/internship projects and, as such, adopts collaborative, experiential, PjBL methodologies. This one-semester program designed by Arvid Andersen (Andersen, 2001) aims to educate engineers with European and international dimensions. The goal of the program is to promote the development of scientific, technical, and soft skills in engineering students through multicultural teamwork and open, multidisciplinary problem-solving (Andersen, 2004). Furthermore, EPS attempts to train conscientious and responsible engineers, which can discover global integrative, innovative, and increasingly sustainable solutions (Andersen, 2001).

The program focuses on teamwork, exposes students to cultural, scientific, and technical diversity, and embraces ethics and sustainability-driven design. Teams of students from diverse scientific and cultural backgrounds collaborate and cooperate to solve real-world problems. The process involves collecting information, finding alternative solutions, choosing their own solution, including methodologies, materials and components, and design, building and evaluating the resulting prototype (Lehman et al., 2008). During this PjBL process, advisers act as coaches, consultants, and facilitators, keeping the team motivated and contributing to the success of the projects. Furthermore, EPS complies with the CDIO directives on design, build and test experiences at the level of capstone projects (Malheiro et al., 2015).

The program is currently offered by nineteen European engineering schools from twelve countries, called the EPS providers, attesting to its success. The EPS providers have discussed, agreed upon and posted on the EPS Providers website the core specification of the European Project Semester framework. The so-called "10 Golden Rules of EPS" are listed in Table 1. Moreover, providers offer EPS program implementations not only compliant with this core framework, but also with diverse "flavors".

Table 1. 10 Golden Rules of EPS

Nº	Rule
I	English is the working language of EPS
II	EPS is multinational with a group size of a minimum of three and maximum of six students, being four or five the ideal number; a minimum of three nationalities must be represented in each EPS group
III	Ideally, but not necessarily, an EPS project is multidisciplinary
IV	An EPS semester is a 30 European Credit Transfer System Units (ECTU) package, the duration of which is not less than 15 weeks
V	An EPS project has a minimum of 20 ECTU and the complementary subjects account for a minimum of 5 ECTU and a maximum of 10 ECTU
VI	The focus of EPS is on teamwork
VII	The subjects included in the EPS must be project supportive; English and a basic crash course in the local language must be offered
VIII	The subjects must include team building at the very beginning and Project Management at the beginning of an EPS semester
IX	Project supervision/coaching must focus on the process as well as the product
X	EPS must have continuous assessment including an interim and a final Report

EPS@ISEP IMPLEMENTATION

The School of Engineering – Instituto Superior de Engenharia do Porto (ISEP) – of the Porto Polytechnic became an EPS provider in the academic year of 2010-2011 (Malheiro et al., 2013). Currently, EPS@ ISEP is in its 11th consecutive year and, during this period, has welcomed 230 participants from Europe, Asia, America, and Africa.

EPS@ISEP is structured as follows: two-thirds (20 ECTU) of the credits are assigned to the project module and one-third (10 ECTU) to complementary modules (EPS rule V). The complementary modules are focused on the development of soft skills considered essential in training for the twenty-first-century engineers, such as communication or team building, project-related activities, such as project management, and transversal topics, such as energy & sustainable development or ethics & deontology (Malheiro et al., 2015). Figure 1 presents the EPS@ISEP 15-week schedule and illustrates the fulfillment of golden rules VIII and X.

Figure 1. EPS@ISEP Schedule

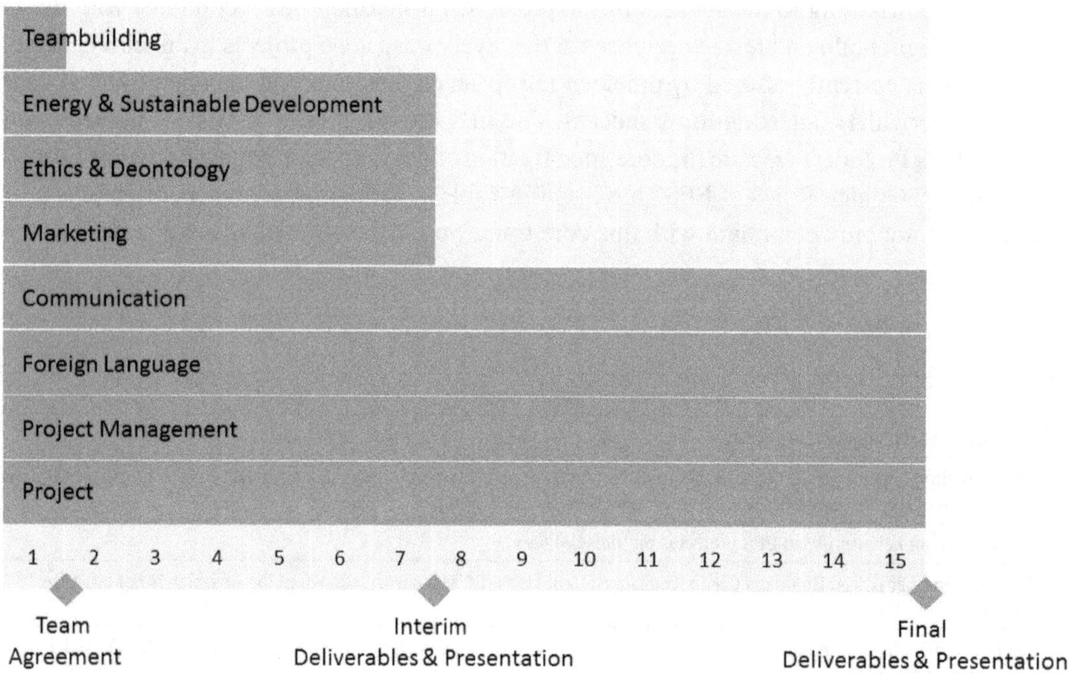

The teams are fully responsible for the project management (task identification, allocation, planning), the solution design (research, brainstorming, weekly meetings), selection and specification of materials and components, as well as assembling and testing of the prototype.

In terms of the learning outcomes, EPS@ISEP aims to develop the following competencies in the participants: (*i*) the ability to effectively work and communicate in multicultural and multidisciplinary teams; (*ii*) the ability to contribute, within their field of expertise, to the solution of a larger multidisciplinary problem; (*iii*) the ability to adopt sustainable development practices; (*iv*) the ability to adopt ethical and deontological engineering habits of mind; and (*v*) the ability to adopt a PjBL lifelong stance.

Proposals

Before the beginning of the semester, the EPS@ISEP staff team gathers project proposals regarding real-world problems. The origin of proposals varies and includes industry, services, R&D institutions, or the school itself. The proposals refer to multidisciplinary problems, i.e., require the integration of multiple technical and scientific competencies, and are open, i.e., specify exclusively top-level requirements such as compliance with the applicable directives/norms and the budget.

A proposal defines the problem/challenge to tackle, the minimal set of requirements, mostly mandatory directives and standards, and the maximum budget for building a proof-of-concept prototype. This type of proposal directs the team towards the conception and design stages and then towards the implementation and test stages of the capstone project/internship. The average cost of an EPS@ISEP project, in terms of materials and components, is approximately 317 €.

While many proposals are inspired on the National Academy of Engineering 14 Grand Challenges for Engineering (National Academy of Engineering 2008), the list of proposals offered tends to be eclectic, ranging from bio-mimetic robotics (Silva et al., 2015) to sustainability (Llauradó et al., 2015; Fountain et al., 2016; Augustyns et al., 2017) and maritime (Ferreira et al., 2015) applications.

Teaming

To accomplish rule II, a Belbin questionnaire is used to determine the individual teamwork profiles (Belbin, 2010), allowing the design of teams with complementary elements from as many diverse fields of studies and as multinational as possible. On the one hand, by identifying the teamwork profile/role of the participants, the Belbin questionnaire helps to assemble teams made of individuals with diverse, complementary teamwork roles. On the other hand, individuals become aware of their teamwork role and associated responsibilities in the group organization, increasing the likelihood of success significantly. The main team roles identified by Belbin are Company Worker, Chairman, Shaper, Plant, Resource Investigator, Monitor/Evaluator, Team Worker, and Finisher. In the case of EPS, since the team size is less than ten (four to six students) there are individuals with overlapping roles as well as assignments. Finally, the teams are assembled so that they include students with different team worker profiles (Belbin), at least three nationalities (EPS rule II) and diverse scientific backgrounds (EPS@ISEP).

This focus on diversity is important to prepare the students to act worldwide as engineers and members of multinational and multidisciplinary teams. One of the first tasks the teams need to accomplish during team building activities (EPS rule VIII) is to define their own organization and conflict resolution rules – the Teamwork Agreement – using two identical questionnaires as proposed by Hansen (2010).

While the first is to be filled individually, the second questionnaire must be completed and signed together by the team. The questionnaire covers operational issues, like task and responsibility assignment, communication methods, task priority and implementation, self-assessment and project monitoring, and conflict resolution approaches. Indirectly, the Teamwork Agreement promotes autonomy and responsibility as well as leads to early awareness and accountability of conflicts.

Coaching

EPS@ISEP adopts a unique model where a panel of multidisciplinary experts acts as a consulting committee regarding project supervision. The teams meet with this advisory panel once a week to discuss the previously agenda posted on their wiki. Each meeting is conducted in turn by a team member, promoting commitment and responsibility. After the meeting, the teams write on their wiki the respective minute, reporting the most prominent issues discussed. This interaction protocol aims to develop the ability to define agendas, leading meetings and write minutes.

The key role of the panel is to promote a free exchange of ideas between team members, privileging original and innovative suggestions and perspectives, which stem from their distinct cultural and scientific backgrounds. Even when suggestions seem implausible, the panel uses them as the starting point of further brainstorming. During this process, the advisers identify problems and challenge the team to find innovative contributions to improve the proposed solution. As far as communication is concerned, the panel is aware that it is interacting with students from diverse scientific and cultural backgrounds.

Another particularly important aspect of the coaching methodology is the prompt feedback given to the students. The communication between the panel and the team involves presentations, discussions, e-mail messages, annotated deliverables, drawings, videos, or cardboard models.

Deliverables

The teams must produce several deliverables, including the project wiki, report, video, paper, manual, flyer, leaflet poster and build a proof-of-concept prototype with the allocated budget. The teams develop a report, video, paper, manual, flyer, leaflet, poster and several presentations with the assistance of the communication seminar teacher and project coaches. Figure 2 displays the main deliverables with darker blocks and partial deliverables with lighter blocks. The main deliverables are located on the left and include prototype and communication related deliverables. The assembling and testing block represents a core process of prototype development, not a deliverable. The outcome of each partial prototype deliverable may impact the precedent block, defining a continuous refinement cycle represented by the light grey flow. All stages of the conception, design, implementation, and operation of the proof-of-concept prototype are reported in the wiki and report.

Figure 2. EPS@ISEP deliverables

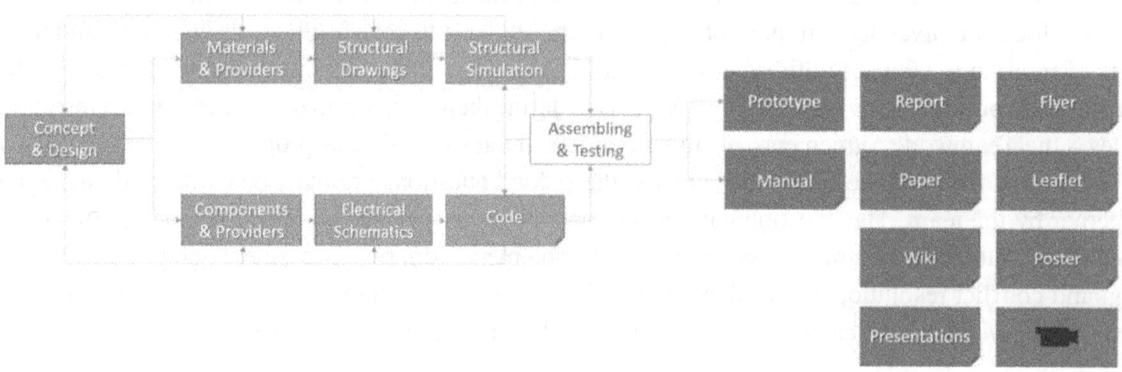

The report structure, which is provided beforehand, closely follows the organization of a scientific report, including as mandatory elements the abstract, introduction, state of the art, project development, marketing, sustainability, ethical concerns, conclusions, glossary, and references. The marketing, eco-efficiency & sustainability and ethics & deontology chapters are produced and refined within the corresponding complementary modules. They address several components of the future engineer profile as suggested by the United Nations Educational, Scientific and Cultural Organization (2010), the Organisation for Economic Co-operation and Development (2018), the European Society for Engineering Education (2016) or the American Society of Engineering Education (2010). Specifically, these chapters focus on the relevant project-related issues identified by the team in each of these areas and the conclusions drawn by the team become project requirements. As project requirements, they then need to be fulfilled during the development of the proof-of-concept prototype.

The paper template and structure are also provided beforehand, and the teams are challenged to prepare the paper for a given call for papers. In this context, the paper is not only a concise version of the report but, since it has a specific focus, requires additional reflection from the team.

Finally, the team wiki is essential to the EPS process since it acts as a collaborative work platform for team members and advisers as well as a project showcase platform. On the one hand, it allows team members to work and prepare together the project plan, logbook, weekly meetings, and the report and, on the other hand, it remains the repository of the deliverables and project-related activities.

Assessment

In terms of assessment, EPS@ISEP follows rule X and uses the scheme proposed by (Hansen, 2010), i.e., the formal assessment takes place twice during the semester and contemplates self & peer as well as teacher and adviser assessment. While the interim assessment is intended to give individuals and teams feedback about their performance so far, from the point of view of their peers and of the advisers, the final assessment is for grading.

The self & peer assessment considers the quality and quantity of the technical contribution, openness to others' ideas, teamwork performance, leadership, attitude, and initiative shown. It is used by advisers to monitor teamwork and to give constructive feedback and advice where needed (Ashworth, 2011). The adviser assessment reflects both team performance as well as individual student performance.

EPS@ISEP OUTCOMES

EPS@ISEP embraces several active learning approaches: (*i*) experiential learning, when students adopt professional engineering practices as, for example, the design, build and test of a prototype (PjBL); (*ii*) collaborative learning, when team members autonomously discover, share and brainstorm findings and commit to solutions; and (*iii*) cooperative learning, when team members work together with teachers, within the project supportive modules, in the ethical, deontological, sustainability, marketing and project management dimensions of their projects.

Table 2 presents citations from several EPS@ISEP reports related to active learning evidence. These testimonials identify several highlights contributing to the EPS vision of active learning, namely multiculturalism, teamwork, collaboration, cooperation and, above all, autonomy.

Table 2. Active Learning in EPS@ISEP – Citations

Project	Citation
UV-based Disinfection	During these four months of work, we have learned many different aspects connected with not only mechanical, electrical and chemical knowledge but also teamwork. We improved our cooperation and communication as an international team. Moreover, we dealt with conflicts and other restrictions and limitations which we encountered on our way. (Bazylinska et al., 2012)
Context-aware Smart Object	This project gave us an opportunity to connect our knowledge from different fields of study with practical skills and teamwork, like in real-life situations. This project gave us a chance to know other cultures, cooperate (...) and practice English. (Harms et al., 2012)
Autonomous Environmental Buoy	"It took an organized and tuned teamwork to arrive to this point. However, we did not always cooperate perfectly; it was especially hard for us in the beginning when we barely knew one another. With time, however, we started displaying better and better teamwork qualities." (Möller et al., 2013)
Bio-Inspired Swimming Robot	Due to our diverse professional backgrounds, experiences and personal interests, we cover a wide range of viewpoints and skills. That is why a high level of quality can be achieved. (Blazejewski et al., 2014)
Travel Logging System	The fact that everyone specializes in different areas adds more diversity to our team. Despite these differences, we have taken a common goal, which is the realization of the project using the skills, experience and knowledge from different disciplines. (Holt et al., 2014)
Bio-inspired flying robot	(...) in a multidisciplinary team (...) it is very important to assign detailed tasks to all group members, (...) but, as people do not know each other backgrounds, it is harder to rely on others to do things than imagined. In short: communication management is the key to success. (Caramin et al., 2014)
Reconfigurable LED Lamp	This experience gave us definitely advantage in future life. We practiced English, studied lighting specifications, learned sustainability and marketing issues as well as improved our communication and cooperation skills. (Aedma et al., 2014)
Aquaponics System	Our main objective was to create a working system that supported both fish and plants and, through long research and development phases, we believe that we created a system that not only fulfills the required objective, but also is aesthetically pleasing. (Docherty et al., 2014)
Solar Food Dehydrator	Every team member has a specific role in this project, according to each person's strengths and weaknesses. However, we value teamwork, working together successfully and in a joint effort, sharing our knowledge, in order to achieve our goal. After a lot of research and brainstorming, we agreed on a final design and on the required features of the product. (Berényl et al., 2015)
Aquaponic Shelter	To make everything work we need a good way to communicate with each other. We have meetings at least twice a week, when every member can tell others what they have done and if there are any problems. We also decide together future plans and set deadlines. (…) A group on Facebook was created for the team to communicate. (Bartosz et al., 2015)
Aquaponic Monitoring	Our team has grown in regard to project management and technical skills in our respective areas of study. The entire experience was focused on how we functioned and overcame challenges both culturally and academically as a team. Considering these categories our overall project was a success. (Davis et al., 2015)
Autonomous Sailboat	(...) during face-to-face meetings, where agendas and targets are clearly highlighted as topics of interest, (...) healthy discussions and debates take place. (…) Our group members have developed project management and ecological rationality. The (...) focus was partially on the functionality as a team and how we collaborated, tasked and controlled while being culturally academically challenged. (Asztalos et al., 2015)

The program engages the students actively in learning by pre-defining a set of deadlines, ensuring timely feedback from coaches and proposing real-world multidisciplinary problems. There are two macro deadlines – the interim and final assessment – and several micro deadlines – proposal selection, task identification and allocation, black box diagrams, detailed schematics, selection of materials and components, as well as the weekly project coaching meetings (where the agenda is defined by the teams). Furthermore, the project supportive modules, the complementary technological crash courses, the collaborative construction of the team wiki (including planning, logbook, report, and deliverables sections) and exciting multidisciplinary problems (aggregating students from different scientific and cultural backgrounds) also contribute to this goal.

Table 3 identifies the active learning tasks and actors. EPS@ISEP is an iterative learning process composed of brainstorm, research, selection, and validation tasks, involving the (T)eam, (C)oaches, (L)ecturers and (E)xperts. The diversity of stakeholders and the imprinted work pace fosters both the autonomous discovery and sharing of knowledge – active learning.

Table 3. Active Learning in EPS@ISEP – Tasks and Actors

	Brainstorm	Research	Selection	Validation
Detailed Requirements	T & S	T	T	C
Related Work	T & S	T	T	E & C
Sustainability	T & L	T	T	L
Ethics & Deontology	T & L	T	T	L
Marketing	T & L	T	T	L
Project Management	T & L	T	T	L
Design	T & E & C	T	T	E & C
List of Materials	T & E & C	T	T	E & C
Prototyping	T & E & C	T	T & E	E & C
Testing	T & E & C	T	T	E & C

Figure 3 summarizes the diversity of EPS@ISEP classes (top row) and teams (bottom row) since 2011. The multicultural experience stems from the daily interaction between individuals from diverse provenance (Figure 3, top left) and fields of study (Figure 3, top center). Moreover, this experience extends far beyond the project, e.g., the initial linguistic uneasiness, once overcome, strengthens the unity between team members. Figure 3 (top right) displays the gender diversity. EPS@ISEP has 36.4% of female participants, which is significantly higher than the default value (23.5%) of ISEP.

Figure 3 (bottom row) illustrates the diversity within the teams, which are composed of students with compatible team worker profiles (Belbin test), a minimum of three nationalities (EPS rule II) and diverse fields of study (a must at EPS@ISEP). The contemplated fields of study are the major fields of civil (CIV), chemical (CHE), electrical (ELE), informatics (IT) and mechanical (MEC) engineering as well as physics (PHY).

Figure 3. EPS@ISEP class (top) and team (bottom) nationality, studies and gender diversity

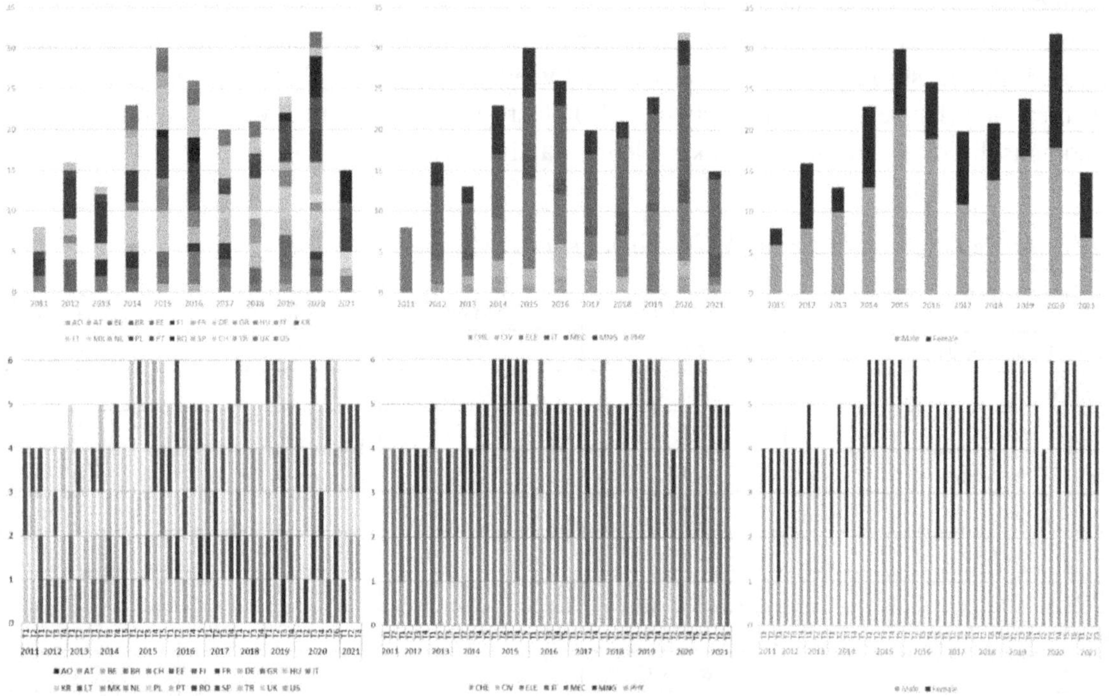

Figure 4 presents the tangible outcomes of the EPS process: the classification and the perceived self & peer evolution. These results exclude the two students, out of 230, who have failed EPS@ISEP so far.

Figure 4. EPS@ISEP classification (left) and self & peer evolution (right)

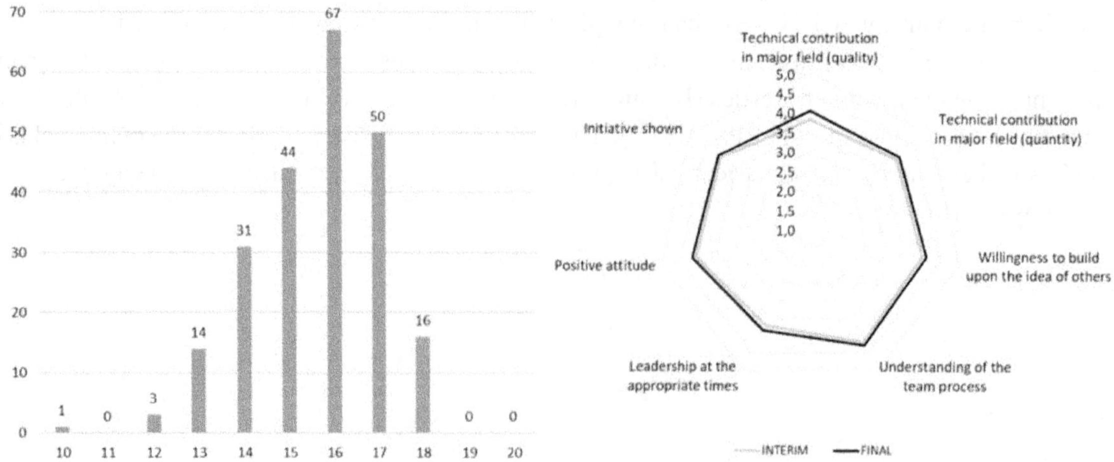

Figure 4 (left) shows the classification histogram with a 0 to 20 grading scale. The most frequent grade is 16 out of 20. However, more important than the grade itself, is the students' perception of the impact of the EPS process. The latter analysis was based on the historical self & peer assessment, which includes the following dimensions: (*i*) technical contribution in major field (quality); (*ii*) technical contribution in major field (quantity); (*iii*) willingness to build upon the idea of others; (*iv*) understanding of the team process; (*v*) leadership at the appropriate times; (*vi*) positive attitude; and (*vii*) initiative shown. The self & peer questionnaire uses a Likert scale (1 to 5). Specifically, Figure 4 (right) depicts the evolution between the intermediate (lighter) and the final stage (darker). In this figure, a perfect heptagon means that all components have been apprehended equally and the size of each component indicates the level attained. The results in Figure 4 (right) show that, according to the students, all assessment dimensions improved from the interim to the final stage. There was an improvement in terms of the quality and quantity of their technical contributions and a greater awareness of how to work in a team and develop a project.

From this analysis, it is possible to conclude that EPS@ISEP is clearly aligned with the general guidelines of Organisation for Economic Co-operation and Development and United Nations Educational, Scientific and Cultural Organization, the engineering education curricula recommendations of the Royal Academy of Engineering, European Network for Accreditation of Engineering Education, European Society for Engineering Education and American Society for Engineering Education, and the accreditation guidelines of European Network for Accreditation of Engineering Education and Accreditation Board for Engineering and Technology. The program is organized around one central module – the "EPS Project" – and a set of complementary supportive modules, among which are "Ethics & Deontology" and "Energy & Sustainable Development". Sustainability is one of the pillars of EPS@ISEP. Not only is it addressed explicitly in the "Ethics & Deontology" and "Energy & Sustainable Development" modules, but it is put into practice during the whole project.

Figure 5 shows that the budget allocated to different teams over the years has been decreasing to encourage reuse, waste reduction and improvisation with the available resources. The teams, in this context, choose to buy few affordable resources, reducing the environmental footprint of the prototypes. Moreover, the budget per team trend in the last years is to stabilize around 100 €. High budgets correspond to projects sponsored by companies and research laboratories.

Figure 5. EPS@ISEP budget per team

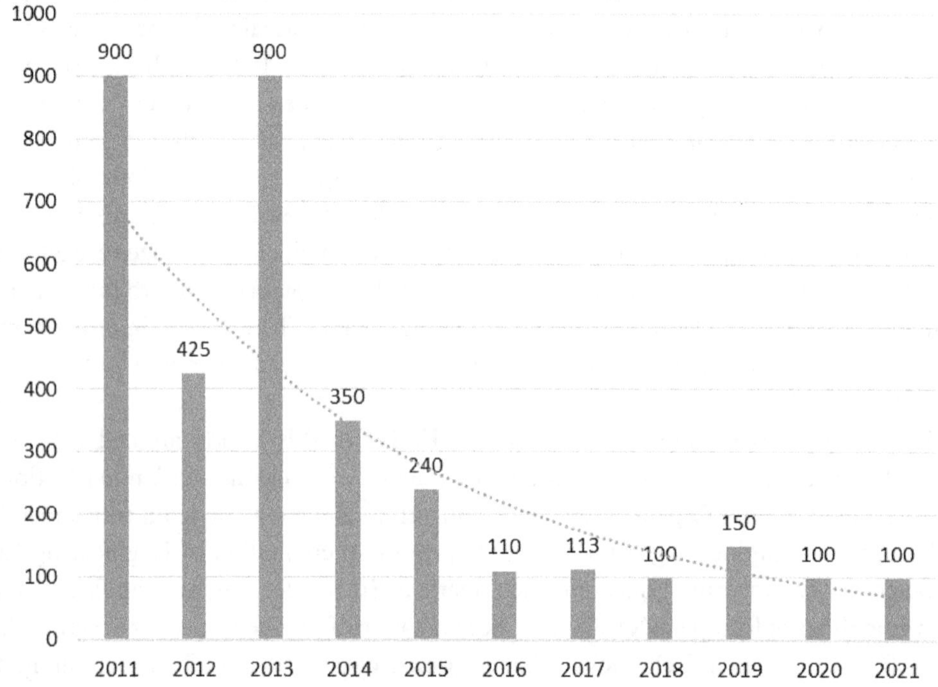

Figure 6 displays the word cloud composed of the most frequent words found in the project reports. It reveals that the teams are highly aware of the ethical, deontological and sustainability concerns.

Figure 6. EPS@ISEP word cloud

In summary, EPS@ISEP promotes teamwork, active learning (lifelong learning), adopts PjBL and addresses ethics & deontology, eco-efficiency & sustainability, and cross-cultural & inter-personal communication.

The program syllabus is in accordance with the following recommendations:

- Engineering curricula must provide a capstone or integrating experience, emphasize the development of professional skills such as solving unstructured problems, communicating effectively and working in teams, and foster awareness of ethical and contextual considerations in engineering (Accreditation Board for Engineering and Technology, 2006).
- Future engineers must be aware of technology's social and environmental impacts, learn to work in complex teams, interact and cooperate with society, and be good communicators. Engineering design must go beyond the traditional objectives of schedule, cost and conventional scope and learn to include broader societal necessities (such as minimizing water, energy and materials use, respecting human and cultural rights, and looking out for health and safety). Engineering education would benefit from less formulaic and more PBL and PjBL and just-in-time approaches, reflecting the problem-based and problem-solving nature of engineering (United Nations Educational, Scientific and Cultural Organization, 2010).
- Engineering professionals need to be educated to consider sustainability issues in all aspects of design and manufacturing and to pursue collaborations with multidisciplinary teams of technical experts. Important attributes for these teams include excellence in communication, an ability to communicate using technology, and understanding the complexities associated with the global market and social context. Furthermore, it is essential for engineers to be flexible, receptive to change and show mutual respect. Attention to ethical issues in engineering will avoid repeating the mistakes of the past and increase the opportunities to emulate "best practice" successes (Committee on the Engineer of 2020, Committee on Engineering Education, and National Academy of Engineering 2005).
- Engineering colleges must promote not only the intellectual and technical development of the students, but prepare future engineers to work in teams, communicate and understand the economic, social, environmental, and international context of their professional activities. Engineering curricula need to incorporate team skills (including collaborative, active learning and communication skills), leadership, a systems perspective, an appreciation of different cultures and business practices (and the understanding that the practice of engineering is now global), a multidisciplinary perspective, a commitment to quality, timeliness and continuous improvement, undergraduate research and engineering work experience, an understanding of the societal, economic and environmental impacts of engineering decisions, and ethics (American Society for Engineering Education, 2010).

In this regard, Duarte et al. (2020) show that Iberian EPS faculty and students are aware of their projects' social, economic, and environmental impact, in terms of quality of life, social responsibility, and the use of resources and environmentally friendly technology.

CONCLUSION

EPS@ISEP is an engineering capstone program intended to prepare future engineering professionals. In recent decades, bodies like the Organisation for Economic Co-operation and Development, the United Nations Educational, Scientific and Cultural Organization, the National Academy of Engineering, the European Society for Engineering Education, the American Society for Engineering Education, the European Network for Accreditation of Engineering Education or the Accreditation Board for Engineering and Technology have emphasized the need, not only, to foster hard and soft skills, but also to embed sustainable and ethical design and development in the engineering curriculum.

In EPS@ISEP this is achieved through PjBL, multi-cultural and multidisciplinary teamwork, and learning autonomy. Thus, it promotes traditional technical-scientific and the desired soft skills required to design solutions for the well-being and advancement of society as well as creativity and life-long learning.

This conclusion is corroborated by the testimonials left on the project reports and by the learning experience improvement perceived by the participants from the interim to the final stage of the EPS process. This learning framework prepares engineers to become agents of positive transformation, contributing to a common, peaceful, and sustainable future on Earth.

ACKNOWLEDGMENT

This research received no specific grant from any funding agency in the public, commercial, or not-for-profit sectors. Nevertheless, the authors thank Jørgen Hansen for the support, the EPS providers for the fruitful exchange of experiences, students, and staff, and ISEP for believing in the program.

REFERENCES

Accreditation Board for Engineering and Technology. (2000). *Viewpoints – Vol. I: Issues of Accreditation in Higher Education*. https://www.abet.org/wp-content/uploads/2015/05/Viewpoints_Vol1.pdf

Accreditation Board for Engineering and Technology. (2006). *Engineering Change – A Study of the Impact of EC2000*. https://www.abet.org/wp-content/uploads/2015/04/EngineeringChange-executive-summary.pdf

Aedma, A., González Alen, D., Norbert, R., Petersen, N., & Rzeznik, P. (2014). *Reconfigurable LED Lamp* (EPS Project Wiki). Instituto Superior de Engenharia do Porto. http://www.eps2014-wiki4.dee.isep.ipp.pt/

American Society for Engineering Education. (2010). The Green Report – Engineering Education for a Changing World. Technical Report. American Society for Engineering Education.

Andersen, A. (2001). Implementation of engineering product design using international student teamwork to comply with future needs. *European Journal of Engineering Education*, *26*(2), 179–186. doi:10.1080/03043790110034438

Andersen, A. (2004). Preparing engineering students to work in a global environment to co-operate, to communicate and to compete. *European Journal of Engineering Education*, *29*(4), 549–558. doi:10.1080/03043790410001711243

Ashworth, D. W. C. (2011). Project Based Learning in International Teams – Monitoring the Effectiveness of Teamwork. *Key Engineering Materials*, *450*, 581–584. doi:10.4028/www.scientific.net/KEM.450.581

Asztalos, I., Giordano, R., Gunther, T., Navarrete Hill, M., & Rabone, G. J. (2015). *Autonomous Sailboat* (EPS Project Wiki). Instituto Superior de Engenharia do Porto. http://www.eps2015-wiki5.dee.isep.ipp.pt/

Augustyns, L., Pogoda, M., Milesi, M., Kang, M., Valls, P., Duarte, A. J., Malheiro, B., Ferreira, F., Ribeiro, M. C., Silva, M. F., Ferreira, P., & Guedes, P. (2017). Sustainable Desalinator – An EPS@ISEP 2016 Project. In *Proceedings of the 45th SEFI Annual Conference* (pp. 491-498). European Society for Engineering Education.

Bartosz, K., Bentin, C., Davies, E., Fernandez de Toro Ronda, P., Fountain, A., & Suits, K. (2015). *Aquaponic Shelter* (EPS Project Wiki). Instituto Superior de Engenharia do Porto. http://www.eps2015-wiki3.dee.isep.ipp.pt/

Bazylinska, A., Jenei, A., Juht, T., Küttis, S., & Walczak, J. (2012). *UV-based Disinfection* (EPS Project Wiki). Instituto Superior de Engenharia do Porto. http://www.eps2012-wiki2.dee.isep.ipp.pt/

Belbin, M. (2010). *Management Teams: Why They Succeed or Fail* (3rd ed.). Butterworth-Heinemann. doi:10.4324/9780080963594

Bell, S. (2010). Project-Based Learning for the 21st Century: Skills for the Future. *The Clearing House: A Journal of Educational Strategies, Issues and Ideas*, *83*(2), 39–43. doi:10.1080/00098650903505415

Berényl, D., Brunke, F., de la Torre, A., Hood, L., & Nogal, H. (2015). *Solar Food Dehydrator* (EPS Project Wiki). Instituto Superior de Engenharia do Porto. http://www.eps2015-wiki1.dee.isep.ipp.pt/

BessonL. (2014). *Lucy*. https://www.lucymovie.com

Blazejewski, M., Heibeck, M., Moreno Ishii, A., & Nybjörk, R. G. (2014). *Bio-Inspired Swimming Robot* (EPS Project Wiki). Instituto Superior de Engenharia do Porto. http://www.eps2014-wiki1.dee.isep.ipp.pt/

Caramin, B. A., Dunn, I., Klawikowski, Y., & Ney, R. (2014). *Bio-inspired Flying Robot* (EPS Project Wiki). Instituto Superior de Engenharia do Porto. http://www.eps2014-wiki3.dee.isep.ipp.pt/

Committee on the Engineer of. (2020). *Committee on Engineering Education, and National Academy of Engineering. In Educating the Engineer of 2020: Adapting Engineering Education to the New Century*. The National Academies Press.

Crawley, E. F., Malmqvist, J., Östlund, S., Brodeur, D. R., & Edström, K. (2014). *Rethinking Engineering Education*. Springer International Publishing., doi:10.1007/978-3-319-05561-9

Davis, A., Dragos, S. V., Rozewski, J., Ruiz Ludeña, F. J., van Tendeloo, K., & von Bröckel, R. (2015). *Aquaponic Monitoring* (EPS Project Wiki). Instituto Superior de Engenharia do Porto. http://www.eps2015-wiki4.dee.isep.ipp.pt/

de Graaff, E., & Christensen, H. P. (2004). Editorial: Theme issue on active learning in engineering education. *European Journal of Engineering Education, 29*(4), 461–463. doi:10.1080/03043790410001716310

Docherty, A., Keane, S., Méry, G., Mesas Llauradó, A., & Sokolowska, N. (2014). *Aquaponics System* (EPS Project Wiki). Instituto Superior de Engenharia do Porto. http://www.eps2014-wiki5.dee.isep.ipp.pt/

Duarte, A. J., Malheiro, B., Arnó, E., Perat, I., Silva, M. F., Fuentes-Durá, P., Guedes, P., & Ferreira, P. (2020). Engineering Education for Sustainable Development: The European Project Semester Approach. *IEEE Transactions on Education, 63*(2), 108–117. doi:10.1109/TE.2019.2926944

Elger, D. F., Beyerlein, S. W., & Budwig, R. S. (2000). Using design, build, and test projects to teach engineering. In *Proceedings of the 30th Annual Frontiers in Education Conference* (*vol. 2*, pp. F3C/9-F3C13). Institute of Electrical and Electronics Engineers. 10.1109/FIE.2000.896572

European Network for Accreditation of Engineering Education. (2015). *EUR-ACE® Framework Standards and Guidelines*. http://www.enaee.eu/wp-assets-enaee/uploads/2012/02/EAFSG_full_nov_vorue-bergehend.pdf

European Society for Engineering Education. (2016). *European Society for Engineering Education*. http://www.sefi.be

Ferreira, P., Malheiro, B., Guedes, P., & Silva, M. (2015). Towards Active Course Marks for Autonomous Sailing Competitions. In F. Morgan & D. Tynan (Eds.), *Robotic Sailing 2014* (pp. 67–75). Springer International Publishing. doi:10.1007/978-3-319-10076-0_6

Fountain, A., Kuron, B., Bentin, C., Davies, E., Suits, K., Toro, P., Duarte, A., Malheiro, B., Ribeiro, C., Ferreira, F., Lima, L., Ferreira, P., & Guedes, P. (2016). Learning Sustainability with EPS@ISEP – Development of an Insectarium. In *Proceedings of the Eight International Symposium on Project Approaches in Engineering Education* (pp. 109-116). Project Approaches in Engineering Education Association.

Hansen, J. (2010). European Project Semester – how engineering students can achieve important competences. In *Proceedings of the Fifth International Conference on Engineering Education* (pp. 1296-1298). Academic Press.

Harms, H., Janaszkiewicz, A., & Valauskaitė, J. (2012). *Context-aware Smart Object* (EPS Project Wiki). Instituto Superior de Engenharia do Porto. http://www.eps2012-wiki3.dee.isep.ipp.pt/

Holt, M., Kosterna, K., Pechero Villota, D., Nyby, E. C. V., & Selçuk, O. (2014). *Travel Logging System* (EPS Project Wiki). Instituto Superior de Engenharia do Porto. http://www.eps2014-wiki2.dee.isep.ipp.pt/

Johnson, D. W., Johnson, R. T., & Smith, K. A. (1991). *Cooperative Learning: Increasing College Faculty Instructional Productivity*. ASHE-ERIC Higher Education Report No. 4 1991. Education Resources Information Center. https://files.eric.ed.gov/fulltext/ED343465.pdf

Kolb, D. A. (2014). *Experiential learning: Experience as the source of learning and development* (2nd ed.). Pearson FT Press.

Lehmann, M., Christensen, P., Du, X., & Thrane, M. (2008). Problem-oriented and project-based learning (POPBL) as an innovative learning strategy for sustainable development in engineering education. *European Journal of Engineering Education, 33*(3), 283–295. doi:10.1080/03043790802088566

Leigh, B., & MacGregor, J. T. (1992). What is Collaborative Learning? In A. S. Goodsell, M. R. Maher, V. Tinto, B. Leigh Smith, & J. MacGregor (Eds.), *Collaborative Learning: A Sourcebook for Higher Education* (pp. 10–29). National Center of Postsecondary Teaching, Learning and Assessment.

Llauradó, A. M., Docherty, A., Méry, G., Sokolowska, N., Keane, S., Duarte, A. J., Malheiro, B., Ribeiro, C., Ferreira, F., Silva, M., Ferreira, P., & Guedes, P. (2015). Aquaponics System – An EPS@ISEP 2014 Spring Project. In *Proceedings of the Third International Conference on Technological Ecosystems for Enhancing Multiculturality* (pp. 537-544). Association for Computing Machinery. 10.1145/2808580.2808662

Lucas, B., Hanson, J., & Claxton, G. (2014). Thinking like an engineer: Implications for the education system. Technical Report. Royal Academy of Engineering.

Malheiro, B., Guedes, P., Silva, M. F., & Ferreira, P. (2019). Fostering Professional Competencies in Engineering Undergraduates with EPS@ISEP. *Education Sciences, 9*(2), 119. doi:10.3390/educsci9020119

Malheiro, B., Silva, M., Ferreira, P., & Guedes, P. (2015). CDIO and the European Project Semester: a match for capstone projects? In *Proceedings of the 11th International CDIO Conference* (pp. 1-14). http://rocketship.cdio.org/knowledge-library/documents/cdio-and-european-project-semester-match-capstone-projects

Malheiro, B., Silva, M., Ribeiro, M. C., Guedes, P., & Ferreira, P. (2013). The European Project Semester at ISEP – Learning to Learn Engineering. In *Proceedings of the First International Conference of the Portuguese Society for Engineering Education* (pp. 1-8). 10.1109/CISPEE.2013.6701977

Malheiro, B., Silva, M., Ribeiro, M. C., Guedes, P., & Ferreira, P. (2015). The European Project Semester at ISEP: The challenge of educating global engineers. *European Journal of Engineering Education, 40*(3), 328–346. doi:10.1080/03043797.2014.960509

Michael, J. (2006). Where's the evidence that active learning works? *Advances in Physiology Education, 30*(4), 159–167. doi:10.1152/advan.00053.2006 PMID:17108243

Möller, B., Tarkowski, M., van den Bleeken, T., & Wlazlo, E. K. (2013). *Autonomous Environmental Buoy* (EPS Project Wiki). Instituto Superior de Engenharia do Porto. http://www.eps2013-wiki3.dee.isep.ipp.pt/

National Academy of Engineering. (2012). *Infusing Real World Experiences into Engineering Education.* The National Academies Press.

Organisation for Economic Co-operation and Development. (2014). OECD Science. *Technology and Industry Outlook, 2014.* Advance online publication. doi:10.1787ti_outlook-2014-en

Organisation for Economic Co-operation and Development. (2018). The future of education and skills. *Education, 2030.* https://www.oecd.org/education/2030/oecd-education-2030-position-paper.pdf

Panitz, T. (1999). *Collaborative versus Cooperative Learning: A Comparison of the Two Concepts Which Will Help Us Understand the Underlying Nature of Interactive Learning*. Education Resources Information Center. https://files.eric.ed.gov/fulltext/ED448443.pdf

Silva, M. F., Malheiro, B., Guedes, P., Ferreira, P., Ribeiro, C., Ferreira, F., & Duarte, A. J. (2015). Development of Biomimetic Robots in the EPS Engineering Programme Capstone Project. In *Proceedings of Third International Conference on Technological Ecosystems for Enhancing Multiculturality* (pp. 227-234). Association for Computing Machinery. 10.1145/2808580.2808614

United Nations Educational, Scientific and Cultural Organization. (2010). *Engineering: Issues, Challenges and Opportunities for Development*. United Nations.

Vos, H., & de Graaff, E. (2004). Developing metacognition: A basis for active learning. *European Journal of Engineering Education, 29*(4), 543–548. doi:10.1080/03043790410001716257

KEY TERMS AND DEFINITIONS

Active Learning: A student-centered learning method that promotes autonomous search for knowledge.

Engineering Education: The noble task of training conscientious and proficient engineers.

European Project Semester: A one-semester engineering program that implements active and project-based learning, multicultural and multidisciplinary teamwork and ethics and sustainability driven design to develop key professional competencies in future engineers.

Project-Based Learning: A learning framework where students are responsible for finding a solution to real-world self-motivating challenges.

Socio-Professional Ethics: A moral compass guiding professional actions and decisions with the goal to improve the well-being of individuals, society, and planet.

Sustainability-Driven Design: A conceptual design process where decisions aim to minimize the economic, environmental, and social impact of the product lifecycle.

Teamwork: A collaborative work method where several people work towards a shared goal (usually unattainable individually) that maximizes the strengths and minimizes the weaknesses of the participants.

Chapter 9
Learning Engineering Ethics:
The EPS Contribution

Roger Nylund
Novia University of Applied Sciences, Finland

Benedita Malheiro
https://orcid.org/0000-0001-9083-4292
Instituto Superior de Engenharia do Porto, Instituto Politécnico do Porto, Portugal

ABSTRACT

Ethics promotes collective respect, well-being, and development. Its values drive individual, social, and environmental conduct and apply to all human activities. Professional ethics or deontology refers to the moral obligations binding professionals to society in the work sphere. These moral obligations are fundamental in professions that, like engineering, are behind technological innovation, given its current daily and worldwide impact. As such, it is essential to train engineers for holistic problem-solving, that is, driven by ethical, sustainability, technical, and economic constraints. This chapter analyses engineering ethics and how it is addressed within the European Project Semester, an undergraduate one-semester international exchange program, focusing on this aspect of engineering education. European Project Semester relies on multicultural teamwork, project-based learning, ethics, and sustainability-oriented design to motivate and prepare global engineers committed to protecting and improving life on the planet.

INTRODUCTION

Ethics, as an academic field, is a branch of practical philosophy. Ethics is so practical that it is omnipresent in our daily life. People make decisions, choices and act based on considerations that have an ethical dimension, regardless of whether they are aware of this fact. As engineering professionals, the ever-changing work environment brings new ethical challenges, such as artificial intelligence and autonomous vehicles, but also in handling questions about the economy, health and safety, and risk management (Felländer, 2019). Students need to be prepared and trained to think critically, identify and

DOI: 10.4018/978-1-6684-2309-7.ch009

Copyright © 2022, IGI Global. Copying or distributing in print or electronic forms without written permission of IGI Global is prohibited.

handle ethical issues autonomously in everyday situations. This requires a "moral compass", a basic idea of ethical conduct.

In 2017, the European Union (EU) Commission renewed its agenda for higher education which bears clear ethical expectations:

[...] the success of the European project depends on the EU's capacity to build a better future for European citizens.

Effective education and training systems are a foundation of **fair, open and democratic societies** *and of sustained growth and employment.*

[...] higher education must play its part in facing up to Europe's **social and democratic challenges**.

[...] academics and students have a crucial role to play in **defending empirical facts** *and* **evidence** *and in* **communicating** *research results effectively and widely. (Publications Office of the European Union, 2021).*

These democratic values are not by any means automatically addressed in ordinary engineering education. However, the European Project Semester (EPS) gives, as will be shown below, many opportunities to do so. This contribution aims to introduce the potential ethical challenges facing students of today and to show how these can be addressed during an EPS-semester.

The current chapter includes four additional sections: the second section addresses ethics in general and specifically professional ethics; the third section presents how EPS contributes to the adoption of ethical practices by the participants; the fourth section illustrates how this was achieved in specific EPS projects; and the last section draws the conclusions.

ETHICS AND PROFESSIONAL ETHICS

Ethics is a branch of philosophy dedicated to the study of morals or the human sense of right and wrong (Crane et al., 2019). It emerges from human interaction and, ultimately, expects individuals to treat others as they would like to be treated themselves. Professional ethics or deontology follows the same spirit but has a narrower scope since it applies to professional acts. This section will first give a brief overview of the concept of ethics, then discuss professional ethics, and lastly, provide examples of how ethics can be incorporated in an EPS-semester.

Ethics

While the term Ethics is usually reserved for more theoretical discussions and broader scope, the concept of morality is used to assess *human actions and conduct.* Still, these two terms are often used as interchangeable, and such a distinction is irrelevant for the purpose of this chapter. The question about morality can be about deliberately *doing* something (actions) or about *failing to do* something (omissions). Dichotomies like good versus bad or better versus worse are used to identify the ethical value of these actions or omissions (Buckeridge, 2011). Morality is sometimes a question about knowing "the right thing" to do, but, more often, about a more profound challenge – the willingness to "walk the talk" and do what one considers right. This can be an inner fight that was documented already some 2000 years ago. "*For I do not do the good I want to do, but the evil I do not want to do, this I keep on doing.*" (St. Paul in Romans 7:19). Ethical theories exclusively connect ethics to behavior, not merely words or thoughts. A person's moral qualities are evaluated based on actions or omissions in all ethical theories. The relevance of this becomes obvious, especially nowadays, when people tend hastily to attribute "good values" to others based on speeches rather than deeds, therefore, rejecting any ethical theory.

If we can agree on the idea that morality is evaluated based on actions and omissions rather than words, it raises the question of whether it includes *all* human actions. There are some limitations. A few criteria can be used for applying moral responsibility to an agent in any given situation. First, to be morally responsible, one needs *freedom to act* (Crane et al., 2019). Of course, it can be objected that there is always a choice, even in extreme cases where "doing the right thing" may have severe consequences for the individual. For example, a whistle-blower can lose his or her job. In everyday situations, moral responsibility is often delegated upwards in organizations. If a boss orders a subordinate to do something, he or she might not question the decision, leaving the moral considerations to the executive.

One needs the *ability for moral reasoning* (ibid, 2019). Usually, this is expected only from humans. Still, small children are humans and are not held morally responsible in the same way as adults. The same goes for elderly people with dementia. The case with the elderly also brings forth one more criterion for being held morally responsible, which is *the power or ability to act* (ibid. 2019). Sometimes one may know what should be done but has no actual power to act. Also, objections can be made here. Many of us may think that global warming is beyond our pay level, that the most we can do is adjust our own ways of living. However, it did not stop the Swedish teenager Greta Thunberg from becoming a worldwide front figure of those who accuse politicians of doing too little to stop climate change (Gerretsen & John, 2019). She displayed an ethos and pathos for our shared future that, in retrospect, anyone could have embraced. The value hierarchy plays out different results for different people – even when sharing similar values. For some, they are a basis for action; for others, less so.

Furthermore, in the evaluation of moral responsibility, the question about *intentionality* is seen as relevant. A person is more responsible for things done intentionally than for honest mistakes. However, intentionality does not apply in engineering – when designing a bridge, the engineer is responsible if it collapses with or without any intention. Omissions are evaluated as seriously as deeds. Then there are questions about *understanding* the situation, i.e., detecting the moral aspects of what is happening or what one is about to do. Nonetheless, it is hard to "do the right thing" if one does not see the ethical dimension. Most ethical questions come creeping gradually. As Bertrand Russell puts it: "going down that road, you don't know when you should start screaming" (Russell in Bauman, 1993). In everyday professional situations, ethical awareness, i.e., the ability to detect the ethical dimension in a decision and its moral imperative for the individual, can be difficult. Too often, in technical fields like engineering, issues are handled as purely technical, ignoring the human and ecological aspects of the application of technology. Studies show that one of the main reasons for not taking moral action is the failure to see the ethical aspects of a situation (Ladkin, 2016). *Moral sensitivity* is needed and can be trained. New generations meet new challenges as mentioned in the introduction of this chapter. Thus, there is a need for an ongoing dialogue about ethical questions. When the ethical aspects in a situation *are,* in fact, evaluated, it is done based on the values of the actor.

As humans, we have diverse sources for the *values* that constitute our "moral compass". Values can be described as objectives that have been accepted as worth striving for by the individual, group, or society. They can be of absolute or relative nature. Moreover, values are not equal since some override others (Brytting, 2005). Economic values may often override ecological values at the grocery shelf in the supermarket. This does not make the other value non-existent. It just loses the competition to another value regarded, at that moment, as more important. Based on different sets of values and courses of action, a few major *ethical theories* have been developed. They can be based on religious beliefs, the strive to follow good principles (deontology), the attempt to avoid causing harm or pain and the endeavor to enhance happiness (utilitarianism), aspiring to be a good person, building a good society (virtue ethics), recognizing universal (human) rights, protecting near and dear ones, etc. (Buckeridge, 2011).

Moral objects are the ones acted upon or otherwise directly and indirectly bearing the consequences of our actions. They can be friends, competitors, enemies, strangers, animals, the ecosystem, etc. From this very incomplete list, it becomes clear that *there is often an ethical evaluation of the mere description of people and situations*. How one describes them shows a lot of the perceived morality of the situation. These evaluations are based on the values of the evaluator.

According to Helkama (2017), the top three Finnish values are honesty, safety for the family, and *health*. Globally, *benevolence* is regarded as particularly important (Schwartz, 2012). When working in teams, honesty and benevolence are essential core values to be fostered since they are the basis for building *trust* within the team. Trust between team members has been identified as one of the cornerstones for top-performing teams (Duhigg, 2016; Wheelan, 2016). According to Putnam (2001), it encompasses the trust in others' goodwill (benevolence), in others' intention to keep agreements, and in others' ability to do so. Values are translated into guidelines for everyday actions through *norms*. Norms in a society or within a profession are commonly accepted behavioral models. There is no need to draw a sharp line between values and norms because these terms are used almost interchangeably for the same concepts. Values are usually used to describe the lays behind the norms, which are more action-oriented guidelines (Schein & Schein, 2017). The set of norms for different professions, known as professional ethics or deontology, is often expressed as guidelines defined by professional associations.

Professional Ethics

Professional skills are vital in building a good and well-functioning society. Professions are self-regulating. Together with governmental authorities, they control the training and evaluation processes that admit new people into the field. Professionals usually have *autonomy* in the workplace. They are expected to utilize their *independent judgment* while carrying out professional responsibilities. Engineers cannot work in the field without a documented engineering education. In most European countries, to perform specific engineering acts, they must be enrolled in their national association of engineers. Since professionals are granted much autonomy at work, it is crucial for professions to be guided by *ethical standards*. Such standards are based on shared *professional values*. For example, the Association of German Engineers has published a set of ethical guidelines for engineers (Verein Deutscher Ingenieure, 2021) that defines as core values and norms:

- *Prudence* – an engineer is expected to work thoroughly, not jump to hasty conclusions, and only take on assignments to which he/she (or the organization) has competence.
- *Responsibility* – an engineer is responsible for his or her own actions towards employers, customers, technology users, and political and societal institutions. This responsibility includes the quality, reliability, and safety of products developed and produced.

- *Objectivity* – an engineer needs to give an objective judgment regardless of whether it favors his/her preferences and opinions.
- *Openness and honesty* – an engineer is expected to discuss all key project decisions with other stakeholders to find a common solution in case of conflicting values.
- *Common sense of ethics* – an engineer follows the *same fundamentals of ethics as everybody else* in society. While this is a vague formulation, it provides a general instruction to work for the good of society and avoid the development of technology that could cause harm or be potentially misused.

While the use of such guidelines is optional, engineers must comply with workplace standards. However, if the unwritten workplace standards are unethical, this can be a significant challenge for integrity-oriented engineers. In such circumstances, an engineer may feel a moral obligation to expose the wrongdoing, often with severe consequences both for the organization and whistleblower. Thus, moral autonomy should be a goal in training all kinds of professionals. For the individual, following rules and recommendations can be a good start, but the ability to decide for oneself what to do in new situations is essential. It requires reflection about one's own values and moral basis to become able to identify moral issues and make moral decisions.

One central dimension of ethics is the distribution of wealth. The western world has been accused, on good grounds, of exploiting people in the developing parts of the world to build our wealth. It is often forgotten that technology has been the main force behind the long-term increases in income in the rich world, not the exploitation of the poor. This means that the entire world has a reasonable hope of reaping the benefits of technological advance (Sachs, 2005: 31). In this context, engineering drives social, economic, and human development and underpins knowledge societies and infrastructures. It is a major factor in innovation and indeed the rise and fall of civilizations (UNESCO 2010). Engineers build technology, and as is well-known, technology can be used for good and for bad. The question is not as much about what *can* be done for a better future, but whether there is a *will* to use technological – and other – advances for the good of humankind and the planet. The ultimate mission for engineers is to take on the challenge to make the world a cleaner, safer, and better place to live in on equal terms and, thus, contribute and support for the achievement of the United Nations' sustainable development goals (United Nations, 2021).

EPS AS A TOOL FOR FOSTERING ETHICAL AWARENESS AND BEHAVIOR

As stated in the introduction, the European Commission has presented higher education institutions with ethical challenges. Some of them are not easily met in traditional classroom education. In addition to the professional content taught in different fields, higher education institutions have a key role in fostering tomorrow's leaders to become the builders of a better society. In a study funded by the European Commission, Brandenburg et al. (2016) found that personal development during an exchange period of six months equals the development during four years of studies at home. Based on the memo© approach, the study measured the level of six selected personality traits of students referred to as "memo© factors": tolerance of ambiguity, curiosity, confidence, serenity, decisiveness, and vigor (problem-solving skills) before and after mobility. Across Europe, 93% of the surveyed employers confirmed that the six traits were indeed important for the recruitment and professional development of their employees (Brandenburg et al., 2016). Obviously, the one-semester international mobility provided by EPS offers such an experience. It helps undergraduate engineering students to become autonomous, which is a crucial basis for becoming a moral agent, i.e., not just following instructions but also reflecting and acting on one's beliefs. The ethical challenges facing young engineering students in EPS can be divided into two parts.

- The ethical challenges of co-operation within a team, namely free-rider and harassment issues.
- The ethical challenges of the project itself.

The challenges of working together in a team are covered in the Teambuilding chapter of this book. The project ethical challenges correspond to the default ethical challenges faced by individuals:

- Ability to identify an ethical question.
- The use of moral imagination.
- The will to choose ethically sustainable solutions.
- The courage to speak up and act ethically.

In an EPS project, the coaches can train the students in these four skills. This is easier since market forces do not fully apply to curricular projects. Students are free to embrace solutions based on the identification of the ethical implications of design, development, and marketing decisions. Once an ethical issue is found, it is seldom straightforward to decide what to do. However, first, engineers must identify and solve potential ethical conflicts. Even if they might know what to do in a perfect world, the real world often blurs this vision, requiring *moral imagination* to find doable solutions. The hardest part is still to find the *willingness* to do what one understands as being the right or better choice. In this context, the role of EPS is to train participants in the detection of such conflicts and to embrace ethical solutions from the first design choices till the final project presentation and demo. EPS, where capstone projects are performed by multicultural and multidisciplinary teams of undergraduate students, is particularly well-suited for this purpose.

One key element is to empower teams, i.e., to make them autonomous and responsible for the definition of their own solutions. This starts by proposing ill-structured and open-ended project briefs, where an interdisciplinary problem is presented together with a set of broad requirements such as the compliance with applicable European Directives and the International System of Units, the adoption of open-source software, the reuse of materials and components, and the specified prototype budget. Ill-structured and open-ended design problems are particularly challenging to solve, especially for novice learners with limited familiarity with this type of problem (Hulls & Rennick, 2020). Moreover, multidisciplinary problems require the contributions of the different scientific backgrounds of team members, strengthening the bond between members, and the willingness to explore unknown fields and acquire new knowledge in the process. The budget is kept low to promote innovation rather than the acquisition of off-the-shelf solutions.

Inevitably, ill-structured and open-ended problems lead the teams to a series of exploratory studies to define the specific requirements and produce the concept of their solution. This stage implies surveying, analyzing, and brainstorming about existing and candidate solutions, considering marketing, ethics, and sustainability. Students develop systematization, critical and design thinking in this process. The derived requirements and concept are then the basis to perform ethically and sustainability-driven design (Duarte et al., 2020).

The design stage can start with multiple draft sketches and cardboard models. It is followed by the selection of materials, considering physical properties, sustainability, and ethics, together with 2D and 3D computer-aided modeling, simulation, and refinement cycle. Whenever the solution encompasses a control system, the initial block diagram is iteratively refined into detailed electrical and electronic schematics. In the case of software development, like mobile or web applications, compliance with the requirements is verified through the predefinition of the functional, performance and load tests to be met. These outcomes are the blueprints of the proposed solution.

The development of a proof-of-concept prototype may involve down-scaling, the reuse of existing materials and components, the building and assembly of parts in workshops, the assembly and test of the control system in laboratories, as well as the development and test of the software.

Finally, to determine whether EPS fosters critical professional competencies in engineering undergraduates, Malheiro et al. (2019) proposed the application of the 4C2S analysis framework. 4C2S stands for the identified set of core professional competencies, ranging from *critical thinking and problem solving*, *effective communication*, *collaboration and team building*, and *creativity and innovation competencies* (the four C) suggested by the American Management Association to *socio-professional ethics* and *sustainable development* (the two S). This evidence-based framework, which analyses the process and timeline of the student learning journey, was then applied to a specific EPS project for the purpose of exemplification. The results showed that the participants were able to incorporate sustainable development and socio-professional ethics in the decision-making process, namely in the project design and development stages.

ILLUSTRATIVE EPS CASES

The main ethical question facing engineers is, then, how to devise sound technical solutions with limited knowledge, budget, and resources, meeting the needs and expectations of people and preserving resources and life on Earth (Silva et al., 2019).

EPS teams must consider the ethical issues surrounding the solution design, the potential product, and the proof-of-concept prototype, namely the engineering, marketing, academics, environmental and liability ethics. It is essential for the development of the projects that the students fully consider each of the topics mentioned when making decisions. Since the first week of the semester, EPS teams have been challenged to find solutions to real problems. The first weeks are dedicated to ideation, preliminary studies, brainstorming, and design thinking. These activities are explored with the help of project support intensive seminars and weekly project meetings. Specifically, seminars focusing on sustainable development, ethics and deontology have a clear impact on the design process (Duarte et al., 2020). As a result, the designed decisions are based not only on state of the art but also on the ethical and sustainability dimensions. The following projects illustrate how three EPS teams achieved this desired goal in the spring of 2021. The extracts provided herein are from the technical-scientific papers and project wikis of the teams. They are focused on the decisions related to the design of the solution and the prospective market attitude. Figure 1 shows the designed solutions.

FREE Desk is an ergonomic reconfigurable smart home desk (Vendoorne-Feys et al., 2021a). The team's vision of the product and purpose was influenced by the applicable laws and technical standards, and the desire to create a professional and trustworthy brand. The design complied with public safety, health, and welfare. To create a strong connection between user and product, the students adopted a sustainable strategy and adhered to an engineering code of ethics from the beginning. Each phase of the decision process considered the engineering, marketing, sales, and environmental perspectives. The main goal was to create a sustainable and eco-friendly solution by reducing plastic and reusing recyclable packaging material. The FREE desk design (Figure 1a) addresses all these concerns, providing excellent usability and fulfills applicable directives, legislation, and regulations. Being a reconfigurable product, the smart desk offers the user three levels of adjustability: bench, sitting desk, and standing desk, and includes a folding light-sensor lamp integrated into the tabletop, multiple device ports, and an integrated battery for a proper working space. The selected materials have a low environmental impact (Vendoorne-Feys et al., 2021b).

ScanGo is a crowd orchestration solution for attendees and organizers of large outdoor events (Fohanno et al., 2021a). The design of the product considered ethical values and legislation. For the manufacturing of the product, the team envisioned a company governed by ethical values and professional deontology. Consequently, the team voted in favor of honest communication with both the customers and the competition, ethical marketing to convey understandable and correct information about the product, and recognition of a liability to build a good reputation and strengthen the product brand. The solution focuses on the well-being of the user and presents an eco-friendly design, balancing the social, economic, and environmental aspects. General ethical and deontological issues were carefully considered throughout the project, as they were key differentiator factors. Moreover, the solution complies with the applicable EU directives. The outcome is ScanGo (Figure1b), a solution integrating a mobile application and a scanning device that allows: (i) attendees to receive real-time information about ongoing concerts, crowd density in designated areas and incentives to attend alternative activities; and (ii) organizers to orchestrate the crowd and enjoy a multi-day scanner rental service, including transport, installation, and set-up (Fohanno et al., 2021b).

Figure 1. EPS examples of design driven by ethics and sustainability
(Vendoorne-Feys et al., 2021b; Fohanno et al., 2021b; Mendes et al.2021b)

Wormify is a smart and modular urban vermiponics food production system for apartment building residents (Mendes et al.2021b). The design, presented in Figure1c, was derived from the four main ethical and deontological concerns related to the project, which are engineering, sales and marketing, environmental, and liability. The team pledged to follow a robust code of ethics and thus act with honesty, impartiality, fairness, and equity, and be committed to the safety and welfare of the public. To avoid unprofessional behavior and dishonest practices, the team planned to implement and monitor an ethical marketing strategy to achieve success and create a solid reputation. Moreover, the team decided to make only true claims and adopt sustainable decisions to create a truly environmentally friendly solution. The design considered the legislation to avoid liability issues and protect the environment and public health. Based on these resolutions, the team opted to create a modular keyhole garden for urban buildings, composed of a raised plant bed with a central composting basket. It relies on soil worms and microbes to reduce waste, enhance soil, and grow food without generating vermin or odors. The proposed system can monitor the most important parameters, ensuring optimal conditions both for the worms and the flora, and provides an effective and low-maintenance solution for transforming organic and compostable waste for urban spaces. Wormify is a product designed to respect all aspects mentioned, promising to follow a path of environmental consciousness that allows for sustainable food production everywhere (Mendes et al., 2021b).

CONCLUSION

This chapter examined the problem of fostering ethical awareness and behavior in engineering undergraduates and how this is achieved by the European Project Semester. As a project-based learning framework, EPS promotes the 4C competencies and the practice of *socio-professional ethics* and *sustainable development* (2S). This is particularly relevant since most EPS participants are third-year engineering students who will soon enter the labor market and become responsible for solving many of the upcoming challenges faced by society.

The analysis of three recent EPS projects highlights the type of ethical issues the teams consider when designing their solutions, which cover engineering, marketing, sustainability, and liability aspects. This leads to the conclusion that the EPS learning process exposes participants to ethical problem solving and imprints ethics and sustainability-driven decision-making in future engineers, an essential feature for the well-being of life on the planet.

REFERENCES

Bauman, Z. (1993). *Postmodern Ethics.* Wiley-Blackwell.

Brandenburg, U., Taboadela, O., & Vancea, M. (2016). Mobility Matters: the ERASMUS Impact Study. In Global and Local Internationalization. Transgressions: Cultural Studies and Education. SensePublishers. doi:10.1007/978-94-6300-301-8_16

Brytting, T. (2005). *Företagsetik andra upplagan.* Liber Ekonomi.

Buckeridge, J. (2011). *4 Es Ethics, Engineering, Economics & Environment.* The Federation Press.

Crane, A., Matten, D., Glozer, S., & Spence, L. (2019). *Business Ethics.* Oxford University Press.

Duarte, A. J., Malheiro, B., Arno, E., Perrat, I., Silva, M. F., Fuentes, P., Guedes, P., & Ferreira, P. (2020). Engineering Education for Sustainable Development: The European Project Semester Approach. *IEEE Transactions on Education, 63*(2), 108–117. doi:10.1109/TE.2019.2926944

Duhigg, C. (2016). *What Google Learned from its Quest to Build the Perfect Team*. https://www.nytimes.com/2016/02/28/

Felländer, A. (2019). Anna Felländer vill rädda oss från AI-undergång. *Digital.*

Fohanno, B., Pires, B., Ionescu, C., Ladka, E., & Perek, M. (2021b). *Crowd Orchestration* (Project Wiki). https://www.eps2021-wiki1.dee.isep.ipp.pt/

Fohanno, B., Pires, B., Ionescu, C., Ladka, E., Perek, M., Malheiro, B., Ribeiro, C., Justo, J., Silva, M. F., Ferreira, P., & Guedes, P. (2021a). Crowd Orchestration – An EPS@ISEP 2021 Project. In *Proceedings of Ninth International Conference on Technological Ecosystems for Enhancing Multiculturality* (pp. 411-416). ACM. 10.1145/3486011.3486481

Gerretsen, I., & John, T. (2019). *UK climate protest: Student walk out inspired by Greta Thunberg*. http://www.cnn.com

Hulls, C. C. W., & Rennick, C. (2020). Use of a Cornerstone Project to Teach Ill-Structured Software Design in First Year. *IEEE Transactions on Education, 63*(2), 98–107. doi:10.1109/TE.2019.2959591

Ladkin, D. (2016). *Mastering the Ethical Dimension of Organizations*. Elgar.

Malheiro, B., Guedes, P., Silva, M. F., & Ferreira, P. (2019). Fostering Professional Competencies in Engineering Undergraduates with EPS@ISEP. *Education Sciences, 9*(2), 119. doi:10.3390/educsci9020119

Mendes, A., Tatuc, E., Joos, F., Wyka, J., & Petrevski, K. (2021b). *Sustainable Food Production Through Vermicomposting* (Project Wiki). https://www.eps2021-wiki2.dee.isep.ipp.pt/

Mendes, A., Tatuc, E., Joos, F., Wyka, J., Petrevski, K., Malheiro, B., Ribeiro, C., Justo, J., Silva, M. F., Ferreira, P., & Guedes, P. (2021a). Sustainable Food Production Through Vermicomposting – An EPS@ISEP 2021 Project. In *Proceedings of Ninth International Conference on Technological Ecosystems for Enhancing Multiculturality* (pp. 553-559). ACM. 10.1145/3486011.3486515

Publications Office of the European Union. (2021). *EU law – EUR-Lex*. https://eur-lex.europa.eu/homepage.html

Putnam, R. D. (2001). *Making democracy work: Civic traditions in modern Italy*. Princeton University Press.

Sachs, J. D. (2005). *The end of Poverty*. Penguin.

Schein, E., & Schein, P. (2017). *Organizational Culture and Leadership* (5th ed.). John Wiley Sons inc.

Schwartz, S. H. (2012). An Overview of the Schwartz Theory of Basic Values. *Online Readings in Psychology and Culture, 2*(1). Advance online publication. doi:10.9707/2307-0919.1116

Silva, M. F., Malheiro, B., Ferreira, P., & Guedes, P. (2019). Promoting Ethical Practices in Engineering Undergraduates. In *Artificial Intelligence, Robots and Ethics, Proceedings of the Fourth International Conference on Robot Ethics and Standards* (pp. 55-63). CLAWAR Association. 10.13180/icres.2019.29-30.07.002

UNESCO. (2010). *Engineering: Issues, Challenges and Opportunities for Development.* UNESCO. https://unesdoc.unesco.org/ark:/48223/pf0000189753

United Nations. (2021). United Nations 17 *Sustainable Development Goals to Transform Our World.* https://www.un.org/sustainabledevelopment/

Vendoorne-Feys, A., Nicoara, G.-G., Carasel, I.-S., Karpiak, M., & Koceski, N. (2021b). *Ergonomic Reconfigurable Home-Desk* (Project Wiki). https://www.eps2021-wiki3.dee.isep.ipp.pt/

Vendoorne-Feys, A., Nicoara, G.-G., Carasel, I.-S., Karpiak, M., Koceski, N., Malheiro, B., Ribeiro, C., Justo, J., Silva, M. F., Ferreira, P., & Guedes, P. (2021a). Reconfigurable and Ergonomic Smart Desk – An EPS@ISEP 2021 Project. In *Proceedings of Ninth International Conference on Technological Ecosystems for Enhancing Multiculturality* (pp. 464-470). ACM. 10.1145/3486011.3486566

Verein Deutscher Ingenieure. (2021). *Ethische Grundsätze des Ingenieurberufs* [Ethical Principles of the Engineering Profession]. https://www.vdi.de/themen/ethische-grundsaetze

Wheelan, S. A. (2016). *Creating Effective Teams – Guide for Members and Leaders* (5th ed.). Sage.

KEY TERMS AND DEFINITIONS

Deontological Ethics: Duty-based professional ethics (see professional ethics below).

Engineering Education: The training of future engineering professionals.

Ethical Awareness: The ability to identify and act based on moral or ethical issues.

Ethical Decision-Making: A decision process guided by ethical concerns.

European Project Semester: A one-semester engineering learning framework supported by project-based learning, multicultural and multidisciplinary teamwork, ethics, deontological and sustainability-driven problem-solving.

Professional Ethics: The moral duties or rules that must be followed in the exercise of a profession.

Sustainability-Driven Solutions: The design of solutions considering the economic, environmental, and social dimensions.

Chapter 10
Towards Human-Oriented Engineering Education:
A Model of Empathetic Engineering in the EPS Program at the International Faculty of Engineering, Lodz

Grazyna Budzinska
Lodz University of Technology, Poland

Anna Karczemska
Lodz University of Technology, Poland

Dariusz Witkowski
Lodz University of Technology, Poland

Malgorzata Klimka-Kolysko
SWPS University of Social Sciences and Humanities, Poland

ABSTRACT

Continuous changes in different spheres of human activity rapidly intensifying in the globalized world pose dramatic challenges for individuals and educational institutions whose mission is to prepare their graduates for professional, lifelong success. The changes trigger a continuous discourse on the nature and form of educational paradigms and pivotal values that should be utilized by schools and universities to fulfill their mission. The chapter presents the human-oriented model of engineering focused on empathy, constructed in the group of academics at Lodz University of Technology. This group has defined a methodological approach that can help to guide students and teachers in finding the optimal approach in project-based study programs, most notably in the European Project Semester. Moreover, the chapter includes the results of interviews with students and project mentors and survey results measuring empathy in both stakeholder groups.

DOI: 10.4018/978-1-6684-2309-7.ch010

Copyright © 2022, IGI Global. Copying or distributing in print or electronic forms without written permission of IGI Global is prohibited.

INTRODUCTION

The precarious times we are living in, characterized by quick and dynamic changes in all spheres of human activity, call for an adequate response from universities and schools in the form of modern methodological concepts, study programs and, even more importantly, philosophical and psychological bases that could direct the future engineers in constructing human-oriented technological solutions for complex problems that societies experience on an every-day basis. The trigger factors for contemporary "wicked problems" (Kolko, 2012) and societal crises are of various nature: economic, political, social, environmental, populational, philosophical and, most recently, of a health-related nature.

Challenges of the globalized world cannot be successfully resolved without engineering technologies that are already deeply "woven into the fabric of our society" (Grasso & Burkins, 2009), and as such may themselves constitute the origin of challenges if not addressed appropriately. The most recent crisis was evoked by the pandemics which caused disruptions in many areas of human activity including education. Remote education resulted in numerous problems; however, the most acute issue at universities was the impossibility to communicate and share through direct contact and develop practical skills in relation with others. Encountering the Other (Levinas, 1969) has again been proven to be a crucial element in the learning process as it conditions human connection, dialogue, the confrontation of ideas and knowledge transference on cognitive, emotional and ethical levels.

The chapter presents the result of a several-year dialogue in a diverse multidisciplinary group of academics representing engineering and non-engineering disciplines, who attempted to identify the most important values for contemporary engineers. These values constitute the pivot of engineering project-organized education aimed at addressing the challenges of dynamically changing environments. The results of this dialogue, enriched with the perceptions of students recorded in reflective writing and collected in student interviews and the opinions of external experts, contributed to the identification of empathy as a core value that may be used in contemporary engineering to enhance its integration with human experience and social contexts. The core value of empathy was, in turn, used as a pivotal value that is to underpin and organize various components of the proposed educational model used in the European Project Semester (EPS) at the International Faculty of Engineering (IFE), Lodz University of Technology (TUL). The model can be characterized as a human-oriented empathy-based model for engineering education exercised in project-organized interdisciplinary and intercultural teams and it constitutes a proposal for an educational paradigm that can help address the challenges of demanding, quickly changing precarious environments, challenges that constitute a real challenge for young people, universities, businesses and societies.

The chapter introduces different theoretical perspectives on empathy[1] that were used as a basis of an academic dialogue at TUL, a dialogue completed with a human oriented empathy-based model for engineering education. The theoretical perspective on empathy originates from literature in humanistic (Levenson & Ruef, 1992), design (Heilighen & Dong, 2019), and engineering studies (Walther et al., 2019; Afroogh et al., 2021). Its presentation explores the ways these perspectives may be utilized in problem solving while keeping the human as the central focus. The theoretical basis is further synthesized in a human-oriented empathy-based model created and utilized in EPS and exemplified in a case study that describes in detail how empathy is implemented in project-organized engineering education using Design Thinking (DT) (Kelley & Kelley, 2013) and Project-Based Learning (PBL) methodology. The concept is enriched with qualitative research results exploring the perceptions of students and project mentors on the role of empathy in project work and quantitative research measuring the levels of affective and cognitive empathy in the group of students.

The chapter includes six sections. Section One presents the concept of empathy describing the roots of empathy meaning, recent development of its conceptualization in social cognitive neuroscience, a concept analysis based on 52 articles on empathy published between 1980 and 2019, which is followed by two subsections subsequently presenting the concept of empathy in engineering and empathy presentation in design. Section Two describes the EPS Human-oriented Model of Engineering, a proposal for project-based engineering education developed by a group of academics who mentored EPS project teams at the IFE. Section Three shows a case study of a student project that reflects the characteristics of the EPS Human-oriented Model of Engineering. Section Four explores students' experience and understanding of the role of empathy in project work based on the survey conducted among 112 students, an in-depth interview with a student who participated in the project realization described in the case study and questions on empathy asked to projects' mentors. Section Five discusses the levels of empathy among students and mentors measured with the Questionnaire of Cognitive and Affective Empathy (Reniers et al., 2011). Section Six concludes the findings of the empathy measurement in cognitive and affective dimensions and explores selected controversies on empathy in education and advocates selected directions for future research.

THE CONCEPT OF EMPATHY

The concept of empathy, used today in many fields such as philosophy, psychology, the neurosciences, sociology, design and engineering, is often understood as a psychological resource relying on numerous "capacities that are thought of as being central for constituting humans", and the resource that allows us to "know what other people are thinking and feeling, to emotionally engage with them, to share their thoughts and feelings, and to care for their well–being" (Stueber, 2019). The roots of its meaning go back to the well-established philosophical writings of Adam Smith and David Hume (Hume, 1739-40), who stated that the "minds of men are mirrors to one another". However, it is a relatively new phenomenon that empathy has been under investigation for scientific exploration in diverse areas of science.

The word empathy was first used in the English lexicon by an American psychologist Edward Titchener (Titchener, 1909), who referred to Theodor Lipps' (Lipps, 1903) concept of 'Einfühlung' as an aesthetic experience that is always an experience of "another human", an inner imitation or kinesthetic sensation experienced by an observer while looking at another person or another object. It lets the observer imagine or learn what the other person feels or thinks (Iacoboni, 2008) but also lets the observer learn to take up roles in life and society (Hoffman, 2000).

Recent developments in social cognitive neuroscience have identified processes that mediate empathy in the brain, which help to confirm previous suppositions that, when observing others, we tend to mimic actions that we see. This process is, to a high extent, an automatic and unconscious experience and it is possible thanks to mirror neurons that "fire" when the action is performed or observed. Then, "the neurons send signals to emotion brain centers in the limbic system to make us feel what other people feel" (Iacoboni, 2008). Brain neural circuits responsible for experiencing affective states have been explored due to the existence of neuroimaging technologies (De Waal & Preston, 2017). Nonetheless, there is also evidence supporting genetic determinants of attachment linked with the role of the oxytocin receptor gene (Smith et al., 2014). Thus, the notion of empathy between people may be related to a state of shared physiology that triggers psychological and mental phenomena. As such, empathy might be a crucial factor that helps us learn through seeing and empathically experiencing model phenomena and behaviors, because the neural networks in our brain can evoke feelings that mirror the experiences of those that we observe or hear. We know we can empathize but cannot be sure about how precise we are in detecting the feelings and thoughts of the Other in the empathy process. The accuracy of the human ability to know through empathy what other people feel or think is still questionable, still being explored and examined (Levenson & Ruef, 1992; Roth & Altman, 2021).

In relation to the new findings in social cognitive neuroscience, empathy was conceptualized by Decety and colleagues (Decety & Lamm, 2006; Decety & Moriguchi, 2007) in an interdisciplinary model combining the interaction of four elements:

Affective sharing understood as experiencing similar emotions by the empathizer and empathee; the concept created with reference to mirror neurons theories and neurocognitive approaches to empathy represented in both *the perception-action model* (De Waal & Preston, 2017) and *simulation theory* (Goleman 1998).

Self-awareness presented as an ability to recognize and differentiate between our own public and private perceptions. Thank to self-awareness, we can understand how we are perceived by others, develop a social perception of the self, and can provide self-insight.

Mental flexibility conceptualized as a cognitive ability to take the subjective perspective of the other while still maintaining the boundaries between an empathizer and an empathee as well as the ability to switch between the two perspectives, which requires a rational and logistical understanding of the experiences of other people and enables role taking.

Emotion regulation shown as a factor due to which it is possible to regulate subjective emotions and feelings while using different cognitive and affective strategies (judgment, affirmation-seeking, mindfulness techniques, etc.).

The model is advocated by its creators to be completed with an important component of action that will go beyond the understanding of empathy function as a mere induction. Empathy would require a conscious decision-making component that completes the process of empathetic induction with "action that is positive, that promotes well-being, that is socially just." (Gerdes & Segal, 2009). Such an understanding and conceptualization of empathy exceeds its meaning beyond affective and cognitive states and allows for its active exploration, which can provoke actions that will have the power to evoke change and have an impact on other people's lives and communities.

An interesting voice in the discussion on empathy belongs to Eklund & Meranius (2021), two Swedish researchers who synthesized thematically a variety of aspects and approaches to empathy in the form of a concept analysis. They used existing multidisciplinary literature that comprised 52 articles published between 1980 and 2019 and attempted to identify similarities and differences among conceptualizations. The thematic analysis can help us synthesize the contemporary understanding of empathy, especially considering the diversity of scientific disciplines involved in its exploration. In spite of the popular opinion that empathy is understood in uncontrollably diverse ways (Kohut, 1959; Rogers, 1957; Gerdes et al., 2010; Iacoboni, 2008; Decety & Moriguchi, 2007; Segal, 2011), the meta-analysis of the concept indicates a thematic consensus in the group of researchers in various fields: medical scientists, psychologists, philosophers, sociologists. Most of them identify four salient components in empathy, the components that stay to a great extent in alignment with the conceptualization presented by Decety and colleagues. Eklund & Meranius (2021) identify four components that constitute a common basis of the majority of the studies on empathy and prove that empathy operates in the realm of understanding, feeling, sharing with the Other, and maintaining distance between people in the process.

Understanding includes knowing, perspective-taking and cognition. The empathizer must have "recognition and understanding of another's emotional state" (Decety & Moriguchi, 2007) and be able to "perceive and understand other people emotions and [...] react appropriately" (Leiberg & Anders, 2006). Smith claims an epistemological role for empathy, that "allows us to know how others feel" (Smith, 2017).

Feeling refers to an effective response to another person's situation and includes affect, emotion and sensation. Empathy is presented as "emotional and mental sensitivity to another's state" (De Waal & Preston, 2017), a "vital emotional force" (Gonzales-Liencrez et al., 2013) or as "an emotional response" (Cuff et al., 2016).

Sharing refers to experiencing similar states to the empathee: sharing experiences, feelings, and representations (Bernhardt & Singer, 2012).

Self-other differentiation refers to recognizing differences between an empathizer and an empathee, and comprises the differentiation of feelings, objectivity and self-distinction. Empathy is conceptualized as "self-other awareness" (Decety & Meyer, 2008), an ability "to recognize others as autonomous agents and to participate imaginatively in their separate lives" (Deigh, 1995).

Both in social cognitive neuroscience and in Eklund and Meranius' models, there is reference to cognition, affect, sharing and self-other awareness, which proves the reliability of conclusions of the two analyses. Eklund and Meranius examined an important issue of relations among the four core themes in empathy studies, and discovered that in some studies, the elements are presented in isolation. However, in the majority of studies, the themes are described as creating a meaningful whole among the four core elements (De Vignemont & Singer, 2006; Decety & Lamm, 2006; Preckel et al., 2018; Cuff et al., 2016) and reveal three important insights: (*i*) "empathy is both closeness and distance; (*ii*) empathy is both emotion and cognition; and (*iii*) empathy is both body and mind." (Eklund & Meranius, 2021). "For it to be empathy, a delicate balance between closeness and distance is required" (Eklund & Meranius, 2021). Looking at an issue or phenomenon from another's perspective requires the involvement of cognitive and affective processes, the involvement of 'mind and heart' to understand and feel how the other person does. The process of empathy also requires awareness that an empathizer will never identify fully with the person he/she empathizes with. Self-other differentiation plays an important role in the process of empathizing, as it helps avoid an overly subjective perspective as well as the appropriation of the Other with the aim of exploiting the knowledge about another person for the sake of one's own interests. The empathizer's identity stays different and intact and so does the identity of the Other: both are needed for empathy to take place.

Another interesting aspect that was discovered in the meta-analysis of empathy is the location of empathy occurrence: neuroscience studies present it as an inner phenomenon that occurs in a single individual, whereas in philosophy and psychotherapy studies, there is a tendency to perceive empathy as an interpersonal phenomenon that needs at least two persons to be involved in communication. In all studies, there is a difference emphasized between empathy and sympathy or empathy and compassion.

Empathy in Engineering

The contemporary discourse about future engineering accentuates empathy as an important factor in professional success. Empathy is presented as a salient professional skill for 21st century engineers, very often classified as a so-called 'soft skill' together with problem solving, communication, teamwork, sustainability, emotional intelligence and social responsibility (Penzenstadler et al., 2009), or as an important foundational component of these 'soft skills,' since it appears impossible to develop 'soft skills' without the important component of empathy. High emotional intelligence, so important for social skills, cannot be developed without emotion regulation (Goleman, 1998). The ability to understand others and recognize their feelings is essential in teamwork and in effective communication. To design solutions customized for the users, it is necessary to explore the users' needs while truly understanding their motives, feelings, pains and aspirations (ideo.org, 2022). In spite of the apparent relationship of empathy and engineering, empathy is still presented in opposition to 'core', professional hard engineering skills, as if the relation between the two was hardly noticeable.

As a consequence of more and more globalized economies and the increasing complexity of problems that humanity is experiencing today, the role and impact of engineering and technology in the way we construct our reality and human existence is continuously increasing. Nonetheless, globalization leads to growing competition in more challenging job markets, competition that young engineers inevitably will face. That is why "soft skills [including empathy] for engineers have become indispensable in this highly competitive global environment" (Balaji & Somashekar, 2009; Rasoal et al., 2012). Empathetic skills can also help to confront the challenges of multicultural communication, languages, international virtual teamwork (Sheppard et al., 2004), and create a window of opportunity for constructing solutions with more ethical considerations. It is especially important today, because researchers state that engineers very often make their decisions without recognizing their responsibility for a potentially negative impact on others (Lathem et al., 2011). Such an attitude is contrary to social and corporate expectations as made visible in an OCED survey resulting in 80.3% of respondent employers who acknowledge the importance of ethics" (OECD, 2015; Casner-Lotto & Barrington, 2006). Directing engineers' way of thinking and decision-making towards the social, environmental and economic needs of communities will increase their assertiveness and positively impact their professionalism and the overall well-being of all the stakeholders (Jones et al., 2017).

The importance of cross-cultural teamwork in the global business world draws attention to empathy as the main predictor for team performance based on an ability to create psychological safety and lower interpersonal risk (Woodley et al., 2010). The other predictor is conversational turn-taking (Edmondson, 1999), being closely connected with empathic components described in this article: affective sharing, mental flexibility or emotion regulation, as well as self -other differentiation. Without these components in place, it would be hard to imagine effective teamwork based on mutual respect for the diversity of perspectives, variety of disciplines, cultures and interests, that are inevitably linked with the nature of contemporary "wicked problems" (Kelley & Kelley, 2013) that engineers are commissioned to tackle.

The existing need to address and use empathy in engineer education and practice is acknowledged beyond question. Nevertheless, it may stand in contrast to its practical applications. The studies by Cech (Cech, 2014) concluded with an image of engineering education as promoting 'the culture of disengagement'. The aim of these longitudinal studies was to investigate the success of the US education system in the effective promotion of responsibility for the welfare of the public. Unfortunately, it proved that young engineering graduates lowered their level of empathy during their engineering education. The results of the studies call for a serious and prompt consideration of the matter, and more profound solutions to be implemented, to combat the risk of educating new generations of engineers who are deprived of prosocial reasoning, orientation and responsibility, while having an overwhelming impact on constructing solutions that can dramatically change our being.

Another important conceptual proposal which addresses the role of engineers with a serious consideration of the challenges of today's globalized world, is the model created by Walther and colleagues (Walther et al., 2019). It conceptualizes empathy as a skill, a practice orientation, and a professional way of being, and seems to utilize and synthesize the aspects acknowledged today as important for empathy (Figure 1).

Figure 1. A model of empathy for engineering
(Walther et al., 2019)

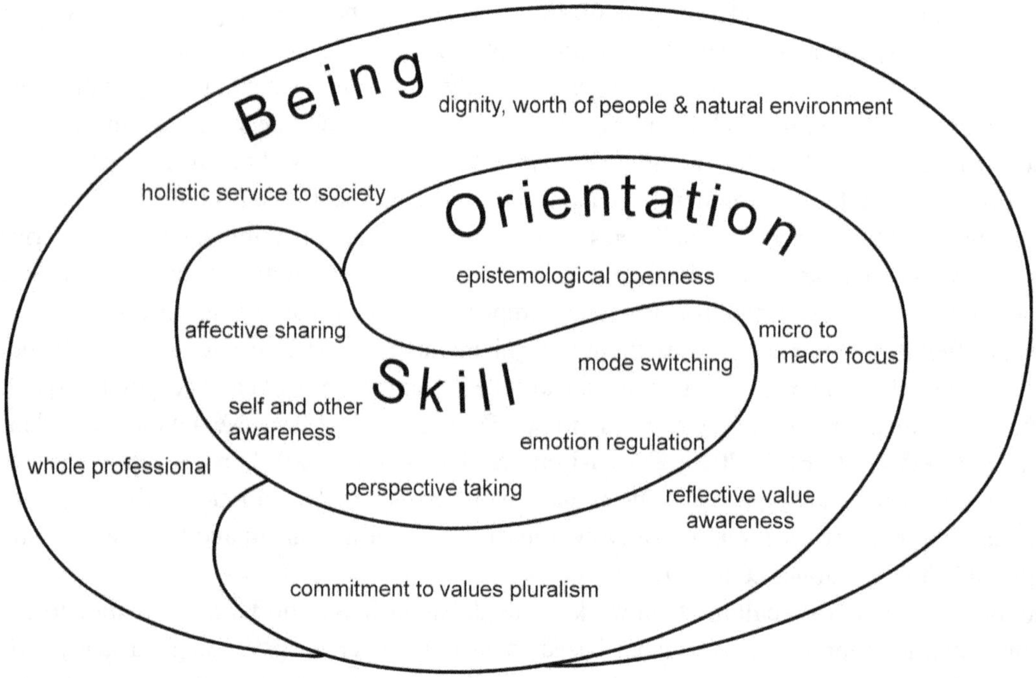

The *skill component* gathers five socio-cognitive processes (affective sharing, self and other awareness, perspective taking, mode switching) that are indispensable for effective communication, building relationships and decision-making. The *practice orientation* guides engineers in the application of empathy in engineering practice and these are openness, micro to macro focus, reflective value awareness and commitment to value pluralism. The *being component* advocates that all the values transformed into a professional way of being can help to educate an engineer - a professional who will be able to realize the mission of serving society while respecting social values, dignity, the worth of people, and the natural environment.

Empathy in Design

Empathy for the end users, although it may be manifested in a variety of design techniques and not always named in a direct way, is naturally an important factor in design. Describing the culture of design, Cross (Cross, 1982) presents its values as follows: practicality, empathy, and concern for appropriateness.

Using empathy in gaining the user's feedback, the design community has made an attempt to restore the connection between designer, user and maker (Jones, 1970). This link was distorted by the rapid development of industry and technology going back to the Industrial Revolution (Jones, 1970). In order to understand human behavior, cognitive models were used predominantly without an emotional "nuisance variable" (Dandavate et al., 1996). Patricia Moore promoted an empathic model of design (Moore & Conn, 1985) when she dressed herself as an elderly woman to experience and better understand the life of the elderly and increase the usability of the product she wanted to design, while embracing the needs and feelings of its primary users. Moore's attitude opened the way to empathy as an established way of data collection and data analysis, or even more as a way of thinking underpinning the design process (Segal & Suri, 1997). The frameworks of universal design, human-centered design, user-centered design and experience-centered design treat empathy as an important core component of the design process (Kouprie & Visser, 2009) and "the powerful tool that designers offer" (Cooper et al., 2014).

A natural consequence of a more inclusive approach to design that accentuates human emotions and feelings was the adoption of the empathy concept as developed in psychology. Kouprie and Visser (2009) ignited a discussion in the design community by accentuating several empathetic relations relevant to design: (*i*) the relation between cognitive and affective components of empathy (Kouprie & Visser, 2009); (*ii*) the relation between closeness and distance, used by the designer towards the user in an endeavor to understand the user's needs and feelings; and (*iii*) the relation between stages of empathizing that are to be coherently integrated into the process of design, important moments at which the designers attempt to enter and share the user's world, and the moments of stepping out of it. There are several empathy frameworks established today that have both common and different elements, but all use empathy as a coherent insight into the user's world through the whole process of design: IDEO, d.school or UX design. They promote certain techniques to be used at different stages of the process (Heylighen & Dong, 2019). Ethnographic quality techniques are often advocated at the research /empathy/ discover stage that helps establish close contact between the designer and the user and enables the designer to achieve an empathic insight into the user's experience. Among others, these techniques are participants' observations, immersion, interviews, and in business, questionnaires and surveys. To communicate the results of research to different stakeholders, storytelling and narrative inquiry have been promoted, as they allow for more insightful and personalized presentations of thoughts, feelings and experiences of others, and as such, are more prone to create empathic solutions. The personas method used in user-centered design (Cooper 1999) presents end users in a more personalized narrative-based way. Ideation tools include techniques that enable a designer to get closer to the experience of the user, e.g., wearable simulations let designers experience what the real user experiences, as they may restrict the movements of the body, and evoke bodily sensations similar to those who feel a certain type of discomfort in real life (Kullman, 2016). Fast prototyping and recurrent testing of the design on the end user to get feedback and improve the design, not only promote contact with these stakeholders, but also, thanks to repeated contact and efforts to understand the experience on the level of mental and physical sensations and emotions, work for increasing empathy levels between designers and users, as well as for establishing such an approach as desired in the process of design. In consequence, management and business literature practically adopted the empathy concept as being at the heart of innovation, because, in times of high competition, achieving customer satisfaction on the level of desire could contribute to the competitive advantage of a company, the company's success in the market and its sustainability. An appropriate perspective taking to understand the customer more deeply, combined with new technologies such as artificial intelligence or augmented reality, can make it possible to attract new customers on an unimaginable scale.

THE MODEL OF HUMAN-ORIENTED ENGINEERING

Following several years of discourse with different stakeholders in engineering education, and acknowledging the concept of holistic engineering (Grasso & Burkins, 2009) that advocates that engineering systems must evolve together with unpredictable and dynamic environments and the unpredictability of the systems themselves, the EPS community was looking for a scheme that would enable young engineers to be better prepared for the complexity of influences, positive opportunities and potential risks, and "better serve the humanity" by attempting "to understand a human condition in all its complexity" (Grasso & Martinelli, 2007). The endeavor resulted in constructing a model that explores the idea of human-focused engineering, whose pivotal basis is constituted by empathy exercised in relation to different stakeholders involved in the educational process and the educational paradigm used in the European Project Semester. The validity concept was additionally strengthened by the latest presence of empathy oriented discourse in different fields including engineering and the importance of soft skills for engineering formation (Walther et al. 2019).

The EPS seemed to be a good choice because it creates a favorable environment for using empathy in various contexts and with various stakeholders. The EPS educational paradigm is constructed around a project conducted in a multicultural team mentored by an engineer/s and a team dynamics mentor, both sides working in close collaboration. The team is supported with a limited number of courses such as: Project Management, which helps students to become more proficient in organizing their project process; Innovation for Project, whose aim is to orient the team towards creative and innovative solutions; and Team Building and Communication for Projects, that explores team dynamics and management, and trains the team in creative problem techniques specific to approaches of DT or PBL, as well as building a platform for inter-team sharing, and providing feedback.

The human-oriented empathy-based model of engineering project-organized education adheres to several premises used and strengthened in practice by academics involved in the EPS program. Empathy is the focal point and the most important value explored and applied in all other dimensions of the program. High engineering competence in one's engineering field is crucial to ensure the high quality of the technological solutions but also allows for empathetic readiness to consider the premises of other disciplines and fields of science and human activity, which is salient in finding solutions to complex problems. The idea of a holistic approach to engineering (Grasso & Martinelli, 2010) enables engineers to operate in a broader context beyond engineering disciplines and integrate empathically in socioeconomic, cultural, and ethical contexts of the 21 century, and calls for social empathetic competence in engineering education for the future. A global mindset (Levy et al., 2015) that comprises empathetic communication across cultures, different meanings, actions, and enables global collaboration, is necessary in a globalized world of work. Interdisciplinary teams and projects ensure successful addressing of complex problems in collaborative and empathetic dialogue across different engineering and non-engineering disciplines. An emphasis on teamwork provides the basis for the encounter with the Other in communication, in using peer learning, exchange and leadership, and as such creates a platform for finding solutions for more complicated issues. Using a growth mindset (Yeager & Dweck, 2020) as an important guideline in learning allows for empathetic openness to communicate across different cultures, disciplines, and personalities. It also ensures readiness to learn new things and accept innovative solutions. Critical and autonomous thinking (Loch, 2020) and reflection secure independent reasoning, readiness to experiment with new solutions and help to develop maturity in thinking and to grow as a professional. Adaptive flexibility is important to allow smooth transference between empathetic and analytical activities, and to help professionals adapt and react to change.

The model treats empathy as a core value of the EPS educational paradigm. It is assumed by the EPS community that empathy creates the foundation of the values presented in the middle circle of the model (see Figure 2). It is to be exercised by or towards all the stakeholders involved in EPS project execution, presented in the outer circle: the team, the mentors, experts, problem owners and solution recipients. The team itself constitutes the environment for experiencing diversity and challenges of the global mindset since it consists of members coming from different cultural contexts and representing different disciplines of expertise. It requires from team members to feel and understand different modes of communication, very often not consistent with their own contexts. Pursuing a common goal in project work, the team members must learn how to develop adaptive skills, without which it would not be possible to communicate effectively. The better they understand and feel other people's thoughts and intentions, the better they will be able to adapt to a changed way of communication. Thus, teamwork in a multicultural and multidisciplinary project group itself creates numerous challenges which, faced by the team members, allow them to start developing competencies favored by the global job market, competencies which cannot be developed without a deeper understanding of the Other in the team, his/her feelings and ways of thinking, or without sharing affective and cognitive states. The challenge of empathetic encounters is increased even more, due to the necessity to communicate with mentors who come from different engineering and non-engineering backgrounds, and represent different generations; but also, experts coming from either scientific or business contexts, and individuals with a problem for whom the team is to create a new solution, and whose context and lifestyles may be extremely diverse, often unknown to students and requiring deep exploration.

Figure 2. EPS Human-Oriented Model of Engineering

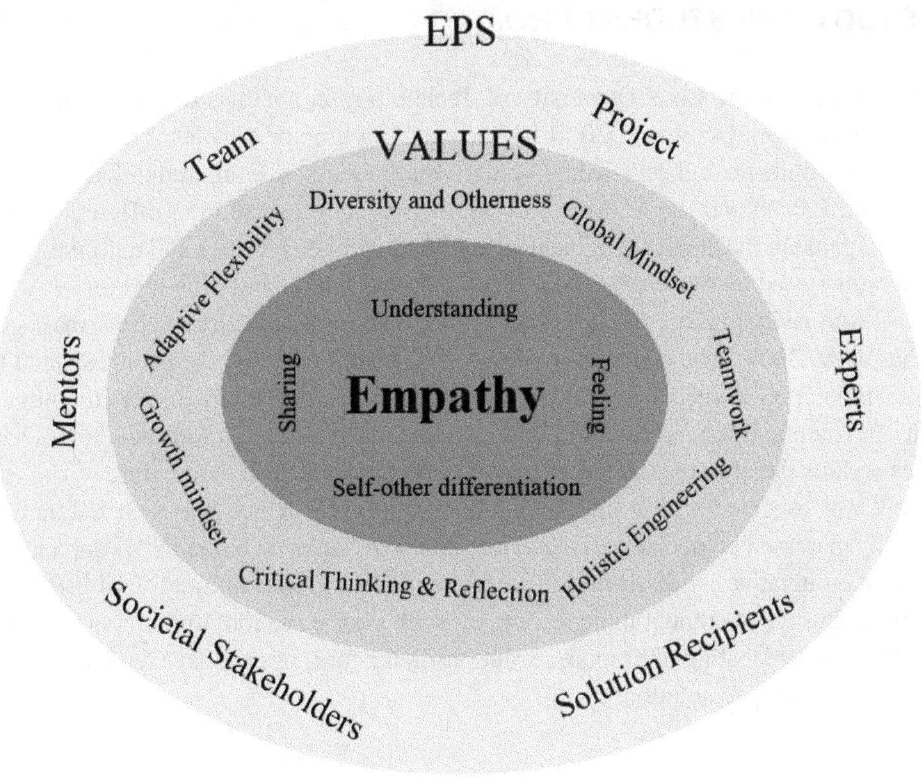

The development of empathy is thus provoked by: (*i*) the nature of EPS projects; (*ii*) the strong emphasis on the ethical and social contexts of the problems; and (*iii*) the PBL and DT approaches that use the component of empathy as the most effective research method. It lets the team get closer to the possessor of the problem, requires a deeper understanding of the user's mentality, emotions, desires as well as of the surrounding social, economic or political contexts. Design thinkers and problem solvers "develop empathy by observing and interviewing users, watching and listening carefully and [...] derive insights based on the user's experience" (Peppler, 2017). Numerous empathy-oriented and human-oriented techniques are introduced during a course of Team Building and Communication for Projects and reinforced later on in meetings of project teams with mentors to be finally applied in the projects. Obviously, the work on solutions that would satisfy problem owners or go beyond the level of satisfaction demands a close empathetic relation with the users but also requires maintaining a necessary distance between the designers and the recipients of the solution at certain stages of the design process, in which the designers must switch to more cognitive modes, and use critical thinking and reflection to use their expertise to create a new solution. Nevertheless, training in DT and PBL approaches increase awareness of empathy importance through provoking closer encounters with all other project stakeholders and more human-oriented attitudes of student-researchers. The teams are given autonomy in empathetic research, problem exploration and definition, building fast prototypes, collecting feedback and designing the final concept or product that is expected to innovatively meet the user's expectations and desires, appealing to the users' functional and emotional needs. Flexibility and autonomy in scheduling can be treated as a specific manifestation of empathy. It reflects mentors' understanding of the project complexity and the student's mental and emotional experience of the process. The scale of the challenge the students have to face and a will to address it in an empathetic way became a basis of the mentors' decision to increase the level of autonomy the project teams should have in an iterative process of the project execution.

A CASE STUDY: THE STUDENT PROJECT

The project "Students of the Lodz University of Technology in a Times of Pandemic" took place in the autumn semester from October 2020 to January 2021, a time of crisis that had a profound impact on people, the environment, and established systems and processes. Four students from three different countries - Spain, Poland and the Netherlands - formed a team representing different fields of study: chemistry, management, mechanical engineering and industrial electronics and automatic engineering. The team was supervised by two engineering mentors and one team building mentor.

Due to the coronavirus crisis, the consumption of disposable plastics such as masks and gloves increased sharply by 300%. The government focused on health care and the economy, and that is why environmental protection seemed to lose its primary status. This resulted in greater amounts of pollution in the oceans and nature. Used masks could be found on the streets and in forests. Plastics have become an even more serious threat to the natural ecosystem and human health than before.

The project was executed with a Design Thinking approach, which was selected as it is strongly connected with "increased prosocial [...] behavior" and a "reliance on empathy" (Peppler, 2017) that is used throughout an iterative process of project execution, e.g., in the exploration of user experience to guide ideation, the usage of ethnographic techniques such as observation, immersion and interviewing, as well as in the repeated testing of proposed solutions in the form of extensive feedback collected from the recipients of a developed solution.

It was noticeable that the team took a very empathic stance in relation to a problem that was having an adverse impact on people and the environment from the very beginning. They empathically decided to respect other team members' opinions and feelings, and they expressed this attitude in the project report, saying:

The most important thing is that we want to help people in the times of pandemic, […] use our knowledge from different areas of studies […] to gain broader understanding of the processes we are facing in our lives and in the project. Together, we want to innovate and find new solutions […] make a difference. We hope that our product could be a foundation for further developments, and in that way our solution […] could be accessible for large groups of people. (Figas et al., 2021).

To get closer to the users of masks, learn about their attitudes, thoughts and feelings, their life, their "pains and gains" in relation to wearing masks, the team collected their observations and synthesized them in the form of a mind-map. The mind-map helped identify more specific issues and select the target user and conduct interviews and a survey with a selected group of users-young people. Young people were selected as the focus for the team interest because it was the audience the students had a good chance to understand more profoundly and access more easily. From the very beginning, the team demonstrated a very empathetic approach: the students decided that one of the most important parts of the project was learning about the feelings and needs of masks users, which was achieved by using such techniques of exploration as interviews, observations, questionnaires, empathy maps, personas. The target audience data was collected and synthesized in a more personalized way, the form of a persona (Figure 3):

Figure 3. Persona

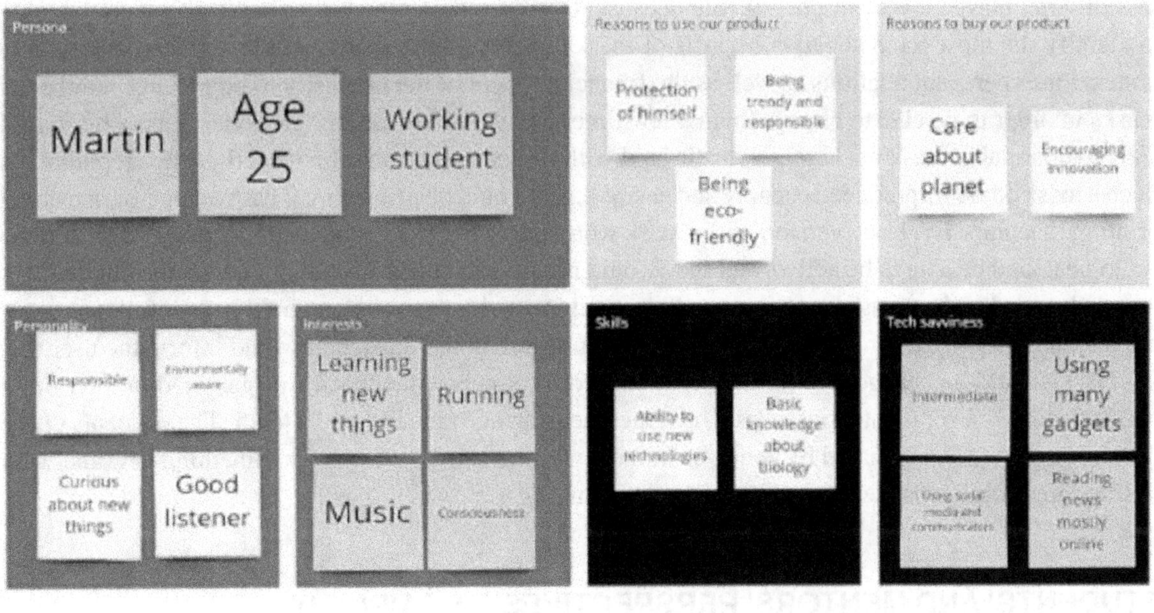

The user is presented as Martin, a 25-year-old working student who lives in the time of the COVID-19 pandemic. He is ready to try new things, has a basic knowledge about technology, and most often uses social media to communicate with others. As he is a responsible person, he wears masks as a protection against Covid-19. He is an ecology-oriented person, tries to be eco-friendly and likes wearing trendy outfits. He likes sports, especially running, so a mask could be a hassle while running or playing other sports. The user was presented in the creative form of a persona with an element of storytelling, which is one of the techniques advocated in Design Thinking as a more empathic tool that enables the designers to understand thoughts, feelings, positive and negative emotions and lifestyle of the target audience for whom the new products are designed. The technique also has the power to evoke empathy when the problem is presented and has the potential to have a more persuasive impact on the recipients of the main message.

The team conducted two surveys among students aged 19 to 27, the identified target audience: the first survey helped them understand what young people know and how they feel about masks. It also allowed the team to diagnose the impact of masks worn throughout the day on people's lives and the environment. The first survey helped the team to construct a more precise and focused questionnaire that led to important conclusions, later used as the basis of a new design concept. The opinions from a hundred responses helped the team understand more deeply people's experiences of mask wearing, their discomforts and emotions, and let them verify the conclusions from their own brainstorming sessions. Empathizing with the user led to very important discoveries: the masks were reported to be uncomfortable, could produce shortage of breath, block facial expression and impair interpersonal communication, and contributed to a high level of dissatisfaction and even frustration. They could be a real problem in communication for people with hearing deficiencies, as these people rely on the interpretation of facial expressions in communicating with others. The masks became the most pervasive form of protection and a number one environmental pollutant.

The discoveries confirmed the assumptions and observations the team had made in their brainstorming sessions, yet played a very important role because, thanks to the empathic research, the team was able to identify the most acute negative impacts of masks on the environment, people, human communication and interpersonal relations as well as the financial aspect of the necessity to buy a huge number of masks in order to be able to change them a few times a day. As a result, the team designed a bio-based and biodegradable mask made of polylactic acid with filters made from polytrimethylene terephthalate. Such a mask could be produced cheaply and easily, e.g., it could be printed on it with a three-dimensional printing machine. The basic version of the mask would meet the requirements of people who want to be ecological, and be able to be active, and they would not have to spend a lot of money on the purchase. A more advanced and expensive version would be equipped with sensors that could be helpful in measuring the humidity of the air and the critical level of moisture in the mask that would inform the user that the mask should be changed. After the production of prototypes, feedback was gathered in interviews conducted with several potential users, young people in an age range from 19 to 25. The concepts of the two products were appreciated by young customers who are active, want to try something new and want to stay safe during a pandemic and buy affordable masks.

STUDENTS' AND MENTORS' PERSPECTIVES ON EMPATHY

Students' perceptions were collected based on open questions added to a questionnaire in the survey conducted in the group of 112 students, out of whom 34 were involved in EPS projects. The students were asked if they have experienced empathy in project work and if they use empathy in project work. The vast majority (98) of the students surveyed admitted using and experiencing empathy in project work;

14 students gave a negative answer. In most cases, the students pointed out the importance of empathy in establishing positive relations in a project team or accentuating the necessity to "understand each other's feelings while working on the project together". They state that "understanding bad emotions" can help in maintaining the team. "Empathizing is crucial to correctly understand different points of view and see the perspective of the others and create the atmosphere of openness" which helps at "the stage of creating ideas and presenting results" and in "designing good solutions and achieving goals in a more effective way". "Empathy used by a supervisor of the project helps in good communication and, thanks to it, it is easier to work on a project.". Empathizing while "communicating with people from other countries" is perceived as especially important by the students, sometimes "empathy influences the results of a discussion and the decisions made, which is helpful" because it may "lead to surprising and fascinating developments". Empathy also helps to "include every member in project work" because it is important "that everyone feels equally engaged and happy about work" in a project. The students see empathy as helpful in conflictual situations or when they experience real differences of opinions either in a project or in social relations with other team members. A small group of students emphasizes the salience of empathy in research as a "creative method in Design Thinking" or Problem-based Learning that is later used in an ideation phase since it "helps to understand the user" and "the core of the problem". It is seen as an effort: it is "hard to always try to put yourself in someone else's shoes and see their point of view". Single individuals were doubtful if empathy should be used or were against using empathy in the project. They accentuated that discussing feelings may be seen as unprofessional, so they mainly focus on "things related to the project". They do not use empathy because they "want to feel good about themselves" or they "do not feel it". Empathy depends on the nature of the project: "some projects are very objective and emotionless, which may not be a good thing, anyway". Using empathy may be risky because "it's frustrating when you use empathy, but you feel negative vibe from another person".

The research on the perceptions of students on empathy includes an interview with one member of the Project "Students of the Lodz University of Technology in a Times of Pandemic". Michał sees empathy as understanding and acceptance of another person's way of thinking or even "a desire to identify with another person". In his opinion, the EPS project was based on empathy because "collaboration in an international context requires division of work and mutual support in achieving a common goal". If one of the team members has a problem, "the other team member looks at the situation from this person's perspective and tries to support this person." He felt empathy "among the team members as well as from the side of mentors", the team was able to develop an interesting solution which proves that there was an empathetic collaboration among team members". "If there is no empathy in a team, people prioritize their own goals and that is why it's difficult to develop a good solution." Another important aspect that Michał emphasizes is using empathy to know the user and the recipient of the solution, looking at things from the user's perspective. He explains that the team was trying to understand the feelings and thoughts of other young people who had to wear masks and wanted to help them in satisfying their needs. "It is important for the customers to believe that the producer tries to take care of their needs and not only treats them as a source of profit, that scientists and producers listen to people and try to help them."

The interview confirmed the importance of team building and communication activities in creating more effective collaboration especially in the times of pandemic when the students had to attend synchronic online classes. The fact that they met face to face at the beginning of the project while going to the museum to learn about the history of Lodz and then spent some time together in an informal setting and learn more about other team members helped a lot in later online communication, which was "difficult" and slowed down communication and progress. The interview stressed the importance of team building and communication classes aimed at developing social skills and learning about cultures of other team members, other personalities, different modes of work, different mentalities of team members as well as the importance of including short informal team building activities in serious project meetings since they "break the ice and help to "create a better environment for work."

It may be concluded that students intuitively see different aspects of empathy as described by researchers. They see it as understanding the feelings and thoughts of another; they perceive the importance of empathy in the efforts to build the team, communicate with mentors and other stakeholders, they acknowledge its importance of empathy in making efforts to explore the needs and expectations of the user in the solution development.

Nine out of thirteen mentors shared their opinions on empathy in open questions added to the questionnaire. Additionally, there were six individual interviews conducted in the group of EPS mentors who agreed to share their perspectives on empathy. The mentors in the majority understand empathy as a skill that can be developed in the process of education, the skill that is used to understand the perspective of another person, a student as well as to understand educational needs and problems that person may have. The mentors strongly emphasize the salience of empathy in understanding the problem owners and in creating solutions, especially in DT projects. Empathy is used "mainly at the *discovery* stage of the project execution when the team is exploring the topic, problem issues and the problem owner, whose needs are crucial and could be explored in an insightful way with an empathetic approach in research" e.g., simulations. Empathy is applied to understand the students in a project team and "consolidate the team". It is especially "important for successful collaboration with a high level of diversity in a team" or with "teams having minority members who may feel isolated", e.g., females at technical universities.

All the mentors believe that project courses "create a better environment for empathetic sharing", especially in small project teams where communication is more personalized, and the project profile has better chances to match intellectual resources of individual team members. Only one mentor admitted not experiencing empathy in project work and not using it since "professional issues should be separated from emotional issues", which guarantees "higher efficiency of the project execution". Additionally, an opinion was expressed that "using empathy in engineering project may be less important", however, this opinion was not fully justified. Answering the question about the type of empathy that contributes to generating knowledge, the EPS mentors state that both cognitive and affective empathy is important, however its application may depend on the type of the project, e.g., numerous IT projects participant concentrate mainly on technical issues and would not have any contact with the user, it is more important in group work than in individual and would be utilized more in small group tutoring and less in traditional classes in large groups. According to mentors, "online synchronous classes diminished empathetic contact and communication, it may seem that it was possible to use new tools to communicate but communication itself was not so rich and empathetic, which paradoxically let us discover the real value of empathy for motivation and engagement". The EPS mentors suggest that "teachers should be trained more extensively in soft skills and empathetic communication and feedback since knowledge sharing and creativity can only happen in the environment that fosters openness in communication without generating fear of reputation loss or fear being ridiculed". "Cognitive and affective empathy should be developed in balance to reach an equilibrium that is optimal for positive intellectual development of young generations that would be able to act and develop new solutions for the good of society." Addressing the aspect of self-other differentiation which allows for treating the others as autonomous agents, the EPS mentors admitted its salience: students should enjoy more autonomy in learning and should be engaged more in creating assessment tools together with mentors and other students.

THE LEVELS AND TYPES OF EMPATHY AMONG STUDENTS AND MENTORS

Inspired by students' and mentors' positive feedback on the importance of empathy in project work, the authors asked a few research questions: What is the level of empathy is among Polish and International students participating in the project courses at Lodz University of Technology, including EPS, for the last two years?; Is there any significant difference in the average empathy levels between Polish and International students?; Is there a significant difference in the levels of empathy between the students involved in EPS projects and other types of projects?; Is there a significant difference in the levels of empathy between the students involved in EPS and the students who had not had project experience before?; Is there a correlation between the levels of empathy among mentors and students?

To measure the levels of empathy, the authors used the scale called: "The QCAE: A Questionnaire of Cognitive and Affective Empathy" constructed by Renate L. E. P. Reniers, Rhiannon Corcoran, Richard Drake, Nick M. Shryane, and Birgit A. Vollm (Reniers et al., 2011) since this scale refers to the categories of affective and cognitive empathy present in the previous empathy studies in engineering (Walther et al., 2019). The QCAE questionnaire consists of 31 items for the respondents to address in a four-degree scale. The instrument includes two subscales: *Affective empathy* and *Cognitive empathy*. Affective empathy consists of Emotion contagion, Proximal responsivity, Peripheral responsivity whereas Cognitive empathy includes Perspective taking and an Online simulation. Perspective taking refers to the ability to put oneself in another person's position; online simulation reflects a will and determination to understand the feelings of another person. In the category of affective empathy, emotional contagion reflects an automatic impact of other people's feelings on us, our mood and feelings, peripheral responsivity refers to affective involvement when observing other people's feelings in a detached context and proximal responsivity refers to the affective involvement when observing another's feelings and being in a close social context. The reliability of the tool, expressed using the Cronbach's alpha coefficient for each of the scales of the questionnaire, was $\alpha < 0.65$, which was considered sufficient confirmation of the reliability of the tool. A high level of internal validity also characterizes the tool. As evidenced by the high value of the correlation coefficient of the two dimensions of empathy. The external validity was verified by means of correlation with Basic Empathy Scale (BES). The cognitive and affective empathy scores on the QCAE and BES showed strong positive correlations: $r = 0.62$, $p < 0.001$ for cognitive empathy; $r = 0.76$, $p < 0.001$ for affective empathy.

The survey was conducted online with the use of MS Forms. There were in all 112 respondents, 51 International students and 61 Polish students; 34 students declared their participation in EPS. The group of 13 EPS mentors who participated in the survey supervised EPS teams in the last three semesters.

The analysis of data in Table 1 allows for evaluating the average level of empathy of the students who took part in the survey as well as the levels in all the subcategories in QCAE scale:

Table 1. Descriptive statistics

Descriptive statistics					
	N	Minimum	Maximum	Average	Standard deviation
Cognitive empathy, Perspective taking	112	21	40	31.78	3.972
Cognitive empathy, Online simulation	112	20	35	27.35	3.979
Cognitive empathy (sum)	112	43	73	59.12	6.993
Affective empathy, Emotion contagion	112	6	16	12.22	2.335
Affective empathy, Proximal responsivity	112	5	16	12.36	2.100
Affective empathy, Peripheral responsivity	112	6	16	10.69	2.177
Affective empathy, sum	112	26	47	35.27	4.619
Empathy sum	112	73	118	94.39	9.715
N Important (observation disabled)	112				

(Reniers et al., 2011)

Table 1 presents the average from the results from the survey of 112 students, the results acquired in a general empathy scale and all the subscales: total empathy, total cognitive empathy broken into perspective taking and online simulation as well as total affective empathy broken into emotion contagion, proximal responsivity and peripheral responsivity. The students reached the average level of 59.12 out of 73 on the general scale of cognitive empathy, where perspective taking was on the level 31.78, and online simulation was on the level of 27.35. In reference to affective empathy, the average reached the level of 35.27 out of a maximum of 47. Emotional contagion of affective empathy was on the average level of 12.22 and proximal responsivity on the level of 12.36. It may be concluded that all the surveyed students demonstrated a high level of general empathy.

Table 2 presents the level of empathy in two groups of International and Polish students.

Table 2. Group statistics

Group statistics					
	Student	N	Average	Standard deviation	Standard mean error
Cognitive empathy, Perspective taking	Intern.	51	31.45	3.987	0.558
	Polish.	61	32.05	3.973	0.509
Cognitive empathy, Online simulation	Intern.	51	27.73	4.080	0.571
	Polish.	61	27.03	3.899	0.499
Cognitive empathy, sum	Intern.	51	59.18	7.213	1.010
	Polish.	61	59.08	6.864	0.879
Affective empathy, Emotion contagion	Intern.	51	12.43	2.211	0.310
	Polish.	61	12.05	2.439	0.312
Affective empathy, Proximal responsivity	Intern.	51	12.27	2.031	0.284
	Polish.	61	12.43	2.171	0.278
Affective empathy, Peripheral responsivity	Intern.	51	10.63	2.332	0.327
	Polish.	61	10.74	2.057	0.263
Affective empathy, sum	Intern.	51	35.33	4.668	0.654
	Polish.	61	35.21	4.616	0.591
Empathy sum	Intern.	51	94.51	10.104	1.415
	Polish.	61	94.30	9.461	1.211

Polish students reached the level of 59.08 in the scale of cognitive empathy broken into Perspective taking on the level of 32.05 and Online simulation reaching 27.03. In contrast, International students reached the levels of 59.18 with Perspective taking being on the level of 32.45 and Online simulation 27.73. In the scale of Affective empathy, the international students were on the level of 35.33 broken into 12.43 in Emotional contagion, Proximal responsivity 12.27 and peripheral responsivity 10.63. General empathy in the group of Polish students stayed on the level of 94.51 and international students reached 94.30. That is why it may be stated that both groups of students are characterized with high general empathy levels.

The survey aimed to answer the question of whether there is a significant difference between Polish and International students in the average levels of empathy. The results are presented in Table 3.

Table 3. Test for independent samples

Test for independent samples										
		Levene's Test for homogeneity of variance		T test of equality of means						
		F	significance	t	df	significance (two-sided)	difference of means	Standard error of the difference	95% confidence interval for the difference of means: lower limit	upper limit
Cognitive empathy, Perspective taking	equality of variance was assumed	0.194	0.660	-0.792	110	0.430	-0.598	0.755	-2.094	0.898
	equality of variance wasn't assumed			-0.792	106.381	0.430	-0.598	0.755	-2.095	0.899
Cognitive empathy, Online simulation	equality of variance was assumed	0.094	0.760	0.917	110	0.361	0.693	0.756	-0.805	2.190
	equality of variance wasn't assumed			0.913	104.657	0.363	0.693	0.759	-0.812	2.197
Cognitive empathy, sum	equality of variance was assumed	0.038	0.847	0.071	110	0.944	0.095	1.333	-2.547	2.736
	equality of variance wasn't assumed			0.071	104.467	0.944	0.095	1.339	-2.560	2.749
Affective empathy, Emotion contagion	equality of variance was assumed	2.169	0.144	0.861	110	0.391	0.382	0.444	-0.497	1.261
	equality of variance wasn't assumed			0.869	109.249	0.387	0.382	0.440	-0.489	1.254
Affective empathy, Proximal responsivity	equality of variance was assumed	0.367	0.546	-0.379	110	0.705	-0.152	0.400	-0.945	0.641
	equality of variance wasn't assumed			-0.382	108.594	0.704	-0.152	0.398	-0.940	0.636

Continued on following page

Table 3. Continued

Test for independent samples										
		Levene's Test for homogeneity of variance		T test of equality of means						
		F	significance	t	df	significance (two-sided)	difference of means	Standard error of the difference	95% confidence interval for the difference of means: lower limit	95% confidence interval for the difference of means: upper limit
Affective empathy, Peripheral responsivity	equality of variance was assumed	0.412	0.522	-0.266	110	0.791	-0.110	0.415	-0.932	0.712
	equality of variance wasn't assumed			-0.263	100.690	0.793	-0.110	0.420	-0.942	0.722
Affective empathy, sum	equality of variance was assumed	0.323	0.571	0.137	110	0.892	0.120	0.880	-1.624	1.865
	equality of variance wasn't assumed			0.136	106.085	0.892	0.120	0.881	-1.627	1.867
Empathy sum	equality of variance was assumed	0.097	0.756	0.116	110	0.908	0.215	1.852	-3.455	3.884
	equality of variance wasn't assumed			0.115	103.724	0.908	0.215	1.863	-3.479	3.908

The analysis with the application of the Student's t-test showed that the difference between the levels of empathy between Polish and International students is not statistically significant.

The authors asked the question if there is a significant difference between the students involved in EPS projects and other types of projects. The results are presented in Table 4.

Table 4. The difference between EPS students and students in other projects

	Test value[a]							
	Cognitive empathy, Perspective taking	**Cognitive empathy, Online simulation**	**Cognitive empathy, sum**	**Affective empathy, Emotion contagion**	**Affective empathy, Proximal responsivity**	**Affective empathy, Peripheral responsivity**	**Affective empathy, sum**	**Empathy sum**
U Manna-Whitney	921.500	973.500	920.500	952.000	952.500	892.500	921.500	976.000
W Wilcoxon	1482.500	1534.500	1481.500	2782.000	2782.500	2722.500	2751.500	2806.000
Z	-0.552	-0.133	-0.559	-0.308	-0.305	-0.792	-0.551	-0.113
two sided asymptotic significance	0.581	0.894	0.576	0.758	0.761	0.428	0.581	0.910
a. Grouping variable: Type of the project you have participated in:								

The analysis based on the test of U Mann-Whitney indicates that the students engaged in EPS projects are not significantly different in the empathy levels from the students involved in other types of projects.

Another interesting question that was posed referred to the difference between the level of empathy between the students in the EPS projects and the students who were not in the project (Table 5).

Table 5. The difference between EPS students and non- project students

	Test value[a]							
	Cognitive empathy, Perspective taking	**Cognitive empathy, Online simulation**	**Cognitive empathy, sum**	**Affective empathy, Emotion contagion**	**Affective empathy, Proximal responsivity**	**Affective empathy, Peripheral responsivity**	**Affective empathy, sum**	**Empathy sum**
U Manna-Whitney	309.000	310.500	311.500	246.000	300.500	244.000	248.500	282.000
W Wilcoxon	499.000	500.500	872.500	436.000	490.500	434.000	438.500	472.000
Z	-0.086	-0.057	-0.038	-1.297	-0.249	-1.343	-1.239	-0.600
two sided asymptotic significance	0.931	0.954	0.970	0.195	0.803	0.179	0.215	0.549
a. Grouping variable: Type of the project you have participated in:								

The analysis based on the U Mann-Whitney test indicates that there is no significant difference between the group of EPS students and those that do not participate in project work at all.

The authors posed another question exploring the difference between the levels of empathy in the groups of the mentors and the students (Table 6).

The rho-Spearman correlation analysis indicated no significant correlation between the empathy of the mentors and the students participating together in the project work.

Student's t-test was used to detect and analyze the average results from the empathy scale (QCAE) measuring the difference between Polish and international students. It was assumed that the distribution of variables is not different from a normal distribution and the groups compared are unequal. Due to the fact that the normality criterion of analyzed variable distributions was not met as well as the parity of the compared groups was not met, it was decided to apply the analyses with the usage of nonparametric tests: the comparison of average frames with U Mann-Whitney Test and correlation analysis with the rho-Spearman correlation coefficient, however, this did not change the situation.

Table 6. Correlations

Correlations									
		Cognitive empathy, Perspective takingS	**Cognitive empathy, Online simulationS**	**Cognitive empathy, sumS**	**Affective empathy, Emotion contagionS**	**Affective empathy, Proximal responsivityS**	**Affective empathy, Peripheral responsivityS**	**Affective empathy, sumS**	**Empathy sumS**
rho-Spearman	Cognitive empathy, Perspective takingM	0.329	-0.171	-0.068	-0.032	-0.048	-0.138	0.058	0.024
		0.273	0.576	0.825	0.917	0.876	0.653	0.850	0.939
		13	13	13	13	13	13	13	13
	Cognitive empathy, Online simulationM	0.393	0.255	0.244	-0.015	-0.006	0.143	0.179	0.263
		0.184	0.401	0.422	0.960	0.985	0.642	0.558	0.386
		13	13	13	13	13	13	13	13
	Cognitive empathy, sumM	0.467	0.044	0.119	-0.081	0.039	0.026	0.130	0.188
		0.108	0.885	0.699	0.792	0.898	0.933	0.672	0.538
		13	13	13	13	13	13	13	13
	Affective empathy, Emotion contagionM	-0.050	-0.270	-0.233	0.108	0.111	0.059	0.117	-0.130
		0.870	0.372	0.444	0.724	0.718	0.849	0.703	0.672
		13	13	13	13	13	13	13	13
	Affective empathy, Proximal responsivityM	0.496	-0.048	0.116	-0.511	0.169	-0.155	-0.105	0.015
		0.085	0.877	0.706	0.074	0.580	0.613	0.733	0.960
		13	13	13	13	13	13	13	13
	Affective empathy, Peripheral responsivityM	0.252	-0.011	0.103	0.171	0.335	-0.337	0.148	0.015
		0.405	0.971	0.737	0.577	0.264	0.260	0.629	0.960
		13	13	13	13	13	13	13	13
	Affective empathy, sumM	0.466	-0.029	0.098	-0.254	0.351	-0.065	0.047	0.055
		0.108	0.925	0.750	0.403	0.240	0.832	0.879	0.858
		13	13	13	13	13	13	13	13
	Empathy sumM	0.437	-0.008	0.094	-0.078	0.111	-0.010	0.150	0.134
		0.135	0.979	0.761	0.800	0.718	0.974	0.624	0.663
		13	13	13	13	13	13	13	13

CONCLUSION

The statistical research conducted on the sample of 112 Polish and International students with 34 EPS students did not show any significant differences between Polish and International students, between the students in EPS projects and other projects, or the students in EPS projects and the students who have never been involved in projects. What is more, there was no correlation discovered between the empathy of mentors and students. The results are not conclusive and did not give any deeper insight into the research on empathy in project work. Surprisingly, all the students and mentors have high levels of empathy in all empathy categories and subcategories, which contradicts the results of the research claiming that engineers very often design solutions without taking responsibility for the adverse influence on others (Lathem et.al., 2011).

It may be hypothesized that the respondents have a high awareness of the importance of empathy in education, science and engineering design. They are also aware of the social expectation that people involved in education and design of solutions for others should demonstrate such a high level of empathy and that is why they are directed by the will to perform in agreement with the social expectations concerning their professionalism. However, it may also mean that they rightly and quite precisely read and categorize components and sub-components of empathy presented in the applied instrument, which indicates the necessity of conducting further research while using different scales. It could also be hypothesized that the results may be caused by a certain degree of inability to self-assess precisely and objectively one's own cognitive and affective states and that may lead to developing a false image of oneself. That is why further research on future EPS project teams is required. However, the explorations of empathy perceptions and their roots should not stop us from admitting that the strong will to be perceived as an empathetic person constitutes a positive basis for future empathy development among engineering students.

It must also be admitted that project-based learning as such may create a different environment for empathy appreciation and development. It is especially true in the European Project Semester at TUL where students representing different cultural backgrounds and different expertise work on a project in small groups throughout the whole semester, develop closed contact with their mentors while meeting them regularly at least once or twice a week and are supported on a regular basis in the development of social group relations, team building activities, interpersonal communication skills and techniques enhancing team decision making and problem solving with the usage of a variety techniques explored in human-oriented approach, PBL approach, universal design within the framework of a Team Building and Communication for Project course. There are usually only four or five projects run each semester, which allows for a group of 20 to 25 EPS students since there are four or five students in each project team. The small size of project teams and the whole EPS group and a long period of 15 weeks of common work create a good environment for developing closer interpersonal relations in their teams, with project mentors, Team Building mentor and the EPS group at large. What is more, project-based education frequently explores more formative methods of assessment, which diminishes internal competition and may redefine the sense of success as more dependent on common work and individual contributions to a common team goal. Students become more self-assured due to a lowered risk connected with mistake making and small failures. They may become more open in expressing their true opinions without the risk of losing their reputation. Project-based learning with a variety of different approaches used in EPS can be a very good environment for developing empathy-based soft skills, so important in engineering at the times of pervasive digitalization and technology developments that very often impoverish human communication and contact, and may potentially deprive it of opportunities to develop empathy. This brings us to other important questions: How dependent is empathy on different assessment methods and techniques? What is the relation between organizational, financial and technological factors that would allow to create environments prone for sustaining empathy in engineering education?

The study presented in the article measured empathy levels at a certain point in time, and the research results reflect a state at this specific moment. Does it mean that empathy is maintained on a similar level in a longer period? Shall sustainability of empathy be treated as a salient issue as well as techniques used to keep it sustainable? This leads us to an interesting question: Is empathy static or a process? If it is a process, is it trainable and sustainable? "Review of studies revealed that empathy can be taught; students [...] can be taught reciprocity and emotional resonance. Once learned, these skills become the part of acquired behavior and enhance the ability to relate across the various situation." (Srivastava & Das,

2016). Currently, empathy towards different stakeholders is taught in EPS based on projects conducted in teams with the usage of Design Thinking or a human-centered approach. The choice of approach in project work allows for creating a structure for empathy development. It triggers the implementation of empathy training. Having started with empathy towards users and clients as an inherent element of DT approach and putting empathy into perspective naturally resulted in attracting students towards this concept and allowed for its further development. Nevertheless, a deeper understanding of empathy and its cognitive and affective components seems to be a necessary requirement for the TUL EPS mentors to complete and extend the training they received at Hasso Plattner Institute D-school in Potsdam. It will direct the way of thinking of these academics and help them utilize factors salient for the development of empathy, e.g., the factors operating in the realm of empathetic objectivity, which seems to rely on self-other differentiation and self-other awareness (Decety & Meyer, 2008) as a basis for recognizing others as autonomous agents (Deigh, 1995). It helps designers avoid a purely subjective perspective.

Considering the popularity of empathy in the public discourse, it may be claimed that the concept of empathy is acknowledged as extremely important and has become one of the ideals that people want to follow and are trying try to achieve. It is considered to be required and highly valued in developing human relations and is perceived as an indispensable basis of communication, knowledge sharing and positive interpersonal relations. What it more, it became an important aspect of design and engineering. That is why it stays in the center of interest of teachers and students who want to empathize and expect to be empathized, teachers who want to be student-centered and human-centered in their approach to another person. It is true that the usage of tools based on self-evaluation would always be prone to bias due to numerous factors such as a desire to match social expectations, create a positive image of yourself or an inability to evaluate yourself impartially. That is why the authors advocate to continue research on empathy on the groups of students and teachers to attempt to find an answer to a few important research questions such as: how does the level of empathy change during the project life cycle when students have to empathize more with a variety of different stakeholders? What is the impact of empathy or the lack of it on an individual team member? How is it perceived in different groups of students and mentors, e.g., do the perceptions and levels differ among men and women, among specific cultures or engineering disciplines?

Considering the fact that empathy is declared to be in the center of interest among academics and students involved in project work and is highly praised by both groups, especially after the period of learning provided exclusively online, which was visible in the answers to open questions and in interviews, the suggested Human-oriented Model of Engineering will be further enhanced and developed. Empathy will remain its core that establishes positive human relations and creates the environment for learning and engineering design that would bring a positive change.

However, it is important to take a reflexive stance towards the positive side of empathy and acknowledge the potential inconclusiveness of empathy impacts. Becoming so pervasively used in design and so popularized in learning, the lack of focus on its potential drawbacks may create a risk that it could become "ideology rather than a principle that is appropriate" to be used on extremely high level in "some situations and inappropriate under other circumstances" (Highlighen & Dong, 2019). The authors believe empathy should stay a means and not an end or a means to improve and create with the full awareness that who "treats a person as 'another I' does not really see a person but only a projected image of himself." (Friedman, 2002).

REFERENCES

Afroogh, S., Esmalian, A., Donaldson, J., & Mostafavi, A. (2021). Empathic Design in Engineering Education and Practice: An Approach for Achieving Inclusive and Effective Community Resilience. *Sustainability, 13*(7), 4060. doi:10.3390u13074060

Balaji, K. V. A., & Somashekar, P. (2009). A comparative study of soft skills among engineers. *The IUP Journal of Soft Skills, 3*(3/4), 50–57.

Bernhardt, B. C., & Singer, T. (2012). The neural basis of empathy. *Annual Review of Neuroscience, 35*(1), 1–23. doi:10.1146/annurev-neuro-062111-150536 PMID:22715878

Casner-Lotto, J., & Barrington, L. (2006). *Are They Really Ready to Work? Employers' Perspectives on the Basic Knowledge and Applied Skills of New Entrants to the 21st Century US Workforce*. The Conference Board, Inc., the Partnership for 21st Century Skills, Corporate Voices for Working Families, and the Society for Human Resource Management. https://www.conference-board.org/pdf_free/BED-06-Workforce.pdf

Cech, E. A. (2014). Culture of Disengagement in Engineering Education? *Science, Technology & Human Values, 39*(1), 42–72. doi:10.1177/0162243913504305

Cooper, A. (1999). *The Inmates Are Running the Asylum: Why High Tech Products Drive Us Crazy and How to Restore the Sanity* (1st ed.). John Wiley & Sons. doi:10.1007/978-3-322-99786-9_1

Cooper, A., Reimann, R., Cronin, D., & Noessel, C. (2014). *About face: The essentials of interaction design* (4th ed.). John Wiley & Sons.

Cross, N. (1982). Designerly ways of knowing. *Design Studies, 3*(4), 221–227. doi:10.1016/0142-694X(82)90040-0

Cuff, B. M. P., Brown, S. J., Taylor, I., & Howat, D. J. (2016). Empathy: A review of concept. *Emotion Review, 8*(2), 144–153. doi:10.1177/1754073914558466

Dandavate, U., Sanders, E. B. N., & Stuart, S. (1996). Emotions Matter: User Empathy in the Product Development Process. *Proceedings of the Human Factors and Ergonomics Society Annual Meeting, 40*(7), 415–418. doi:10.1177/154193129604000709

De Vignemont, F., & Singer, T. (2006). The empathic brain: How, when, why? *Trends in Cognitive Sciences, 10*(10), 435–441. doi:10.1016/j.tics.2006.08.008 PMID:16949331

De Waal, F. B. M., & Preston, S D. (n.d.). Mammalian empathy: behavioral manifestations and neural basis. *Nature Reviews Neuroscience, 18*(8), 498–509. doi:10.1038/nrn.2017.72

Decety, J., & Lamm, C. (2006). Human empathy through the lens of social neuroscience. *TheScientificWorldJournal, 6*, 1146–1163. doi:10.1100/tsw.2006.221 PMID:16998603

Decety, J., & Meyer, M. (2008). From emotion resonance to empathic understanding: A Social Developmental Neuroscience Account. *Development and Psychopathology, 20*(4), 1053–1080. doi:10.1017/S0954579408000503 PMID:18838031

Decety, J., & Moriguchi, Y. (2007). The empathic brain and its dysfunction in psychiatric populations: Implications for intervention across different clinical conditions. *BioPsychoSocial Medicine*, *1*(1), 22. doi:10.1186/1751-0759-1-22 PMID:18021398

Deigh, J. (1995). Empathy and universalizability. *Ethics*, *105*(4), 743–763. doi:10.1086/293751

Edmondson, A. (1999). Psychological Safety and Learning Behavior in Work Teams. *Administrative Science Quarterly*, *44*(2), 350–383. doi:10.2307/2666999

Eklund, H. J., & Meranius, M. S. (2021). Towards a consensus on the nature of empathy: A review of reviews. *Patient Education and Counseling*, *104*(2), 300–307. doi:10.1016/j.pec.2020.08.022 PMID:32888755

Figas, M., Lejarza, A., Rappard, R., & Sarachaga, E. (2021). Students in the times of pandemic. European Project Semester, Lodz University of Technology.

Friedman, M. S. (2002). *Martin Buber: The Life of Dialogue*. Routledge.

Gerdes, K. E., & Segal, E. A. (2009). A social work model of empathy. *Advances in Social Work*, *10*(2), 114–127. doi:10.18060/235

Gerdes, K. E., Segal, E. A., & Lietz, C. A. (2010). Conceptualising and measuring empathy. *British Journal of Social Work*, *40*(7), 2326–2343. doi:10.1093/bjsw/bcq048

Goleman, D. (1998). *Working with emotional intelligence*. Bloomsbury.

Gonzales-Liencres, C., Shamaay-Tsoory, S. G., & Brüne, M. (2013). Towards a neuroscience of empathy: Ontogeny, phylogeny, brain mechanism, context and psychopathology. *Neuroscience and Biobehavioral Reviews*, *37*(8), 1537–1538. doi:10.1016/j.neubiorev.2013.05.001 PMID:23680700

Grasso, D., & Burkins, M. (Eds.). (2010). *Holistic Engineering Education: Beyond Technology*. Springer. doi:10.1007/978-1-4419-1393-7

Grasso, D., & Martinelli, D. (2007). Holistic engineering. *The Chronicle of Higher Education*, *53*, B8.

Grasso, D., & Martinelli, D. (2010). Holistic engineering education: Beyond technology. *Choice Reviews Online*, *47*(10), 1–10. doi:10.5860/CHOICE.47-5674

Heylighen, A., & Dong, A. (2019). To empathise or not to empathise? Empathy and its limits in design. *Design Studies*, *65*, 107–124. doi:10.1016/j.destud.2019.10.007

Hoffman, M. L. (2000). *Empathy and moral development: Implications for caring and justice*. Cambridge University Press. doi:10.1017/CBO9780511805851

Hume, D. (1978). *(1739–40 [1978]). A Treatise of Human Nature*. Oxford University Press.

Iacoboni, M. (2008). *Mirroring people: The science of empathy and how we connect with others*. Picador.

ideo.org. (2022). *IDEO.org is a nonprofit design studio | IDEO*. https://www.ideo.org/

Jones, J. C. (1970). *Design Methods: Seeds of Human Futures*. Wiley-Interscience.

Jones, S. A., Michelfelder, D., & Nair, I. (2017). Engineering Managers and Sustainable Systems: The Need for and Challenges of Using an Ethical Framework for Transformative Leadership. *Journal of Cleaner Production, 140*, 205–212. doi:10.1016/j.jclepro.2015.02.009

Kelley, T., & Kelley, D. (2013). *Creative confidence*. Crown Business.

Kohut, H. (1959). Introspection, empathy, and psychoanalysis: An examination of the relationship between the mode of observation and therapy. *Journal of the American Psychoanalytic Association, 7*(3), 459–483. doi:10.1177/000306515900700304 PMID:13672863

Kolko, J. (2012). *Wicked Problems: Problems Worth Solving*. Austin Center for Design.

Kouprie, M., & Visser, F. S. (2009). A framework for empathy in design: Stepping into and out of the user's life. *Journal of Engineering Design, 20*(5), 437–448. doi:10.1080/09544820902875033

Kullman, K. (2016). Prototyping bodies: A post-phenomenology of wearable simulations. *Design Studies, 47*, 73–90. doi:10.1016/j.destud.2016.08.004

Lathem, S. A., Neumann, M. D., & Hayden, N. (2011). The Socially Responsible Engineer: Assessing Student Attitudes of Roles and Responsibilities. *Journal of Engineering Education, 100*(3), 444–474. doi:10.1002/j.2168-9830.2011.tb00022.x

Leiberg, S., & Anders, S. (2006). The multiple facets of empathy: A survey of theory and evidence. *Progress in Brain Research, 156*, 419–440. doi:10.1016/S0079-6123(06)56023-6 PMID:17015094

Levenson, R., & Ruef, A. (1992). Empathy: A physiological substrate. *Journal of Personality and Social Psychology, 63*(2), 234–246. doi:10.1037/0022-3514.63.2.234 PMID:1403614

Levinas, E. (1969). *Totality and Infinity*. Duquesne University Press.

Levy, O., Beechler, S., Taylor, S., & Boyacigiller, N. (2015). Global Mindset. Wiley Encyclopedia of Management, 1-4. doi:10.1002/9781118785317.weom060094

Lipps, T. (1903). Einfühlung, Innere Nachahmung und Organempfindung [Empathy, Inner Imitation and Sense-Feelings]. *Archiv fur die Gesamte Psychologie, 1*, 465–519.

Loch, W. (2020). Critical Thinking in Problem Exploration in Design and Technology Design Project. *Design and Technology Education: An International Journal, 25*(1), 35–58.

Moore, P., & Conn, C. P. (1985). *Disguised: A true story*. Word Books.

OECD. (2015). Skills for Social Progress: The Power of Social and Emotional Skills. OECD Skills Studies. Paris: OECD Publishing. doi:10.1787/9789264226159-en

Penzenstadler, B., Haller, G., Schlosser, T., & Frenzel, G. (2009). Soft skills required: A practical approach for empowering soft skills in the engineering world. *Proceedings of the 2009 Collaboration and Intercultural Issues on Requirements: Communication, Understanding and Softskills*. 10.1109/CIRCUS.2009.5

Peppler, K. (2017). *The SAGE Encyclopedia of Out-of-School Learning*. doi:10.4135/9781483385198

Preckel, K., Kanske, P., & Singer, T. (2018). On the interaction of social affect and cognition: Empathy, compassion and theory of mind. *Current Opinion in Behavioral Sciences*, *19*, 1–6. doi:10.1016/j.cobeha.2017.07.010

Rasoal, C., Danielsson, H., & Tomas, J. (2012). Empathy among students in engineering programmes. *European Journal of Engineering Education*, *37*(5), 427–435. doi:10.1080/03043797.2012.708720

Reniers, R. L. E. P., Corcoran, R., Drake, R., Shryane, N. M., & Vollm, B. A. (2011). The QCAE: A Questionnaire of Cognitive and Affective Empathy. *Journal of Personality Assessment*, *93*(1), 84–95. doi:10.1080/00223891.2010.528484 PMID:21184334

Rogers, C. (1957). The necessary and sufficient conditions for therapeutic personality change. *Journal of Consulting Psychology*, *21*(2), 95–103. doi:10.1037/h0045357 PMID:13416422

Roth, L., & Altman, T. (2021). The self-other agreement of multiple informants on empathy measures and its relation to empathic accuracy. *Personality and Individual Differences*, *171*, 110499. doi:10.1016/j.paid.2020.110499

Segal, E. A. (2011). Social empathy: A model built on empathy, contextual understanding, and social responsibility that promotes social justice. *Journal of Social Service Research*, *37*(3), 266–277. doi:10.1080/01488376.2011.564040

Segal, L. D., & Suri, J. F. (1997). The Empathic Practitioner: Measurement and Interpretation of User Experience. *Proceedings of the Human Factors and Ergonomics Society Annual Meeting*, *41*(1), 451–454. doi:10.1177/107118139704100199

Sheppard, K., Dominick, P., & Aronson, Z. (2004). Preparing engineering students for the new business paradigm of international teamwork and global orientation. *International Journal of Engineering Education*, *20*(3), 475–483. https://www.ijee.ie/articles/Vol20-3/IJEE1486.pdf

Smith, J. (2017). What is empathy for? *Synthese*, *194*(3), 709–722. doi:10.100711229-015-0771-8 PMID:32226174

Smith, K. E., Porges, E. C., Norman, G. J., Connelly, J. J., & Decety, J. (2014). Oxytocin receptor gene variation predicts empathic concern and autonomic arousal while perceiving harm to others. *Social Neuroscience*, *9*(1), 1–9. doi:10.1080/17470919.2013.863223 PMID:24295535

Stueber, K. (2019). Empathy. In E. N. Zalta (Ed.), *The Stanford Encyclopedia of Philosophy* (Fall 2019 Edition). https://plato.stanford.edu/archives/fall2019/entries/empathy/

Titchener, E. B. (1909). *Lectures on the Experimental Psychology of Thought-Processes*. Macmillan. doi:10.1037/10877-000

Walther, J., Brewer, M., Sochacka, N., & Miller, S. (2019). Empathy and engineering formation. *Journal of Engineering Education*, *109*(1), 1–23. doi:10.1002/jee.20301

Woodley, A. W., Chabris, C. F., Pentland, A., Hashmi, N., & Malone, T. W. (2010). Evidence for a collective intelligence factor in the performance of human groups. *Science*, *330*(6004), 686–688. doi:10.1126cience.1193147 PMID:20929725

Yeager, D., & Dweck, C. (2020). What can be learned from growth mindset controversies? *The American Psychologist, 75*(9), 1269–1284. doi:10.1037/amp0000794 PMID:33382294

ADDITIONAL READING

Bandura, A., & Walters, R. H. (1963). *Social learning and personality development.* Holt, Rinehart, and Winston.

Decety, J., & Ickes, W. (2011). *The social neuroscience of empathy*. The MIT Press.

Håkansson, J., & Montgomery, H. (2002). The role of action in empathy from the perspectives of the empathizer and the target. *Current Research in Social Psychology, 8*, 50–62. https://crisp.org.uiowa.edu/sites/crisp.org.uiowa.edu/files/2020-04/8.4.pdf

Håkansson, J., & Montgomery, H. (2003). Empathy as an Interpersonal Phenomenon. *Journal of Social and Personal Relationships, 20*(3), 267–284. doi:10.1177/02654075030200003001

Kiverstein, J. (2015). Empathy and the responsiveness to social affordances. *Consciousness and Cognition, 36*, 532–542. doi:10.1016/j.concog.2015.05.002 PMID:26021190

Neumann, M., Bensing, J., Mercer, S., Ernstmann, N., Ommen, O., & Pfaff, H. (2009). Analyzing the "nature" and "specific effectiveness" of clinical empathy: A theoretical overview and contribution towards a theory-based research agenda. *Patient Education and Counseling, 74*(3), 339–346. doi:10.1016/j.pec.2008.11.013 PMID:19124216

Pedersen, R. (2007). Empathy: A wolf in sheep's clothing? *Medicine, Health Care, and Philosophy, 11*(3), 325–335. doi:10.100711019-007-9104-0 PMID:17926143

Rader, M. M. (Ed.). (1979). *A Modern Book of Esthetics*. Holt, Rinehart and Winston.

Shamay-Tsoory, S. G. (2010). The Neural Bases for Empathy. *The Neuroscientist, 17*(1), 18–24. doi:10.1177/1073858410379268 PMID:21071616

Singer, T., & Lamm, C. (2009). The social neuroscience of empathy. *Annals of the New York Academy of Sciences, 1156*(1), 81–96. doi:10.1111/j.1749-6632.2009.04418.x PMID:19338504

KEY TERMS AND DEFINITIONS

Affective Sharing: An ability to experience similar emotions by the empathizer and empathee.

Emotion Regulation: An ability to regulate subjective emotions and feelings while using different cognitive and affective strategies (judgement, affirmation-seeking, mindfulness techniques etc.).

Feeling: An effective response to another person's situation and includes affect, emotion, and sensation.

Mental Flexibility: A cognitive ability to take the subjective perspective of the other while still maintaining the boundaries between an empathizer and an empathee, as well as the ability to switch between the two perspectives.

Mirror Neurons: A type of neurons that modulate their activity when a person executes a motor act and when the person observes the same or similar act performed by another person.

Self-Awareness: An ability to recognize and differentiate between our own public and private perceptions.

Self-Other Differentiation: An ability to recognize differences between an empathizer and an empathee, and comprises the differentiation of feelings, objectivity and self-distinction.

ENDNOTE

1 The authors acknowledge the importance of differentiation and congruence between empathy, sympathy and compassion and their research exploration, however, the detailed analysis of the differentiation is beyond the scope of this chapter. The number of perspectives of the topic would require separate detailed research. For further discussion, see, e.g., Hoffman, 2000 or Nichols, 2004.

Chapter 11
Robotics and the European Project Semester

Manuel Fernando Silva
https://orcid.org/0000-0002-0593-2865
Instituto Superior de Engenharia do Porto, Instituto Politécnico do Porto, Portugal

Abel José Duarte
Instituto Superior de Engenharia do Porto, Instituto Politécnico do Porto, Portugal

Paulo Duarte Ferreira
https://orcid.org/0000-0001-6432-6366
Instituto Superior de Engenharia do Porto, Instituto Politécnico do Porto, Portugal

Pedro Barbosa Guedes
https://orcid.org/0000-0001-9254-7692
Instituto Superior de Engenharia do Porto, Instituto Politécnico do Porto, Portugal

ABSTRACT

Robotics is a multidisciplinary subject that typically involves mechanics, electronics, and computer science concepts. For this reason, robotic projects are particularly well suited to the European Project Semester framework since they allow students with different backgrounds to contribute to the overall team objective in their specific knowledge areas. This chapter briefly presents illustrative examples of robotic projects that have been developed by teams of students participating in the European Project Semester at the School of Engineering of the Polytechnic Institute of Porto. It concludes by presenting and discussing student feedback, namely on the program and the projects developed.

INTRODUCTION

The EPS@ISEP is the EPS implementation at Instituto Superior the Engenharia do Porto (ISEP) – the School of Engineering of the Porto Polytechnic. This one-semester program welcomes engineering, business, and product design students since the academic year or 2010-2011 and, naturally, follows

DOI: 10.4018/978-1-6684-2309-7.ch011

Copyright © 2022, IGI Global. Copying or distributing in print or electronic forms without written permission of IGI Global is prohibited.

the European Project Semester concept: prepare future engineers to think and act globally, by adopting project-based learning and teamwork methodologies, fostering the development of complementary skills, and addressing ethics, sustainability and multiculturalism. Multidisciplinary collaborative teamwork (Duarte *et al.*, 2015) as well as ethics and sustainability-driven design (Duarte *et al.*, 2020) are pervasive concerns within EPS@ISEP projects. EPS creates a diverse environment within each team, where each member brings his/her own specific set of skills, knowledge, and culture. All projects developed within the EPS@ISEP are based on the application of scientific, economic, social, and practical knowledge to design and invent solutions to improve the quality of life on Earth.

EPS@ISEP is a 30 European Credit Transfer System Units (ECTU) package structured in six modules: Project (20 ECTU), Energy & Sustainable Development (2 ECTU), Ethics & Deontology (2 ECTU), Marketing & Communication (2 ECTU), Project Management & Team Building (2 ECTU), Portuguese Culture & Language (2 ECTU). Apart from Portuguese, the remaining 2 ECTU are project support modules. They allow students to analyze essential aspects with impact in their solution's design and development, such as sustainability, marketing, ethics, and teamwork. The faculty involved in project coaching and module teaching are from seven departments of the School of Engineering (chemical, electrical, informatics, mechanics, management, mathematics, and physics).

Before the beginning of the semester, a set of project proposals regarding multidisciplinary problems affecting society are collected. The origin of proposals varies and includes industry, services, R&D institutions, or the school itself. The proposals tend to be multidisciplinary problems, *i.e.*, require the integration of multiple technical and scientific competences. A proposal defines the problem/challenge to address, the minimal set of requirements, mostly mandatory directives and standards, and the maximum budget. Depending on the complexity of the projects, the average cost of an EPS@ISEP project in terms of materials is approximately 200 €. Each team is expected to choose, design, build, test, and deliver a prototype of the solution to the chosen problem together with several other deliverables, ranging from reports (interim and final), presentations (interim and final), a paper, poster, video, leaflet, user manual as well as the drawings, 3D model, code developed and demo of the prototype. A strong effort is made to ensure the team elements are from different countries and have a diversified technical background. This aims to fulfill all technical, scientific, and commercial demands of a successful prototype evolution while maximizing the cultural and knowledge exchanges among the students.

These projects share another ulterior goal in the context of engineering education: to help the students develop personal, teamwork and problem-solving skills while applying and enriching their technical-scientific knowledge. Gopakumar has coined this ability as public leadership, *i.e.*, engineers who excel not only in technical problem solving but also have the understanding and skills to operate within a complex, fast-changing social, political, and cultural environment (Gopakumar, 2014).

The above-stated objectives can be fulfilled by developing biomimetic or artistic robots. Furthermore, the development of biomimetic robots has already been used for attracting K–12 students to pursue careers in science, technology, engineering, and mathematics (Laut *et al.*, 2015). This type of project directs teams towards the conceptualization, design, implementation, and operation stages of the capstone project/internship (Malheiro *et al.*, 2015).

Given these ideas, the four projects analyzed in this paper are: (*i*) Bro-Fish, an educational construction kit, directed to children between eight to twelve years old; (*ii*) Bubbles, an educational robotic toy for teaching young children the basics of programming through play; (*iii*) MyBird, a bioinspired flying robot; and (*iv*) GraphBot, a color drawing robot for children, commanded by voice.

Following this introduction, the remainder of this chapter is organized according to the following structure: Section 2 describes four examples of robotics projects developed within EPS@ISEP. Section 3 analyses the impact of robotics projects, including student testimonials, and, finally, Section 4 presents the conclusions.

EXAMPLES OF ROBOTIC PROJECTS DEVELOPED WITHIN EPS@ISEP

The following subsections present four examples of different robots developed by teams of students participating in different editions of EPS@ISEP. First, it describes three examples of biologically inspired robots, mimicking fish (BroFish and Bubbles) and birds (MyBird), followed by a more conventional drawing robot (GraphBot).

Bro-Fish – An Educational Construction Kit

The Bro-Fish robot was developed by a gender-balanced team composed of Spanish, Finnish, German and Polish undergraduates enrolled in Civil Engineering, Electrical Engineering, Naval Architecture and Maritime Engineering, and Mechanical Engineering and Computer Science.

According to the team, their motivation for choosing this project was based on the following main reasons: (*i*) the unquestionable relevance of oceans to Earth, which derives from their cumulative dimension (70% of the surface of the planet) and amount and variety of life forms supported (99% of existing habitats); and (*ii*) the challenge of designing a biomimetic swimming robot, a design choice which, when compared to thrust propelled vehicles, grants higher efficiency and increased maneuverability (Ishii *et al.*, 2014).

With this motivation, their main goal was to provide an innovative and engaging opportunity for children to learn. The idea is that children who are early exposed to technology will hopefully be able to contribute to future technological innovation and development. To stem children´s curiosity and enthusiasm for technology, the team wants to provide kids the opportunity to experiment with laws of nature and state-of-the-art technologies rather than having them spend more time in traditional classroom activities. Therefore, they decided to develop a toy for children, which is a construction kit for a swimming robot with biomimetic features. According to the team's opinion, by self-assembling the product and being able to program and configure certain features, the user naturally learns about robotics and the physics of floating objects.

Based on these ideas, the team embraced the design and development of a biologically inspired swimming robot prototype with a 12-hour power autonomy. The literature review performed on bioinspired mechanisms of displacement led the team to adopt the movements of the body and/or caudal fin, the undulation of the middle or pectoral fins and oscillations of the middle or pectoral (Lindsey, 1978) (Sfakiotakis *et al.*, 1999). The key features of the anguilliform-inspired design include: a waterproof body, forward propulsion, horizontal steering, depth control and exchangeable parts to allow children to conduct physical experiments. The considered changeable parts were the tail and fins (shape, size, and motion pattern) as well as the mass (number and distribution of mass modules).

The students, within the Marketing and Communication module, defined the market plan for their product. They started by conducting a survey among twenty-one twelve years old kids from a school in Porto, Portugal, to better understand the market and the customer's requirements, so that the product fits their needs exactly. This allowed them to clearly identify their product and the target market (EU and, in particular, Germany) (Ishii *et al.*, 2014). According to the team, "Bro-Fish" should be an educational construction kit, directed to children between eight to twelve years old, presenting the following features: (*i*) fish-like movement, (*ii*) remotely controlled (left/right and up/down), (*iii*) easy to program, and (*iv*) allow learning about physics of motion in the water by changing parts. Children should: (*i*) assemble the toy following the user manual, (*ii*) write programs based on the provided templates, corresponding to the various levels of expertise, and (*iii*) play and experiment by changing the customizable parts like the fins and the tail. The idea is that "Bro-Fish" is not only a toy but also a laboratory to experiment with physics and robotics. Therefore, to make their brand known, they developed the logo presented in Figure 1 (left).

Figure 1. Logo (left), anguilliform design (center) and prototype (right)
(Ishii et al., 2014)

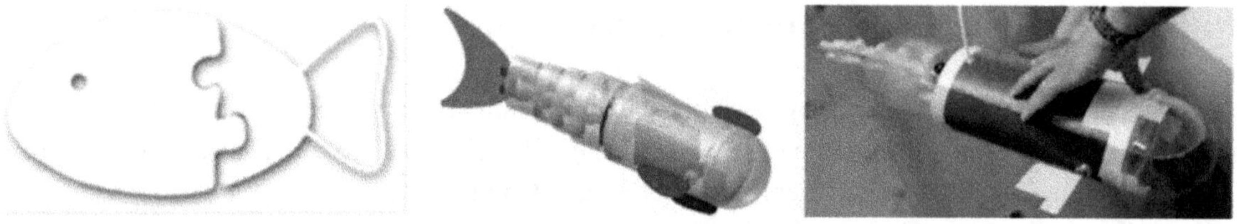

In summary, "Bro-Fish" was designed as a construction kit targeted to kids of an age range between eight and twelve. Not only does it develop competences in three different fields (mechanics, robotics, and programming), but continued learning through customizable parts (tail, fins, mass) and different programming levels.

In the Energy and Sustainable Development module, the team addressed the set of eco-efficiency measures for sustainability. Since the beginning of this project that sustainability engineering was in the team's minds since this toy is going to operate in water and is made for kids, and in the Marketing Plan, was already mentioned the development of an "Eco-friendly plan". This plan considers the inclusion, in each toy of their future product line-up, of an environmental leaflet for kids. In the case of the "Swimming Robot", kids will find a leaflet called "Why are our oceans important?".

Supported by these background studies, the next step was to design the structure of the bioinspired swimming robot meeting the identified requirements. First, it was necessary to define the number of actuated segments and fins to propel and steer the robot. Considering it was intended for children, the team decided to use motor-based propulsion. The approach adopted was aimed at creating a command-response pattern that is easier to handle. Therefore, the designed robot has one back fin and two pectoral fins at (one per side of the body), and a steering fin at the tail. Unlike the prototype proposed by Silva *et al.* (2014), BroFish uses the back fin for propulsion instead of the pectoral fins – an approach identical to the natural locomotion of fish. To propel the robot with an oscillating back fin, the designed solution, rather than using a servomotor, transfers the continuous rotation of a DC motor into a swinging tail movement. For this, a rotating shaft was bent and placed inside vertical slots of the body, leading the body segments to move in accordance with the shaft´s shape in the horizontal plane. The final design of the bioinspired robot was simulated using 3D design software (Figure 1, center).

After selecting suitable components (Ishii *et al.*, 2014), the BroFish prototype was successfully assembled and tested (Figure 1, right). During the experimental evaluation, the pool tests revealed the potential to improve the robot forward motion (refinement of the tail design) and the longitudinal mass distribution. A list of adaptations was proposed for the tail refinement, namely, to increase its dimensions and, thus, improve force transfer.

A detailed description of this project can be found on the team's final report (Ishii *et al.*, 2014) and accompanying paper (Ishii *et al.*, 2015).

Bubbles – An Educational Robotic Toy

The didactic robotic fish was selected by a team of five members comprising a Media and Technology student from Germany, an Engineering and Architecture student from Spain, a Logistics Management student from Poland, an Electrical Engineering student from Scotland and a Mechanical Engineering student from the U.S.A.

The specified goal of the robotic fish was to create a robot able to swim in the water which had at least a single degree of freedom. To provide this movement, the design team looked for inspiration in nature. Millions of years of evolution have filled the oceans with the perfect swimming machines: fish. Most fish (85%) utilize as main source of propulsion body and/or caudal (back) fin (BCF) locomotion (Reinhardt, *et al.*, 2016a). Inspired by this finding, the team decided early on to mimic BCF propulsion. However, this form of movement falls on a spectrum from undulatory to oscillatory. These classifications concern how the propulsion wave propagates from the tail through the body (Lindsey, 1978). With the imposed design restrictions of this project, ostraciiform locomotion made the most sense. In accordance with the project charter, the fish should be inexpensive, controlled by servo motor(s), and have a 3D printed body, and there is no straightforward way to create undulatory movement without sacrificing one or more of these tenets.

Given this, the team decided that the body shape should be based on that of a typical wider body of ostraciiform fish, like the box fish and the puffer fish. It was designed to be 150 mm long and with a compartmentalized dry interior, designed to house all the fish's electronic components, as depicted in Figure 2 (left and center). The main body consists of two pieces, top and bottom. These are connected via bolts and separated by a waterproof gasket.

As dictated by the project directives, the body is 3D printed (Figure 2, right). This print was completed using polylactic acid (PLA) due to its easy printability, nontoxicity, and biodegradability. However, the body can be created using any 3D printable material.

The motion of the fish is provided by a single servo motor connected to the rear fin vertical drive axle via springs. The springs dampen the motion and help to prevent component damage from repetitive stress. The axle, which is split into two sections to allow for simple disassembly, leaves the fish's interior through holes waterproofed with O-rings. The movement is controlled by an Arduino Nano, which is custom programmable by the user. To start an input program, the user removes a magnet from the exterior surface of the fish. This magnet opens a reed switch, and the Arduino begins to run its code. Finally, the team chose a 7.4 V / 900 mAh battery due to budget constraints. This battery grants power autonomy of over one hour and has a charging time of approximately two hours.

Figure 2. Exploded (left), vertical cut (center) and prototype (right)
(Reinhardt, et al., 2016a)

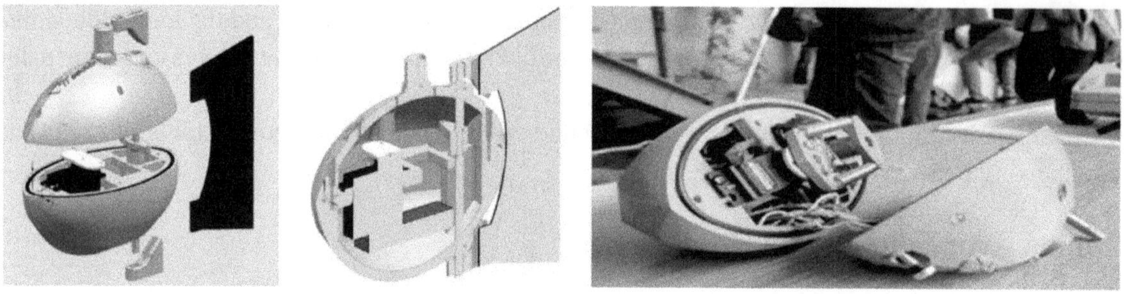

The tail is attached to the vertical drive axle, using small 3D printed parts placed on the top and bottom parts of the body. These parts slide over a flattened portion of the axle and are secured via setscrews. This design allows experimenting with several types of tails, namely, in terms of dimensions and materials.

To provide a programming environment suitable for children's cognitive abilities, the team considered both block and text-based development environments. Different from the text-based programming concept, the block-based concept relies on blocks that can be arranged by the user to create computing processes. The advantage of this workflow is that the correct syntax and terminology do not have significance because the user can only use pre-built elements. On the other side, block-based programming does not support the same level of complexity and number of functions as text-based programming. Since no complex functionalities are required for this project, the team decided to use the concept of block-based programming (Figure 3, left).

Figure 3. Ardublock program (left) and corresponding Arduino code automatically generated (right)
(Reinhardt et al., 2016a)

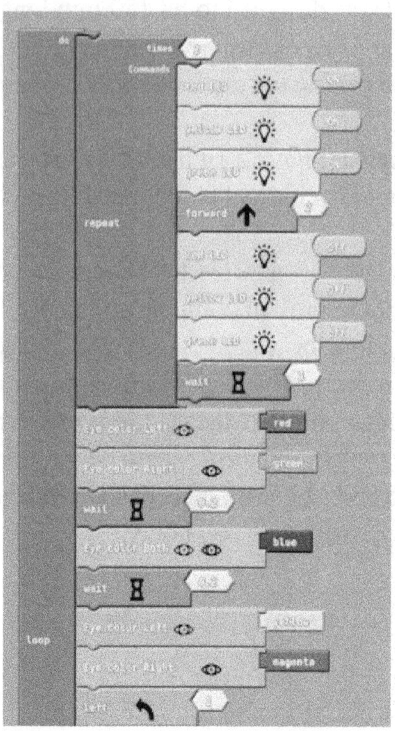

```
  pinMode(button, INPUT_PULLUP);
  pinMode(12, OUTPUT);
}

void loop()
{
  buttonWait();
  for (_ABVAR_1_a=1; _ABVAR_1_a<= ( 2 ); ++_ABVAR_1_a )
  {
    LED(true,"red");
    LED(true,"yellow");
    LED(true,"green");
    forward(2);
    LED(false,"red");
    LED(false,"yellow");
    LED(false,"green");
    delay(1*1000);
  }
  eyeLED("red",'l');
  eyeLED("green",'r');
  delay(0.2*1000);
  eyeLED("blue",'b');
  delay(0.2*1000);
  eyeLED("yellow",'l');
  eyeLED("magenta",'r');
  left(1);
  delay(0.2*1000);
  eyeLED("white",'b');
  delay(0.2*1000);
  eyeLED("cyan",'b');
  delay(0.2*1000);
  right(1);
  eyeLED("white",'l');
  eyeLED("black",'r');
  delay(0.2*1000);
  eyeLED("black",'l');
  eyeLED("white",'r');
  delay(0.2*1000);
  eyeLED("white",'b');
}
```

The team developed an illustrative program using the Ardublock (Figure 3). The code was uploaded to the Arduino, and successfully tested. The Arduino Nano recognized the software, and the program ran smoothly. Bubbles operated as expected, becoming a fully functioning robot.

A detailed description of this project can be found in the final report (Reinhardt *et al.*, 2016a) and accompanying paper (Reinhardt *et al.*, 2016b) produced by the team.

MyBird – A Bioinspired Flying Robot

The biologically inspired flying robot project was developed by a team of four including a Mechanical Engineer student from the United Kingdom, a Product Designer student from Belgium, a student of Sales and Purchasing Engineering from Germany and a student of Materials Engineering from Estonia.

The main goal was to create a robot that could fly up, and land safely but without any kind of propeller, within a budget of 250 €, and their motivation for choosing this project is in their statement: "This project appealed to us because it would be challenging, yet possible. Furthermore, we found it to be an interesting topic since none of us has any experience with robotics, but all are interested in the field of mechanics. Besides our interest in robotics, we liked the bio-inspired aspect of the proposal, since nature creates the most elegant and intelligent solutions to its problems" (Dunn *et al.*, 2014).

The team started their work by making a state-of-the art study, including some historical aspects of crewed flight (in particular, some preliminary models of human-powered flight), and analyzing several academic and research prototypes as well as commercially available ornithopters. During this phase, they also got acquainted with the characteristics of some electrical components and materials used in the construction of radio-controlled (RC) flying toys, namely power sources, actuation systems, communication systems, structural and wing materials. In the end, they were able to choose the materials and components for their prototype.

In parallel, in the Marketing and Communication module, the students tried first to understand the market and the customers' requirements. Next, they determined the features matching these requirements. With this knowledge, they were able to create a customer-orientated marketing strategy and develop an integrated marketing program. Finally, they clearly identified their product – a remotely controlled bioinspired flying robot that mimics the movement of an ornithopter – and the target market – Australia since it displays a sustained toy sales growth in the last years (Dunn *et al.*, 2014). Furthermore, the flying bird should have replaceable wings and a camera to further engage children.

At the end of these two stages, the team decided to model the robot after a white dove (Figure 4).

Figure 4. Planned wing beat of the ornithopter
(Body, 2009)

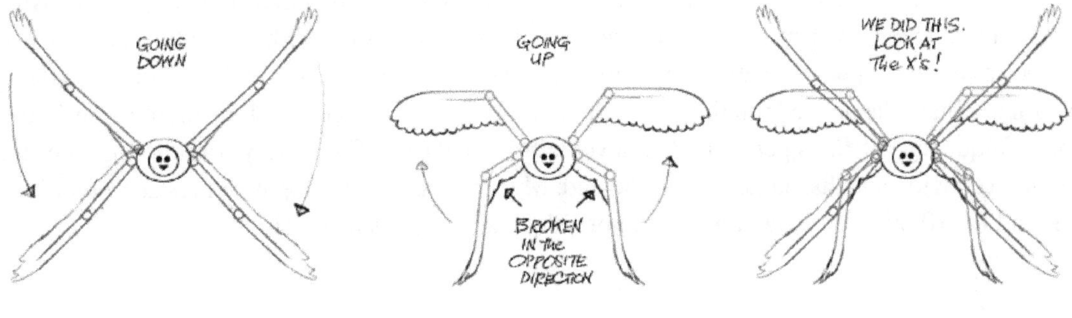

In the Energy and Sustainable Development module, the team addressed the set of eco-efficiency measures for sustainability considered during the project development and latter industrialization and commercialization. The team concluded that sustainability adds value to toys and influences buyer attitudes and purchasing behavior. According to the results of the consumer and trade study "Toys go green – Sustainability in Toys" (Spielwarenmesse, 2014), not only quality and the game concept convince the consumers to buy a toy, but it is also the sustainability which counts to their three most important purchasing criteria for toys.

During the state-of-the-art study, the team identified several different solutions to make the ornithopter fly and decided to implement a biplane. The reasons behind this decision are presented and justified in Dunn *et al.* (2014). After some conceptual studies, the complete ornithopter (including the wings, mechanical structure, and detailed actuation system) was designed in SolidWorks™, as depicted in Figure 5 (left), and the lift and drag forces involved in flight were determined (Dunn *et al.*, 2014). These studies lead to the design of a 30 cm long and 7 cm wide body (at the thickest points), a wingspan of 50 cm (when the wings are positioned horizontal) and a weight around 100 g (due to the materials chosen). MyBird flaps the wings at a frequency of 11 flaps per second and has an autonomy of 10 minutes (Dunn *et al.*, 2014; Caramin *et al.*, 2015).

Figure 5. Flying robot design (left) and prototype (right)
(Dunn et al., 2014)

The body was built in polystyrene and the wings in Mylar (as depicted in Figure 5). Polystyrene is a lightweight, shock absorbing material which can be easily shaped without any specific tools or skills. Mylar is the most common material for ornithopter wings due to its low density and mechanical properties. Leading edges and connection rods were made from carbon fiber rods. The flight is controlled by an Arduino Pro Mini together with a brushed inrunner motor and a motor driver, to amplify the Arduino output current. The chosen motor already had a 10:1 gear reduction. With this ratio, the output shaft would be spinning at approximately 2000 rpm at 4 V. However, the team's design required further gear reduction to obtain the desired rotating speed. The final reduction stage of 3:1 ensured the desired wing flapping frequency (11 flaps per second) at a velocity of 660 rpm. The battery was chosen according to its weight, capacity, and discharge rate. In the end, MyBird was able to flap its wings at around 11 flaps per second for 10 minutes. However, it was not able to achieve sustained flight.

A detailed description of this project can be found in the team's final report (Dunn *et al.*, 2014) and accompanying paper (Caramin *et al.*, 2015).

GraphBot – A Color Drawing Robot

A team of five undergraduates chose this proposal. It was composed of an Information Technology student from Poland, an Electrical, Electronic and Energy Engineering student from Scotland, a student of Power Systems Engineering from Finland, a student of Mechatronics from Romania and a student of Industrial Design from Belgium.

According to the team, the development of this robot was an opportunity to learn more about robotics and programming. The team had limited experience with Arduino and had never embraced a project of this size. As future engineers, they considered this project interesting since it builds upon the multidisciplinary skills pre-acquired by the team members and develops new skills in robotics, which are in high market demand. The development of a drawing robot, considering project management, marketing sustainability and ethics, while working in a team composed of people with distinct cultural and study backgrounds was a challenge and an advantage. The team had the chance to learn from each other and, by sharing knowledge, achieve a common goal: bringing this project to a good end.

After conducting a marketing study on drawing robots, the team decided to build a drawing robot that targeted children between the ages of 5 and 12. It was named Graphic Robot (GraphBot). The robot should be able to draw in assorted colors and move in different directions, the area of drawing being defined by the paper size. The robot is intended to introduce children, in particular girls, to technology from an early age, motivating girls to study and embrace later a career in Science Technology Engineering and Mathematics (STEM). Since girls are traditionally directed towards other fields than STEM, few tend to choose an engineering related field for further education (Royal Academy of Engineering, 2016). The team goal was to change this perception and familiarize girls with technology and engineering at an early age. According to the team's view, the robot should support creativeness (create art), and be versatile by allowing the reconfiguration and upload of new code to the GraphBot.

The specified robot requirements were the following: (*i*) move on a plane; (*ii*) work with assorted colors; (*iii*) be aesthetically pleasing; (*iv*) allow changing the drawing paper; (*v*) maximum dimensions of 1.0 m × 1.0 m × 0.8 m; (*vi*) reuse provided materials; (*vii*) use low-cost hardware solutions and open-source software; (*viii*) comply with the applicable EU Directives; and (*ix*) a maximum budget of 150 €.

After performing a literature review, the team compared the candidate parts, including cost, to decide the type of structure and user interface to be used with the GraphBot. The decision was to adopt a wheeled robot approach, since this solution allows the device to move and draw over a large surface and the structure of the robot is easy to build with low-cost components. The robot should be able to hold a pen and move it over a surface to draw freely or follow instructions previously stored or provided in real-time by a person. The team decided to implement a user voice interface. This way, the robot is controlled in real-time through voice or sound commands, which can be used by people of all ages and be fun.

Based on the performed marketing, sustainability and ethical analyses and multiple discussions, the team specified that, in addition to the initial requirements, the GraphBot should: (*i*) recognize simple voice commands and respond accordingly in a timely manner; (*ii*) be user-friendly, easy to use and robust; (*iii*) detect the drawing area automatically; (*iv*) change the drawing color easily; and (*v*) provide access to the micro-controller interface, allowing the refinement of the code for more advanced users. With this set of requirements clearly defined, the team focused on addressing the following open issues: how to change the drawing color; which components to choose; how to assemble the electronic and structural components; and how to control the system.

Figure 6 (left) presents the preliminary sketch of the GraphBot, a two-wheeled robot equipped with four assorted color pens (Dziomdziora *et al.*, 2016a), with the casing 3D printed in recyclable plastic. Figure 6 (right) displays the placement of the electronic components inside the robot. Two direct current (DC) motors implement a differential wheel motion system and are connected to the micro-controller through the L293D motor driver. A caster wheel is placed underneath the rear of the robot for extra support. The stepper motor, which controls the color pen change, rotates the penholder. The GraphBot detects the edge of the paper using an infrared (IR) sensor placed on the bottom front. The microphone, speaker and complementary components are assembled on a dedicated printed circuit board (PCB) attached to the upper part of the casing. The microphone is used to receive the user commands and the speaker provides feedback, *i.e.*, informs whether the voice command was recognized. The microphone and the speaker are connected to an EasyVR shield. A 7.2 V lithium-polymer ion battery, placed in the center of the robot, powers all electronic components. The battery is accessible for charging through the bottom. A DC jack is used to connect the Arduino to the battery.

Figure 6. GraphBot design (left), internal placement of components (center) and prototype (right)
(Dziomdziora et al., 2016a)

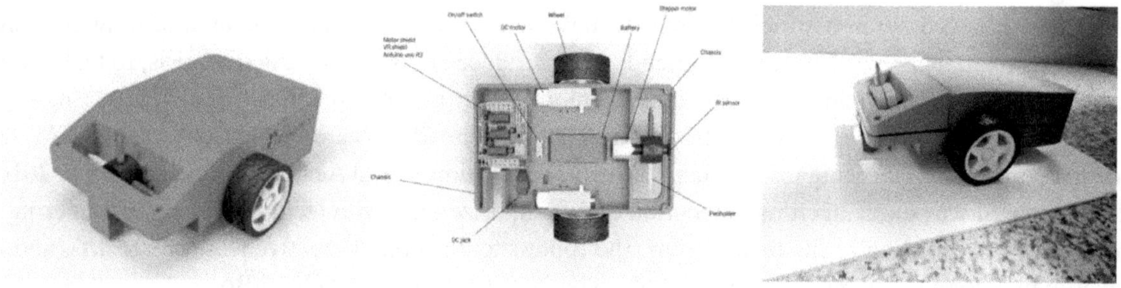

After assembling all parts and updating the code, the GraphBot was ready for the final tests (Figure 6, right). The final tests identified some problems: (*i*) the wheels touched the chassis, requiring a higher torque and drawing imperfect straight lines; (*ii*) the microphone of the EasyVR proved to have low sensitivity, ignoring voice commands at times; (*iii*) the markers occasionally got stuck or did not touch the paper; and (*iv*) the robot failed to detect the end of the drawing area when moving backward. The team solved the first issue by rearranging the positioning of the wheels in relation to the chassis. Regarding the second issue, the team suggests it can be solved with a better microphone, and, regarding the last issue, the team believes it can be addressed by positioning a second IR sensor at the back.

The detailed description of this project can be found in the final report (Dziomdziora *et al.*, 2016a) and accompanying paper (Dziomdziora *et al.*, 2016b) produced by the team.

DISCUSSION

EPS offers students the experience of being integrated in a multicultural and multidisciplinary team, *i.e.*, with people of diverse engineering fields and nationalities. This exposure to the multicultural perspective develops cross cultural communication skills essential to the global market. Furthermore, it involves students in all the phases of the engineering process, ranging from the initial open problem statement to the prototype. Project proposals refer to open multidisciplinary real-world problems. Its purpose is to expose students to problems of a greater dimension and complexity than those faced throughout the degree program as well as to put them in contact with the so-called real world, in opposition to the academic world. The teams are free to find their own solution to the problem, taking into consideration not only technological and financial constraints, but also the sustainability, ethical, deontological, and marketing perspectives. This autonomy directs students to independent thinking and comprehension of the engineering process, essential to fostering future entrepreneurship.

As the EPS participants are undergraduate students, their adaptation to an unfamiliar environment of total independence, at their age, may provoke social retraction, leading to isolation, or in reverse, an exaggeration of "social life". This has only been seen in a small number of cases, that have been solved with the help of the other students and/or the student support office, both providing discrete and efficient assistance on these cases.

We claim that the development of robots, whether using wheels, legs, wings, or fins for displacement, as capstone projects, help EPS students to achieve the above-stated goals more effectively at the capstone project/internship level. Since the number of bioinspired robots found in the literature is low, there are no definite solutions available. Students, constrained by the project budget and duration, need to think out of the box, fostering their creativity. First, they define the robot's purpose or application, and then devise how to transform it into an attractive product and which marketing strategies to adopt. The same applies to ethical and deontological concerns. The fact that robots are new machines raises several questions, especially concerning their use (one of the most cited questions is their use for military purposes), on which the students must think and reflect. Finally, to be able to assemble and test the prototype they need to choose and purchase components in an unknown country, sometimes facing delivery delays and sold-out materials (promoting their autonomy and decision-making abilities).

The building of small robots, using commonly available modules, places the students with a variety of issues (mechanical, electrical, chemical, programming, etc.) that are representative of the issues faced when developing a "real product". When compared with a pure simulation-based study, the development of a physical prototype is more rewarding, providing physical tactile feedback to the students. This physical interaction is many times completed with the integration of the robot in an Internet of Things framework, to solve more complex problems.

The diversity of project themes requires a multidisciplinary, dynamic, and motivated team of supervisors. EPS coaching may need a period of adaptation since the students define the agenda of the weekly meetings and the professors ensue, i.e., the role of teachers changes from leader to consultant.

In terms of the personal and learning outcomes of this process, the students gain autonomy, self-confidence and develop technical, scientific, and complementary skills with this multidisciplinary, multicultural teamwork experience – the design and development of a prototype.

As a partial conclusion of their learning process, the team involved in the development of the biologically inspired flying robot project stated in their final report:

In the first few months of the project, we learned a lot about teamwork. All of us have worked in teams before, but it is more difficult in a multidisciplinary team. In short: communication management is the key to success, and this is not only the case in your own team but also out with the team. If the communication management is not going well, the time management fails too because meeting the deadlines depends on everybody (Dunn et al., 2014).

In the words of the students involved in the development of Bubbles:

The team encountered many obstacles and, without teamwork and the support provided, it would not have been possible to finalize Bubbles. It was vitally important to utilize planning tools to remain up to date with all deadlines and to ensure that the budget was not exceeded. The EPS@ISEP allowed the team to grow stronger and it was important to learn and move on from any mistakes. It also enabled the experience for any future projects. Every team member has gained a much deeper insight into the engineering world, including those without previous engineering experience. It was a thoroughly enjoyable experience for the team as not only did people get to work and support people from across the world, bonds and friendships were also formed. Although the creation of Bubbles was a success, there is always room for improvement and the team could have potentially tackled certain situations differently. Overall, the project was a success, and the creation of bubbles was one that was enjoyable, challenging and life-changing (Reinhardt et al., 2016a).

Finally, the GraphBot team members shared that:

The European Project Semester at ISEP was a great life experience. I very much enjoyed working on and developing the GraphBot with my teammates from all over Europe. While a multidisciplinary team member, I was able to learn from other members when working on the parts of the project in which I was less experienced.

The European Project Semester has been very good practice on engineering and project work in general. Working as a group on a problem for a full semester is an effective way to learn important skills that are needed in the future. I have done a lot of different group work before, but never anything like what the EPS has to offer. Working on a practical project in a multinational group is a great experience for anyone.

Participating in an EPS is great preparation to start your own company and shock the world with a breathtaking innovative solution. We, with our different skills and origins, were able to convert our ideas into a working prototype. We learned the importance of cooperation, i.e., that acting alone does not achieve the same success as a team! Trusting a team member does not only allow you to work properly, but creates friendships, which may even last after EPS is ended. Facing problems and difficulties during project development was an opportunity to learn to find solutions for unexpected situations. Being an engineer is being able to design a solution for a problem, and this is exactly what EPS teaches you.

EPS is a very good way to learn how to work in a team. It teaches people with quite different views on how to work together as a team, namely, how to overcome differences. There is support from the teachers, but problems must be sorted out by the team alone. It is hard to anticipate the knowledge and capabilities of the other team members, which often leads to misunderstandings. EPS taught me a lot about teamwork, different cultures, and study fields.

The EPS Program has been a good experience. I had never done anything like this before. To work with different people from different countries and cultures is wonderful since I met eager students who enjoyed this project with me. I am glad I am a part of the best team I have ever worked with. Together we have made some good memories. I visited beautiful places in Portugal with wonderful views and I recommend everyone to come here and join this program. I learned how cooperation is essential and how important it is to work as a team. Five people from different countries in one team can realize extraordinary things (Dziomdziora et al., 2016b).

CONCLUSION

EPS prepares undergraduate students for their professional life. The diversified engineering and cultural backgrounds of the team members, which are difficult to achieve in a traditional engineering degree, and the freedom to direct the project promotes creativity. The initial specification, procurement, and marketing phases expose the students to engineering tasks usually absent from project-based learning, contributing to the development of standard entrepreneurship-related tasks. As stated by Dunn *et al.*, (2014), the project suffered setbacks caused by the team itself, namely late purchasing of components and delayed decision-making and planning. However, the students claimed they learned a lot from the project and the teamwork experience. In this context, robotics provides a challenging, motivational, multidisciplinary project backdrop to foster ethics and sustainability-driven conceptualization and design.

As a result, we claim that the design and development of robots within EPS promotes the development of highly appreciated professional skills, and, thus, we recommend the adoption of this combination of engineering capstone projects and learning framework for the satisfaction of the students and the benefit of the engineering profession.

ACKNOWLEDGMENT

This work was partially financed by National Funds through the FCT – Fundação para a Ciência e a Tecnologia (Portuguese Foundation for Science and Technology) as part of project UIDB/50014/2020. Moreover, the authors thank the board of direction, international relations office, departments, laboratories, colleagues, and technicians of ISEP; the lecturers, supervisors, consultants, and students of EPS@ISEP; and project sponsors IT sector – Sistemas de Informação and Autonomous Systems Laboratory.

REFERENCES

Body, B. (2009). *Tips and observations of bird flight.* https://www.brendanbody.co.uk/flight_tutorial/index.html

Caramin, B., Dunn, I., Ney, R., Klawikowski, Y., Duarte, A., Malheiro, B., Ribeiro, C., Ferreira, F., Silva, M., Ferreira, P., & Guedes, P. (2015). Design and Implementation of a Biologically Inspired Flying Robot – An EPS@ISEP 2014 Spring Edition Project. In *Proceedings of TEEM – 3rd Technological Ecosystems for Enhancing Multiculturality* (pp. 211-218). doi:10.1145/2808580.280861

Duarte, A. J., Malheiro, B., Arnó, E., Perat, I., Silva, M. F., Fuentes-Durá, P., Guedes, P., & Ferreira, P. (2020). Engineering Education for Sustainable Development: The European Project Semester Approach. *IEEE Transactions on Education*, *63*(2), 108–117. doi:10.1109/TE.2019.2926944

Dunn, I., Caramin, B., Klawikowski, Y., & Ney, R. (2014). *Bio-inspired Flying Robot.* https://drive.google.com/file/d/0B_4jquZNwPGmTjJUbVJpby1ob3M/edit?pli=1

Dziomdziora, A., Sin, D. N., Robertson, F., Mänysalo, M., & Pattiselano, N. (2016a). *Robotic Art – GraphBot.* http://www.eps2016-wiki1.dee.isep.ipp.pt/doku.php?id=report

Dziomdziora, A., Sin, D. N., Robertson, F., Mänysalo, M., Pattiselano, N., Duarte, A., Malheiro, B., Ribeiro, C., Ferreira, F., Silva, M. F., Ferreira, P., & Guedes, P. (2016b). Artistic Robot – An EPS@ISEP 2016 Project. In *Proceedings of ICL2016 – 19th International Conference on Collaborative Learning* (pp. 225-238). 10.1007/978-3-319-50337-0_20

Gopakumar, G. (2014). Public Leadership Framework: Studying Approaches to Diversify Engineering Education. *International Journal of Engineering Pedagogy*, *4*(1), 43–48. doi:10.3991/ijep.v4i1.3269

Ishii, A., Heibeck, M., Blazejewski, M., & Nybjörk, R. (2014). *Design and Implementation of a Biologically Inspired Swimming Robot – Final Report.* http://www.eps2014-wiki1.dee.isep.ipp.pt/lib/exe/fetch.php?media=final_report_team1.pdf

Ishii, A., Heibeck, M., Blazejewski, M., Nybjörk, R., Duarte, A., Malheiro, B., Ribeiro, C., Ferreira, F., Silva, M., Ferreira, P., & Guedes, P. (2015). Design and Implementation of a Biologically Inspired Swimming Robot – An EPS@ISEP 2014 Spring Edition Project. In *Proceedings of TEEM – 3rd Technological Ecosystems for Enhancing Multiculturality* (pp. 219-226). 10.1145/2808580.2808613

Laut, J., Bartolini, T., & Porfiri, M. (2015). Bioinspiring an Interest in STEM. *IEEE Transactions on Education*, *58*(1), 48–55. doi:10.1109/TE.2014.2324533

Lindsey, C. C. (1978). Form, Function, and Locomotory Habits in Fish. In W.S. Hoar & D.J. Randall (Eds.), Fish Physiology. Academic Press. doi:10.1016/S1546-5098(08)60163-6

Malheiro, B., Silva, M., Duarte, P., & Guedes, P. (2015). CDIO and the European Project Semester: a Match for Capstone Projects? In *Proceedings of the 11th International CDIO Conference* (pp. 1-14). 10.13140/RG.2.1.3621.9042

Reinhardt, A., Chousa Esteban, A., Urbanska, J., McPhee, M., & Greene, T. (2016a). *Didactic Robotic Fish Kit – Final Report.* http://www.eps2016-wiki3.dee.isep.ipp.pt/lib/exe/fetch.php?media=final_report.pdf

Reinhardt, A., Chousa Esteban, A., Urbanska, J., McPhee, M., Greene, T., Duarte, A., Malheiro, B., Ribeiro, C., Ferreira, F., Silva, M. F., Ferreira, P., & Guedes, P. (2016b). Didactic Robotic Fish – An EPS@ISEP 2016 Project. In *Proceedings of ICL2016 – 19th International Conference on Collaborative Learning* (pp. 239-253). 10.1007/978-3-319-50337-0_21

Royal Academy of Engineering. (2016). *Thinking like an engineer – Implications for the education system*. https://www.raeng.org.uk/publications/reports/thinking-like-an-engineer-implications-full-report

Sfakiotakis, M., Lane, D. M., & Davies, J. B. C. (1999). Review of fish swimming modes for aquatic locomotion. *IEEE Journal of Oceanic Engineering*, *24*(2), 237–252. doi:10.1109/48.757275

Silva, J. A. M., Silva, M. F., & Barbosa, R. S. (2014). Design and implementation of a biological inspired swimming robot. *Advances in Robotics Research International Journal*, *1*(3), 239–250. doi:10.12989/arr.2015.1.3.239

Spielwarenmesse. (2014). *Toys go green*. https://www.spielwarenmesse.de/uploads/media/Toys_go_green_Wissensbroschuere_de.pdf

KEY TERMS AND DEFINITIONS

Collaborative Learning: A learning method where students work together on a problem, task, or product.

Engineering Education: An area of higher education dedicated to the training of engineering professionals.

European Project Semester: A one-semester learning framework based on project-based, multicultural, and multidisciplinary teamwork as well as ethics and sustainability driven practices.

Multicultural and Multidisciplinary Teamwork: A work method where teams, composed of members from distinct cultural and knowledge backgrounds, address a problem, task or product spanning multiple disciplines, usually too complex for a single individual.

Project-Based Learning: A learning method where students work together on a challenging problem, task, or product for several weeks to gain new knowledge and skills.

Robotics: A multidisciplinary field, involving biology as well as computing, electrical and mechanical engineering, dedicated to the design and implementation of autonomous or supervised aerial, terrestrial, and aquatic platforms.

Student-Centered Learning: A method of teaching where students are responsible for their own knowledge and skill acquisition, fostering lifelong learning and independent problem-solving.

Section 3
Case Studies

Chapter 12
Project-Based Learning With EPS@UPB:
Development of a Robotic Arm With IoT Capabilities

Anca Greculescu
University Politehnica of Bucharest, Romania

Gina Florica Stoica
University Politehnica of Bucharest, Romania

Andrei Opris
University Politehnica of Bucharest, Romania

ABSTRACT

The implementation of the European Project Semester at the University Politehnica of Bucharest has outlined the importance of developing soft skills through project-based learning for 21st century engineering learners. This chapter describes the project conducted by a team of four multinational students in 2018. It aimed to develop soft skills (language, written and oral communication, organization and time-management, research and inquiry, self-assessment and reflection, group participation, and leadership) as well as to foster the necessary professional tools for a smooth insertion into the labor market. The development of a 3D-printed robotic arm with IoT capabilities has proved a useful didactic approach in this sense and, at the same time, an efficient way to meet the future economic and technological demands of our changing world. The main goal was to design an affordable robot arm so that people could buy for any application desired.

INTRODUCTION

The European Project Semester (EPS) at University POLITEHNICA of Bucharest (UPB) was first launched and implemented within the Faculty of Mechanical Engineering and Mechatronics, the De-

DOI: 10.4018/978-1-6684-2309-7.ch012

Copyright © 2022, IGI Global. Copying or distributing in print or electronic forms without written permission of IGI Global is prohibited.

partment of Industrial Design in 2014. Since then, it has welcomed and integrated international students from diverse scientific backgrounds in EPS teams, dedicated to finding solutions to real-world problems.

The goal of this chapter is to describe the development at UPB by an EPS team of a Robotic arm with Internet of Things (IoT) capabilities for a hospital in Bucharest. The arm is expected to have four degrees of freedom, operate on its own, communicate with other devices, and be 3D-printed. Working on such a project, the students had to adapt to different learning styles and cultural backgrounds and join their efforts to achieve the intended outcomes of the project.

The remaining of this manuscript is organized in four sections. They are dedicated to the EPS; the project-related studies leading to the design, printing and assembly of the robotic arm; the implementation of the control, actuation, sensing and communication modules; and the conclusion.

BACKGROUND

It is worth mentioning that promotion of international employability among engineering graduates is one of the key objectives of European Higher Education. Moreover, there is a growing need to prevent youth unemployment and foster educational methods likely to facilitate their transition from education to work. Likewise, Technical Higher Education needs to link engineering education to the needs of industry by means of a learning-by-doing approach. Hence, the European Project Semester (EPS) proves an innovative training program that addresses the new professional demands for the future engineers. The program has been designed in compliance with the new learning outcomes established by the Bologna Process and the European Higher Education Area.

EPS is a program offered by a consortium of European universities in 12 countries throughout Europe (Romania, Spain, United Kingdom, the Netherlands, France, Germany, Austria, Poland, Norway, Belgium, Portugal, Finland) to students who have completed at least two years of study. EPS is crafted to address the design requirements of the degree and prepare engineering students with all the necessary skills to face the challenges of today's world economic and technological advancements. Furthermore, EPS is a mixture of Project Related Courses and Project organized/problem-based Learning.

Students work in international and usually interdisciplinary teams of three to six students and most of the projects are also cross-disciplinary. Some projects are done in cooperation with business partners and industries, and others more academic. Hence, the main objective is to train students from different countries and different disciplines to work together in multi-cultural and multi-disciplinary groups. The students work together to execute an integrated engineering-design-and-business project, focusing on: the development of personal competences, especially the ability to work and communicate within cross-cultural groups and the interrelated work of several educational disciplines like mechanical & electrical engineering, design engineering, information technology, business & management etc.

The main competences modern engineering must develop are: teamwork, interpersonal skills, the ability to work in an international team with students of different disciplines and nationalities. It is widely acceptable that an engineer today must be able to cope with a broad scope of disciplines such as: economics, management, communication, languages and a solid training in interdisciplinary and international teams (Andersen, 2004).

The University POLITEHNICA of Bucharest (UPB) became an EPS provider in 2013, at St Polten -Austria, when it was admitted in the consortium. UPB has agreed to implement the EPS program in compliance with the "10 Golden Rules" established by the consortium (European Project Semester, 2021).

The European Project Semester was first launched and implemented within the Faculty of Mechanical Engineering and Mechatronics, the Department of Industrial Design in 2014. The EPS program at UPB is international, cross-disciplinary and multicultural; the working language is English; it addresses the real needs of companies/business partners; it is an intensive, one-semester program and it is worth 30 European Credit Transfer and Accumulation System (ECTS) credits; it consists of a project and supporting courses: Professional & Intercultural Communication/ Introduction to Romanian Language & Culture (2 ECTS); Project Management (2 ECTS); Technical English (2 ECTS); Eco-Design (2 ECTS) and Product Design and Development (2 ECTS). The technical courses may differ each year in order to support the respective projects.

A project supervisor is also assigned from the company and the team keeps the company supervisor updated on the project progress by handing in a weekly progress report. In addition, throughout the semester, students periodically fix up tutorials on Project Management and Professional Communication that will assist them in refining their project delivery and oral assessments. Project-based learning (PBL) represents an instructional process that reflects a learner-centered environment and that concentrates on students' use of disciplinary concepts, tools, experiences and technologies to answer questions and solve real-world problems (Krajcik & Blumenfeld, 2006; Markham, Larmer, & Ravitz, 2003).

Likewise, project-based learning is defined as "a systematic teaching method that engages students in learning knowledge and skills through an extended inquiry process structured around complex, authentic questions and carefully designed projects and tasks" (Markham et al., 2003, p. 4).

PBL is synonymous with learning in depth. A well-designed project helps students to encounter (and struggle with) the central concepts and principles of a discipline. PBL teaches students twenty-first century skills and content. These skills include communication and presentation, organization and time management, research and inquiry, self-assessment and reflection, and group participation and leadership.

PBL is generally carried out by groups of students working together towards a common goal. Performance is assessed on an individual basis, and considers the quality of the product produced, the depth of content understanding demonstrated, and the contributions made to the project. Finally, PBL allows students to reflect upon their own ideas and opinions, exercise voice and choice, and make decisions that affect project outcomes and the learning process in general.

Nevertheless, it deems fit to address the challenges of assessing student work with a PBL model. The research in this area also described the use of peer assessments and new assessment strategies (Frank & Barzilai, 2004); rubric creation in conjunction with student feedback; individual grades vs. group grades (Thomas & Mergendoller, 2000); and the danger of merely assessing superficial project work vs. deeper content understanding (Marx et al., 1997).

ROBOTIC ARM WITH IOT CAPABILITIES

The current project was devised as a case study likely to meet the principles of Project-based learning and it was assigned to an international team of students from the Netherlands, Norway, France and Romania. The project is a design study into printing off a robotic arm by means of 3D printing with actuation and Internet of Things (IoT) capabilities. The students were supervised by a tutor and most of their work was done in the faculty laboratory. The research supporting the project was carried out at ELIAS Hospital in Bucharest. Figure 1 and Figure 2 display the fully assembled robot, equipped with a supporting base and actuators to facilitate grabbing objects from the ground.

The robotic arm may be used for both industrial and medical purposes. The aim was to design and manufacture such a robotic arm that would be more cost-effective, smaller and simpler as compared to other robotic arms on the market. It will nevertheless be equipped with structural adjustments (4 degrees of freedom – the three basic ones plus one extra gripper); IoT capabilities; and be fully 3D-printed.

Figure 1. Fully assembled robot arm

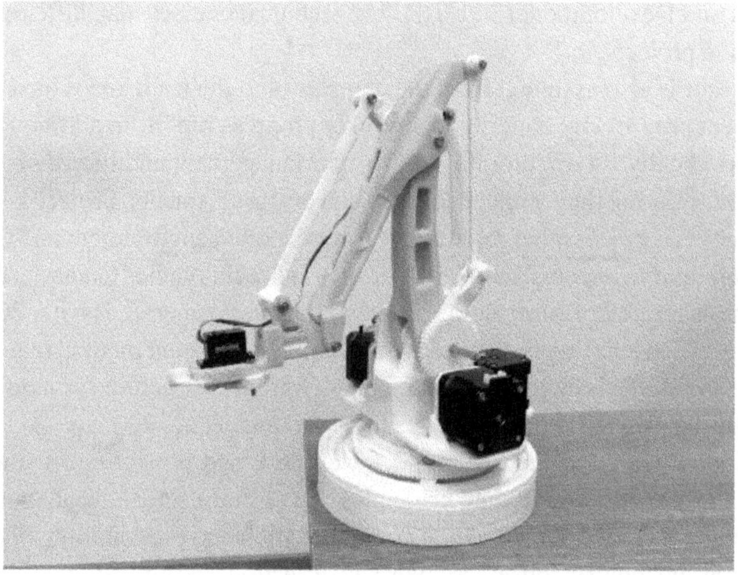

Figure 2. Hardware enabling IoT capabilities

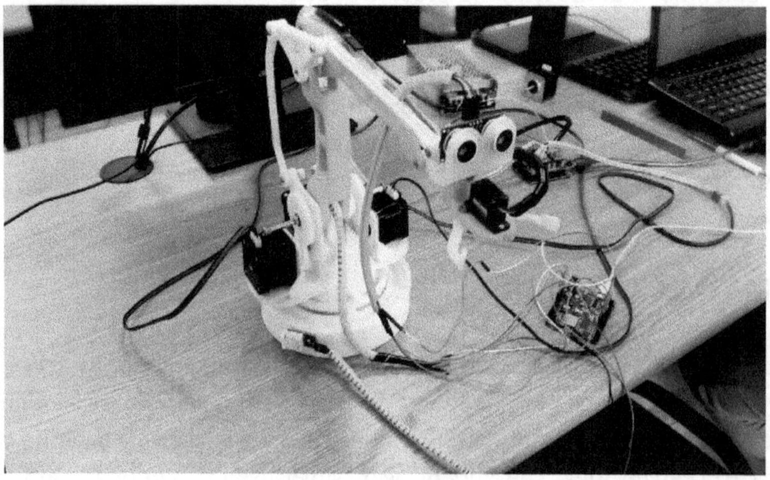

The initial phase of the project was conducted by a research of the market in order to find the most suitable Robotic Arm to be used and improved (Figure 3, Figure 4, Figure 5 and Figure 6).

The students analyzed different models, considering the Degrees of Freedom, components required and the manufacturing process. The starting point model is the second option in Table 1 (Figure 4).

Table 1. Analyzing Robotic Arm Models

Model	Link	Info	Actuators
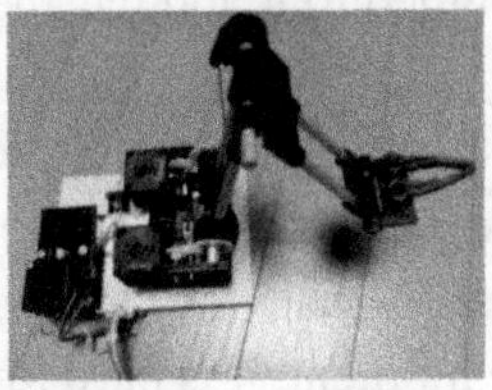	https://www.thingiverse.com/thing:541997	3 DoF Gripper 3D-printed + Aluminum bars	5 motors Model Mabuchi RE 140
Figure 3. Robot Arm RA80			
	https://www.thingiverse.com/thing:2520572	3 DoF Gripper Fully 3D-printed	3 stepper motors 1 servo motor (gripper)
Figure 4. Robot Arm MK2+			
	https://www.thingiverse.com/thing:1015238	3 DoF Gripper Plastic components	3 servo motors 1 servo motor (gripper)
Figure 5. EEZY Bot Arm			
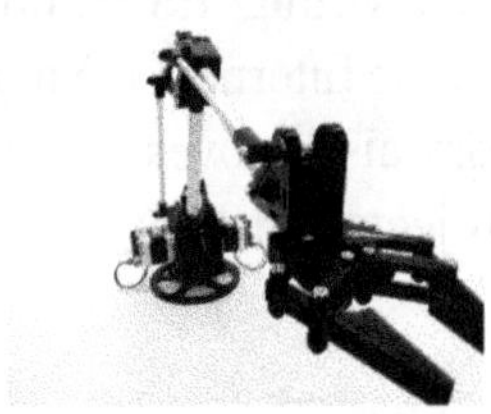	http://www.instructables.com/id/Robot-Arm-UStepper	4 DoF Gripper 5D-printed + Aluminum bars	3 stepper motors 1 servo motor (gripper)
Figure 6. Robot Arm UStepper			

The students identified three target groups for their product: technical high schools and universities –such a 3D-printed robotic arm would be an excellent device for hands-on teaching and learning robotics and mechatronics. It may serve as a showcase model for engineering students, so that they in turn would make their own version of a robot or modify the existing model. Another possible customer would be the retail market if only the production of the robot arm be cheap and affordable (100 € – 200 €).

For example, the owner of a goldfish, too busy or unmindful of feeding it, may set the robot arm to grab a can of food and feed the fish on a pre-set time basis. As prospective customers, factories, big and small, may also need robotic arms to operate fully or semi-automated assembly lines.

IoT or Internet of Things is a new concept about connecting physical objects to the internet and offering them the capability of communicating or exchanging data remotely. It is a relatively new idea and a lot of companies are investing in IoT development because of its huge potential.

The devices that can be connected to IoT are from all domains of activity – cars, watches, smartphones, sensors, actuators, microcontrollers, medical equipment etc. People often refer to these objects as "things". Basically, a "thing" is a physical device which has an IP address assigned to it and has the ability to communicate with other "things" over the internet infrastructure.

Consumers seem to be very satisfied with their devices connected to IoT platforms and the unique features that they benefit of. Although IoT is becoming more and more popular and accessible to people, it is not the safest environment. Cyber-attacks and cyber-criminality (or generically – hacking) have a crucial impact on IoT because of the steady growth in the number of users. At the moment, it is hard for this industry to offer a secure and safe place since everything is developing so fast.

It is worth mentioning that IoT consists of three principles called "the three C": Control, Connection and Cost saving. Control means that the devices work on their own and are isolated. Connected refers to a device that can communicate with other different ones and Cooperation-Coordination or cost saving stands for analysis of data to reveal new business, engineering or scientific insights (Haboud, 2016).

A research was carried out in order to find the most suitable IoT platform with the best attributes both for the user and the developers of the current project. Platforms such as Thinger.io, Blynk.cc, MyDevice-Cayenne, Thingworx, ThingsBoard etc. are accessible for free for an unlimited or a limited period of time and these platforms are basically cloud platforms ready to store, exchange and expound data. Likewise, each platform offers the possibility of creating an application or a web-page with a graphical interface where the collected information can be presented with the help of different tools.

The team found that the convenient solution was Blynk due to its capability of developing an Android or IOS application to show the results directly to phone using their application. Basically, Blynk app offers a unique chance for the developers to create an application for phones inside their main application. Blynk recommends itself as "The most popular mobile app for the IoT. Works with anything: ESP8266, Arduino, Raspberry Pi, SparkFun and others. Blynk will get you online and ready for the Internet Of Your Things." according to their official website. (Blynk, n.d.) Nevertheless, Thingworx also proves a very interesting IoT platform that offers solutions for industrial equipment and small projects and Thingworx was a very good source of documentation for this project.

Design and Assembly of the Robotic Arm

The robot arm had to be made of pieces no longer than about 200 mm (3D printer not able to handle bigger parts); of mostly printable parts (except for motors, screws etc.); closed chain link and stepper motors.

The base is made up of five parts: two bearings holders; main base; base-low and a base-low gear.

The assembly process implied that a nut be inserted into a hole in the main base. Then two screws allowed to fix the base-low with the main base. After that, one bearing holder was placed under the base-low and the other bearing holder on the base-low gear and the ten bearing balls between the two bearings holders alike. Finally, the base-low and the base-low gear together with all the parts between them were fixed with a socket screw. When the assembly was finished, the base did not prove stable because of the bolt sticking outside at the bottom. The team decided to design a stand to support the base by means of CATIA V5, as shown in Figure 7.

Figure 7. CATIA base design

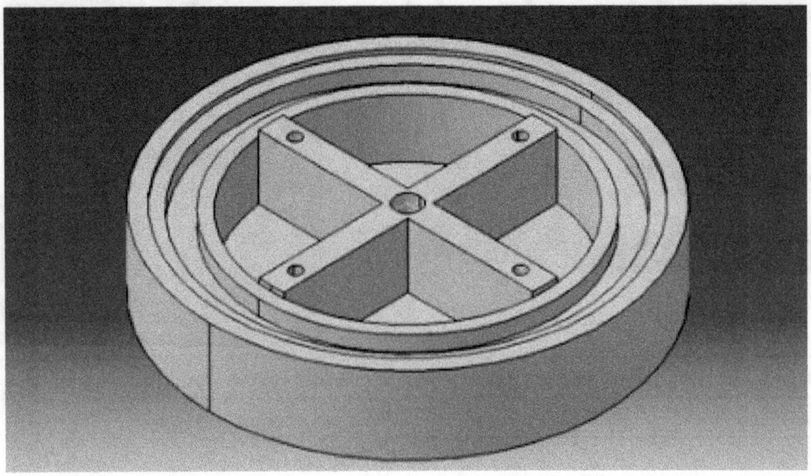

3D Printing Process

The team mainly used two printers (Figure 8 and Figure 9), as shown in Table 2. After all the parts were 3D-printed, it became obvious that some of the parts were not accurate enough. Small parts that were meant to fit together, simply did not fit without having to cut out plastic in certain places. This was solved by using another more accurate printer with a smaller nozzle. The parts that needed to be of higher quality were now printed again with a 0.4 mm nozzle instead of 0.6 mm nozzle.

This gave better results, but on the other hand, it took longer time to print. This printer used green plastic, which is why the robot consists of both white and green parts. The most difficult parts to print were the ten bearing balls for the bearing at the base of the robot. These balls have a diameter of 5 mm, and the spherical shape must be very accurate in order to serve as a moving part in a bearing system. After printing the bearing balls on both printers, the results were far from satisfying as the surface of the balls were not accurate enough. The team also had access to a third 3D printer that was using Acrylonitrile Butadiene Styrene (ABS) material. This printer gave a much better result, but the bearing balls still did not have the desired perfect spherical shape at the surface. By polishing each of the ten balls with sand-paper, the result became slightly better, and when assembling the bearing, the result proved satisfying.

Table 2. Printers used for the 3D-printed robotic arm

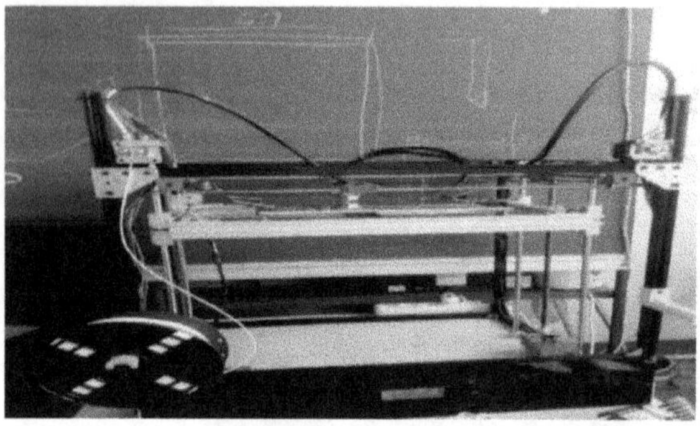	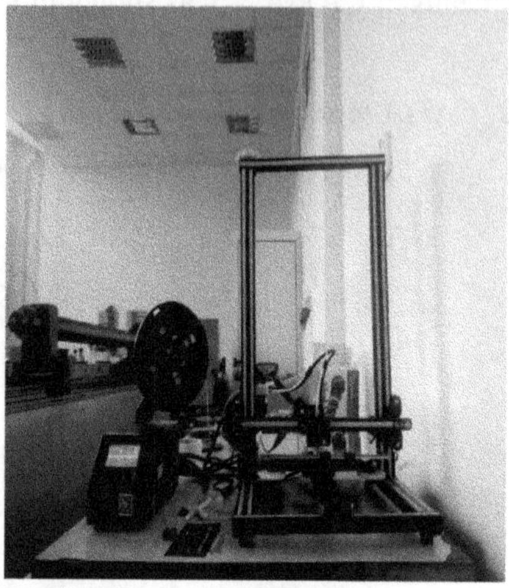
Nozzle Diameter: 0.6 mm	Nozzle Diameter: 0.4 mm
Filament Diameter: 1.75 mm	Filament Diameter: 1.75 mm
Bed Temperature: 75 °C	Bed Temperature: 70 °C
Extruder Temperature: 205 °C	Extruder Temperature: 200 °C
Material Type: Polylactic acid (PLA) Filament	Material Type: Polylactic Acid (PLA) Filament
Printing Surface: Glass	Printing Surface: Glass
Figure 8. Custom-Made Printer, 2018	*Figure 9. Creality CR-10, 2018*

ACTUATION, CONTROL AND INTERNET OF THINGS CAPABILITIES

The team started this process by creating a bill of materials with all the components required. The initial hardware components were: one Arduino UNO microcontroller (Mega2560 as an alternative), one Computer Numerical Control (CNC) Arduino Shield, one Wemos WiFi shield, one servomotor and three stepper motors, including drivers, one 12 V 15 A power supply and cables for different kind of connections.

The bill of materials was created by researching different possibilities of actuating a robot - the result also represented a good relation between the quality and the price of the chosen items. The next phase was to wire everything for the initial tests of actuation. A wiring scheme was used as shown in Figure 10.

The next step was the software process once the hardware parts were connected. There were two major software tools used – Arduino Integrated Development Environment (IDE) and Universal G-code sender, which led to the possibility of actuating and controlling the prototype. The libraries and the code, including standard and custom code sequences developed by the team, were flashed to the microcontroller by using Arduino IDE; the actuation and control part was made through Universal G-code sender which allowed the robot to be controlled from a computer keyboard. One of the main requirements of the project was to include IoT capabilities. This was achieved by adding three sensors on the robot's architecture – sensors for humidity and temperature, vibration and distance measurement.

The Internet of Things feature was fully enabled once the team developed a user-interface for an application displaying the values of the sensors.

Figure 10. Wiring Diagram

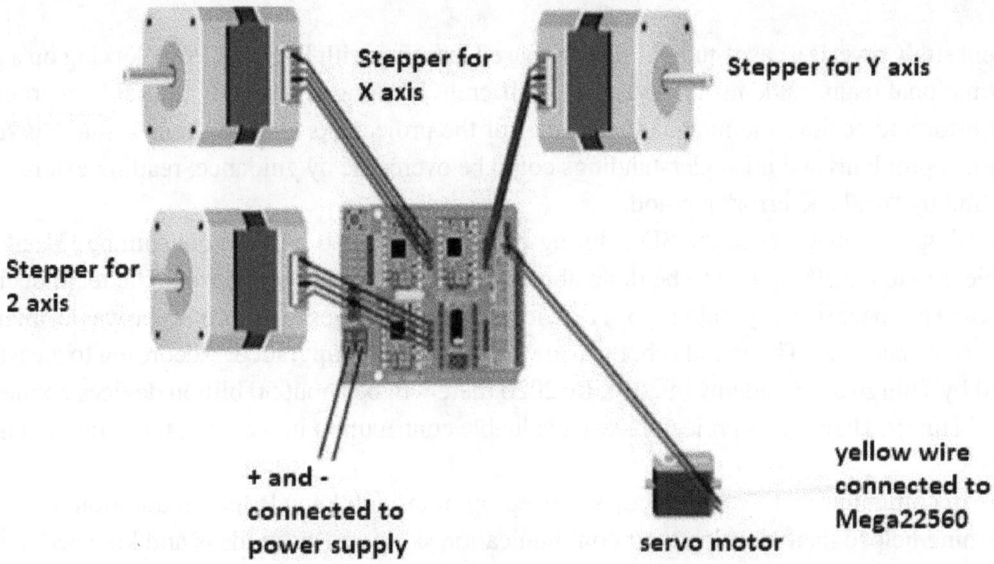

The interface was created using Blynk.io developer tool and it can be seen in Figure 11. The application consisted of four interactive gauge shapes displaying real-time data. The application could be used on both major smartphone operating systems (Android/iOS) - the sensors were sending data to the app through a wireless internet connection. In this phase of the project, the prototype is assembled and it can be controlled and actuated – moreover, the IoT capabilities were enabled by adding sensors and creating and internet-connected application to display data.

Figure 11. User Interface of the application

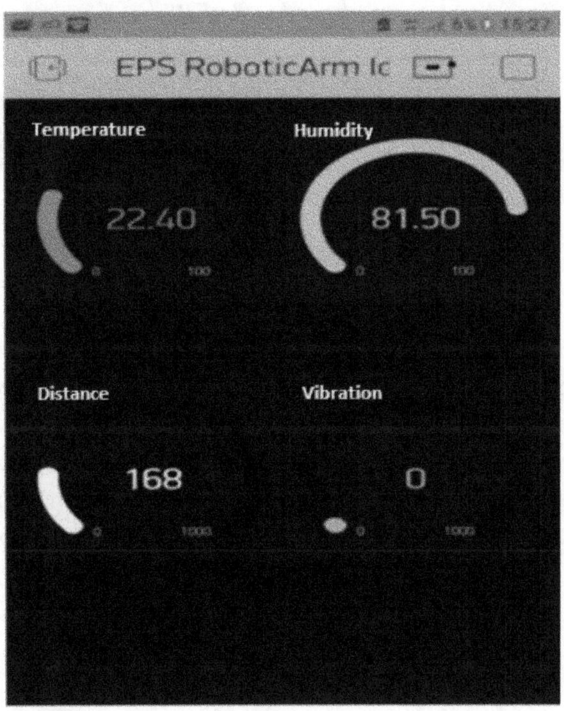

CONCLUSION

The present study reveals a case-study of project-based learning with EPS at UPB. Working on a project as a multinational team, students had to adapt to different learning styles and cultural background and join their efforts to achieve the intended outcomes of the project. As a consequence, students realized that technical problems and misunderstandings could be overcome by guidance, reading extensively on the topic and by "trial and error" method.

Concerning the robotic arm, the 3D printing process showed to be time-consuming. Moreover, it became clear that something had to be done about the stability of the robot arm. The team decided to design a stand for more stability and support. Providing IoT capabilities to the prototype was an interesting idea and a real challenge. The initial robotic arm was changed and upgraded. According to a case study conducted by Thingworx Academy in 2016, by 2020 there will be about 50 billion devices connected to Internet of Things. Hence, this project makes a valuable contribution in reaching the estimated number of IoT devices.

The team communicated well and they swapped opinions and knowledge. In addition, the project-based learning helped them develop their communication skills and swap ideas and knowledge.

The whole experience proved challenging and instructional at the same time. All in all, this EPS project-based learning experience showed that:

At the beginning, when I arrived here I had some apprehensions about this semester because I didn't know the country and I didn't know anyone. But now I can assure that it was one of the best experiences in my life, it permitted to me to meet new people from different country (Remi).

This EPS journey has taught me a lot of things. It has been a really nice journey as well, since I've been able to meet a lot of new people who are all from different origins, meaning profession and country. I have made sustainable friendships. The project itself was really educative, since I didn't have any experience with 3D printing or robotics, so that was nice. Now I know how to operate a 3D printer and how to make a robot work. The choice to go abroad to do EPS has been one of the best of my life so far, I'm very satisfied (Floris).

Working in a multicultural group was a great opportunity for me. This experience helped me a lot to improve some personal and technical skills. The communication was good and we managed to create an interesting project. It was very nice that we could share experiences and ideas and through this we helped each other in one way or another. All in all, the EPS experience was a good one and I'm glad that I had the opportunity to meet and work with foreign people (Andrei).

Since our world is getting smaller and smaller, international companies can hugely benefit from the experience gained by exchange students, from their skills acquired and developed through innovative hands-on learning activities such as the project-based approach. Hence, the development of soft skills through PBL, provided by an EPS program, lies at the very heart of integrated engineering. Such integrated engineering projects are learning and teaching methods by which new materials and competencies may be introduced, it is more than just an opportunity to practice what has been taught.

REFERENCES

Alluwar, N. (2017). *What is IoT architecture?* https://www.quora.com/What-is-IoT-architecture

Andersen, A. (2004). Preparing engineering students to work in a global environment to co-operate, to communicate and to compete. *European Journal of Engineering Education*, *29*(4), 549–558. doi:10.1080/03043790410001711243

Blynk. (n.d.). *How Blynk Works.* http://docs.blynk.cc/#getting-started

Chakravorty, D. (2017). *STL File Format for 3D printing – Simply Explained.* all3dp. https://all3dp.com/what-is-stl-file-format-extension-3d-printing/

European Project Semester. (2021). *Concept | European Project Semester.* http://europeanprojectsemester.eu/concept

Frank, M., & Barzilai, A. (2004). Integrating alternative assessment in a project-based learning course for preservice science and technology teachers. *Assessment & Evaluation in Higher Education*, *29*(1), 41–61. doi:10.1080/0260293042000160401

Haboud, D. (2016). *IoT and The Three C's.* https://resources.altium.com/pcb-design-blog/iot-and-the-three-c-s

Jazar, P. R. (2006). *Theory of Applied Robotics.* Springer.

Krajcik, J. S., & Blumenfeld, P. (2006). Project-based learning. In R. K. Sawyer (Ed.), The Cambridge Handbook of the Learning Sciences (pp. 317–333). Cambridge.

Markham, T., Larmer, J., & Ravitz, J. (2003). *Project based learning handbook: A guide to standards-focused project based learning* (2nd ed.). Buck Institute for Education.

Marx, R. W., Blumenfeld, P. C., Krajcik, J. S., & Soloway, E. (1997). Enacting project-based science: Challenges for practice and policy. *The Elementary School Journal*, *97*(4), 341–358. doi:10.1086/461870

Poole, H. H. (1989). Fundamentals of Robotics Engineering. Van Nostrand Reinhold. doi:10.1007/978-94-011-7050-5

Thomas, J. W., & Mergendoller, J. R. (2000). *Managing project based learning: Principles from the field.* Paper presented at the Annual Meetings of the American Educational Research Association, New Orleans, LA.

KEY TERMS AND DEFINITIONS

3D Printing: Manufacturing process which enables the creation of physical 3D solids of objects using a series of layered development framework, where layers are laid down in succession to create a complete 3D object by a 3D printer.

Degree of Freedom (DoF): The number of independent parameters, in a mechanical system, that can define its configuration – parameters that determines the state of a physical system.

Forward Kinematics: It is a branch of kinematics that studies and analyses where the exact position of the end-effector would be given by the certain angles provided to the robot actuators.

Internet of Things: Different electronic devices embedded in everyday objects, enabling them to send and receive data – these devices can communicate and interact with others over the Internet, and they can be remotely monitored and controlled.

Microcontroller: Compact electronic device made from an integrated circuit including a processor, memory, and input/output peripherals.

Chapter 13
Bicycle Speedometer Plus

Thomas Mechlinski
Osnabrück University of Applied Sciences, Germany

Jürgen Wübbelmann
Osnabrück University of Applied Sciences, Germany

Ralf Hagemann
Osnabrück University of Applied Sciences, Germany

Nikita Korovin
https://orcid.org/0000-0002-4670-6733
Moscow Polytechnic University, Russia

Amanda Abós Sanchez
Universitat Politècnica de Catalunya, Spain

Nichiket Goyal
Chitkara University, India

ABSTRACT

The European Project Semester program at the University of Applied Science Osnabrück deals mainly with engineering projects in different engineering domains. Developing a product based on complex technical requirements and economic concerns, such as cost-sensible development and marketing and environmentally-friendly design, are goals of a successful project during the one-semester internship at Osnabrück for a multicultural project team. In 2018, a team of four students received the task of developing a speedometer for bicycles with a brake light to enhance the rider's safety, especially in an urban environment. Besides the technical challenges, the team developed a marketing plan for the product. To the students, it was an excellent opportunity to understand the benefits and risks of working in a multicultural team and the necessity of following strict project management guidelines to finish a project as a team successfully.

DOI: 10.4018/978-1-6684-2309-7.ch013

Copyright © 2022, IGI Global. Copying or distributing in print or electronic forms without written permission of IGI Global is prohibited.

INTRODUCTION

The Osnabrück University of Applied Sciences (OUAS) offers the European Project Semester (EPS) since 2015. EPS projects address scientific and technical problems related to local laboratories and companies. In this context, technological projects like smart objects are attractive to engineering students, fostering autonomous research and learning.

The objective of this chapter is then twofold: (*i*) present a smart object developed by a team of EPS students at OUAS; and (*ii*) illustrate the type of EPS projects conducted at OUAS. In this case study, the team was challenged to develop a smart bicycle speedometer enhanced with a light, data storage and a Bluetooth communication interface.

This chapter has four additional sections dedicated to: project research and requirement analysis; market research of competitors; implementation of the smart speedometer; and conclusion.

PROJECT CHARTER AND REQUIREMENT ANALYSIS

The objective of the project was the development of a speedometer for a bicycle with some enhanced features such as a bike light, and the preparation of a Bluetooth connection with mobile devices so that navigational information could be displayed on the speedometer. The researchers (student team) were tasked with the challenge of integrating the results of a previous project into their conceptual design of the speedometer. The basic requirements given to the team at the beginning of the project are listed in Table 1.

Table 1. Specific requirements of the project

Feature	Explanation
General features	
Subject	To develop a speedometer for a bicycle with some enhanced features
Team	The members of the Speedometer Plus team
Timeframe	Four months
Location	Osnabrück University of Applied Sciences
Specific features	
Energy supply	A hub dynamo delivers the energy for the product. When it stops, an energy buffer must be able to provide energy to display data for 5 minutes without using any kind of battery.
Test bench	A test bench has to be developed, which allows the usage of a hub dynamo during the development phase of the speedometer in the lab.
Electrical characteristics	All required distinguishing characteristics of a hub dynamo, such as voltage over speed, voltage overload, etc., must be measured and documented.
Protection	The speedometer must be protected against overvoltage under every operating condition. The protection circuit shall be a standalone circuit and permanently fixed to the bicycle. As the dynamo provides dynamic voltage, the speedometer should work under every condition, and it should be protected against overloading.
Sensor	The accelerating voltage of the hub dynamo shall be used to determine the speed and acceleration of the bicycle by detecting the frequency. No additional sensors shall be used.
Configuration	The user must (at the least) be able to configure the wheel diameter, the number of pole pairs of the hub dynamo, and the unit of speed (mph or km/h) and distance (miles or km). The configuration of the odometer and trip counters is optional.

Continued on following page

Table 1. Continued

Feature	Explanation
Basic features	The speedometer must (at the least) be able to display the speed, the total distance (odometer), and one resettable trip recorder. Displaying average speed, trip time, and other information shall be foreseen
Plus features	
Brake lights	The "Bicycle Speedometer Plus" must (at the least) be able to control the rear light. In case of negative acceleration (as detected by the hub dynamo), the rear light shall flash. SAFETY: If the product is not mounted on the bike, the rear light must be supplied with power by the hub dynamo.
Connections to a smart device	The product shall provide the hardware for a Bluetooth LE interface to be able to connect the product with a smart device, e. g., to transfer navigational information (further enhancements).
Data retention	Configuration data (see Feature "Configuration") and operational data such as total distance (see Feature "Basic features") must be stored persistently. Configuration data and total distance must be protected against corruption and the user should be able to reset the data anytime they wish to.
Housing	The product must have a housing.
Mechanics	During operation, the product must be fixed on the handlebar of the bicycle. The housing must be removable. The protection class shall be at least IP65 (dust protected and water jet protected).
Electrical connections	When the housing is fixed to the handlebar, the Speedometer Plus must be able to control the rear light. Without a Speedometer Plus being connected, the front and rear lights must be connected to the hub dynamo. SAFETY: In case of a malfunction, the lights shall turn on.

A previous project team already developed a schematic for overvoltage protection and basic software based on an Arduino hardware. This software didn't fulfill the real-time requirements of the speed measurement.

Before a definition of the target group was determined, market segmentation was made of the group of people who would likely use the speedometer. This step was necessary in order to know the characteristics of the different groups and to then choose which of them could be final customers for the product.

Germany has a population of 83 million inhabitants, and people are using bikes more and more. Germans own 72 million bicycles (FAZIT Communication GmbH, 2018, April 1). Therefore, in purely statistical terms, almost every one of Germany's inhabitants has a bike.

As an example, the city of Münster (Westfalen) has twice as many bicycles as its inhabitants (City of Münster Presse- und Informationsamt, 2018). Over 100 000 people there travel by bike every day. The bike users were divided into three groups:

- **Competition / professional bikers**: Professional clothing and accessories for cycling. They usually spend a lot of money on this because they care about quality.
- **Urban bikers**: These ones are the people who use the bike for transport in the street. Maybe they do not care so much about the quality. Normally they prefer good-looking and cheap devices; and additionally, they do not typically need any professional accessories. However, they might need the speedometer in order to know the speed at which they are moving.
- **Recreational bikers**: People who only use the bike for fun or only on specific days. Usually, they are not going to need a speedometer or any other technical device.

Regarding these groups of users, it is necessary to comment that the different classes can be distinguished according to purchasing level. This is an important point – because depending on the purchasing level of each group, the criteria for the functions of the product can differ.

After considering these different target groups, it was decided to build a speedometer for urban bikers, who will thus be enabled to control their speed and get some additional information during their rides.

ENVIRONMENT

Bikes and bicycle sports are very popular in Germany. In many cities, separate roads and traffic lights for bikers are to be found. It is a developing means of transport. Nowadays, there are many different types of bicycles available, such as electric bikes, motorized bikes, and many other kinds.

The main benefits of cycling are:

- Enabling driving without a license.
- Enabling saving the cost of fuel.
- Being a good source of exercise.
- Doing no harm to the environment.
- Being affordable by most.
- Requiring only a little maintenance.

Cycling is an effective and environmentally friendly transport but riding a two-wheeler in the city is not as safe as using a car there. Germany and The Netherlands are pioneers in developing laws with an aim to protect cyclists and people on foot. For example, it is forbidden to use a phone while riding a bicycle except if the person is using the phone in hands-free mode.

MARKET RESEARCH

This is a sample report of the data and analysis from up-to-date market research conducted during this EPS Project.

The target market and specific companies analyzed in this report have not changed.

Case Study and Background Information

Based on competitive research, the new speedometer (device) should have some characteristics that are not available from current products on the market. The new idea of a "Bicycle Speedometer Plus" is that of a dynamo-powered speedometer with some enhanced features such as brake lights and the preparation of a Bluetooth connection with mobile devices, where navigation information can be displayed on the speedometer. The development is based on the results of a previous project and meets all current laws as well as all announced federal standards in Germany.

Methodology

Sources of information include:

- Public information.
- Company marketing information; advertising.
- Local news sources.
- Media sources and reviews.

Every effort was made to provide accurate information. The information provided in Table 2 was the "best available" data at development time.

Table 2. Market comparison

Material	Connectivity	Battery autonomy	Screen	Weight (g)	Works with	Price (€)
Garmin Edge 25 (Garmin, 2022b)						
Plastic (IPX7 resistant)	ANT+ Bluetooth	8 h	0.9'' Monochrome LCD display	25	GPS GLONASS	127
Garmin Edge 1030 (Garmin, 2022a)						
Plastic (IPX7 resistant)	ANT+ Bluetooth Wi-Fi	20 h	3.5'' Color Touchscreen	123	Android iOS GPS	499
Wahoo Elemnt Bolt (Wahoo Fitness, 2022)						
Plastic (IPX7 resistant)	ANT+ Bluetooth	17 h	2.2'' Monochrome LCD display	60	Android iOS GPS GLONASS	329
Lezyne Super GPS Enhanced (Smith, C., 2016)						
Weather resistant	ANT+ Bluetooth	24 h	2'' Monochrome LCD display	76	Android iOS GPS GLONASS	135
CatEye Velo 9 (Cateye, 2022)						
Plastic (water resistant)	–	3 year	2.3'' Monochrome LCD display	28	No app support	14
Beeline navi + speedometer (Beeline, 2022)						
Rubber (IP66 resistant)	Bluetooth	10 h	1.18'' Monochrome LCD display		Android iOS	90
Omata (Omata, 2022)						
Aluminum / plastic	ANT+ Bluetooth	17 h	B&W Analog display	79	iOS GPS	620
SmartHalo (Hemsing, R., 2018)						
Rubber / plastic (IP66 resistant)	Bluetooth	20 h	Monochrome LCD display	195	Android iOS GPS	170

Market Summary

There are no companies that definitively dominate this market segment. The most famous company researched is a company with one of the biggest lines of products for bicycles and bicyclists, such as Global Positioning System (GPS) bike computers, turn-by-turn navigation, cycling awareness accessories, and smart lights along with personal apps for them.

Industry Trends

Bicycle accessory product development continues to increase, and price trends continue to rise. The startups are especially strong in sport and bicycle devices.

SPEEDOMETER IMPLEMENTATION

The engineering part of the project consisted of four major topics:

i. A test bench had to be developed to measure the electrical characteristics of a hub dynamo. The hub dynamo would provide the speedometer with energy, and the frequency of the dynamo voltage would be used to calculate the bicycle's speed. The dynamo effective voltage might vary across a wide range (between 70 V and less than 6 V) depending on the electrical load and the bicycle speed. The test bench was to be used to characterize different hub dynamos and to provide reliable laboratory equipment for testing the speedometer electronics.
ii. The electronic circuitry had to be designed for the speedometer. The circuitry had to consist of an overvoltage protection, a signal shaping to retrieve a measurable signal from the dynamo voltage, a buffer energy supply, and a microcontroller with a display.
iii. The software for the microcontroller had to be written.
iv. The speedometer housing had to be designed.

Test Bench

Based on the requirements, a test bench plan was made. Figure 1 depicts the design decision flow. The dynamo or a complete wheel with a spoked hub dynamo can be driven by a motor or a drilling machine connected to a friction wheel. Based on the fact that a drilling machine with a friction wheel makes it possible to change the transmission ratio of the test bench quite simply and therefore makes the test bench adaptable to quite a variety of hub dynamos with different diameters, the team decided to drive the hub dynamo directly by a drilling machine.

Figure 1. Test bench plan, beginning the process by choosing between a drill or a motor

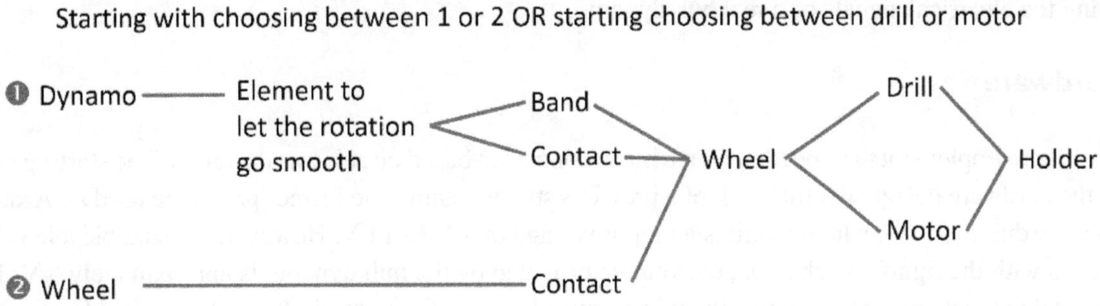

The final test bench was used in order to derive experimental data for the relevant points of the bicycle's operational conditions. Presuming the hub dynamo is mounted in a 28 '' wheel, the rotational speed was varied in a range that can be converted in a translational speed of the bicycle between 0 km/h and 70 km/h, and the electrical load was varied between 0 Ω (short circuit) and ∞ Ω (bicycle light switched off). These parameter variations were done for two different hub dynamos – Shimano DH-3D30 and Shimano HB-NX60 (Fahrradbeleuchtung, 2012). Therefore, approximately 200 measurements were done. The analysis of the obtained data showed that the dynamo in use is a multi-pole alternating voltage machine. Without load, the voltage peaks can reach up to 100 V. Figure 2 depicts the voltage signal with a load of 20 Ω (representing the rear light being) at a bicycle speed of 30 km/h.

Figure 2. Measurement of a Shimano DH 3D30 hub dynamo with a load of 20 Ω (the resistance of a rear light), the frequency is corresponding to a bicycle speed of 30 km/h

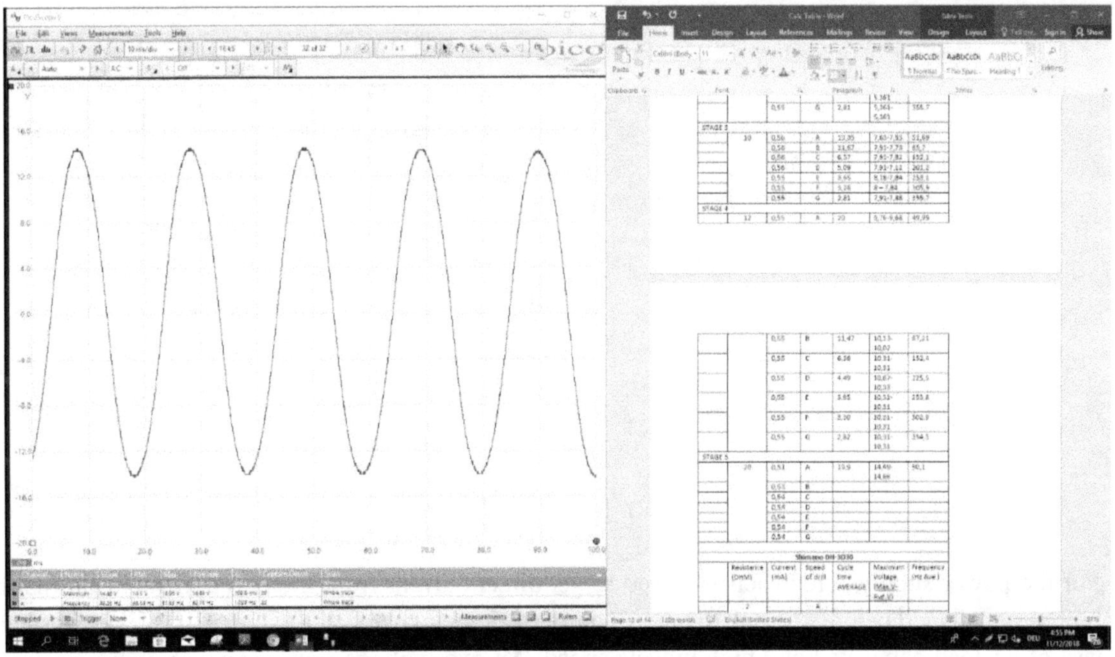

In this operation point, the peak voltage is already 15 V. Besides its usage to determine the characteristics of two different hub dynamos, the test bench was used in the software development for the testing using the electrical signals of a real hub dynamo.

Hardware

In order to implement the Speedometer, a microcontroller-based circuit was designed. The starting point of the hardware design was the work of a previous student team. The former prototype used an Arduino Uno (Arduino, 2019), which requires an input voltage of 7 V to 12 V. However, at usual bicycle riding speeds with the lights switched on, the voltage provided by the hub dynamo is approximately 6 V. This resulted in the decision to redesign the microcontroller part of the circuit. Instead, an ARM Cortex M3-based (Yiu, 2010) microcontroller – which only requires a 3.3 V input voltage – was chosen: the NXP LPC1769 (NXP, 2016). Additionally, OUAS provided an object-orientated software library for hardware-related programming. This proprietary library is used in the OUAS module "Embedded Systems".

Figure 3. Speedometer schematic

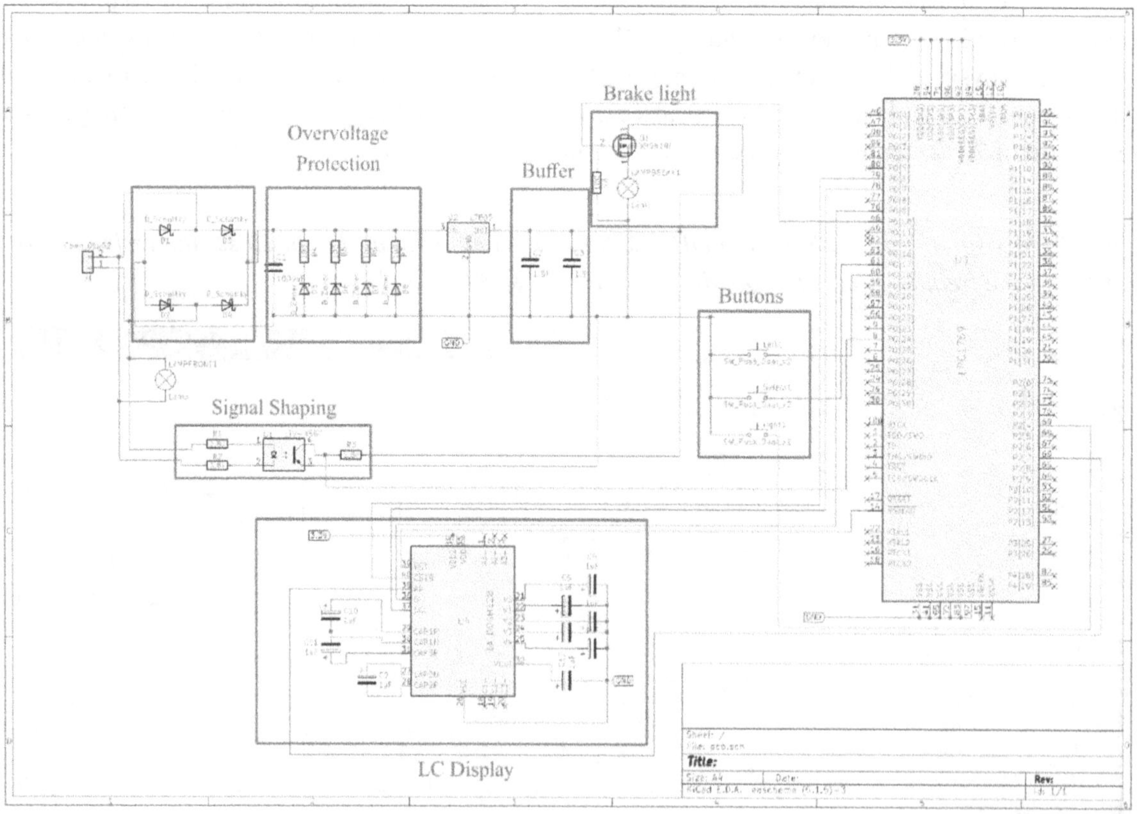

For the power supply, a rectifier was used to get direct current from the hub dynamo's alternating current. As explained before, the hub dynamo's output voltage may reach a value of 100 V peak voltage in case of no-load connected and high bicycle speed. An over-voltage protection circuit was designed to limit the input voltage to a value that is uncritical for the connected electronic devices. As an energy buffer, the need for two supercapacitors (Conway, 1999) with a capacity of 1.5 F each was foreseen. The tacho signal is shaped by an optocoupler. For presenting the information to the rider, a Liquid Crystal Display (LCD) was chosen and connected via an SPI Bus to the microcontroller (Figure 3, block "LC Display"). The rear light is controlled by the microcontroller. In the case of negative acceleration, the software can use the rear light as a brake light.

For the hardware tests, the circuit was built on a breadboard, as there was no electrical engineering student in the team to do a Printed Circuit Board (PCB) layout. The tests of the electric circuitry showed that the hardware design fulfilled the project requirements.

Software

Software development was done using the C++ programming language and GNU development tools (Breymann, 2016; Barr & Massa, 2006).

The main requirements for the Speedometer are that no additional sensor should be used to determine the speed; it should be configurable by the user; it should store data; and the rear light should blink whenever the brakes are applied. Thus, to determine the speed, the frequency generated by the dynamo hub is used. The speed is calculated by calling an interrupt service routine at every rising edge of the shaped voltage signal (see Figure 3, block "Signal shaping"). The time between two rising edges of the signal is measured in system ticks and used to calculate the bicycle speed. As the circumference of the bicycle wheel is stored in mm, the calculated speed is calculated and stored in mm/s (Listing 1).

```
// get bicycle speed in mm / s
uint32_t MeasurementTask::calcSpeed(uint32_t ticks, const Config &config){
uint32_t v = UINT32_MAX;
if (0 != ticks){
v =(uint64_t)config.getCircumference() / config.getPoles() * ticksPerSecond /
ticks;
        }
return v;
}
```

Listing 1: Method to calculate bicycle speed

The traveled distance and the trip counters are updated at every rotation of the wheel. When displaying the speed on the LCD, according to the chosen units, the stored speed and distances are converted to either km/h (kilometers per hour) or mph (miles per hour). All calculations are done in fixed-point arithmetic.

Operation data such as traveled distance, trip times, and distances are stored in nonvolatile memory whenever the bike stops. This is detected by a compare interrupt, which occurs 20 seconds after the last rising edge interrupt. With the given energy buffer, there should be no data loss due to the bike stopping in normal traffic situations. Even when the bike stops for a longer time period and the microcontroller is not powered any longer, the microcontroller can recover the data of the last operation after reset.

Besides the overall data, two trip counters are implemented. The trip counters store the trip distance, the riding time, and the maximum trip speed. These counters can be reset individually.

For the chosen LCD, no appropriate font and driver software were available. Therefore, the font and the display driver had to be programmed within the project. Figure 4 shows the display content for speed and travel distance and for one trip counter.

Figure 4. LCD showing speed and one of two trip counters

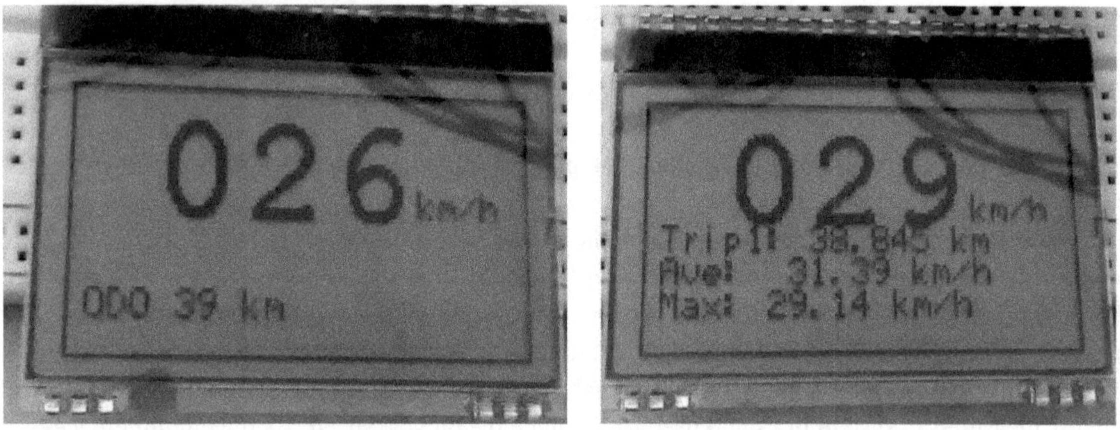

Housing Design

The housing is an important part of the product, as it holds the technical part together and protects it against water and other forces from outside.
The requirements for the housing are as follows:

- Must show the screen to the user.
- Must be able to be attached to the handlebar (via a holder).
- Must be removable.
- Must protect the technical parts needed for the speedometer against weather (storm) and forces (there being no damage when it falls).
- Must feel good and natural to use.

During the research on possible materials for 3D printing (Coward, 2015) the housing, the team decided that the best material would be polylactide because it fits all the requirements perfectly and because it is a material currently in use. It is not the best choice regarding electrical resistivity ($\approx 5 \times 10^{17}$ ¼ Ω cm); but its conductance is low enough for the speedometer application.

The design started with a (three-day) sketching phase. During the sketching phase, the size of the LC display was determined between the hardware and housing team. The sketch in Figure 5 was also made to decide the shape of the housing.

After sketching, the team looked for the best aspects in each of them and determined that they wanted to place the buttons next to each other and to place the screen above the buttons.

Figure 5. Initial sketches of the housing

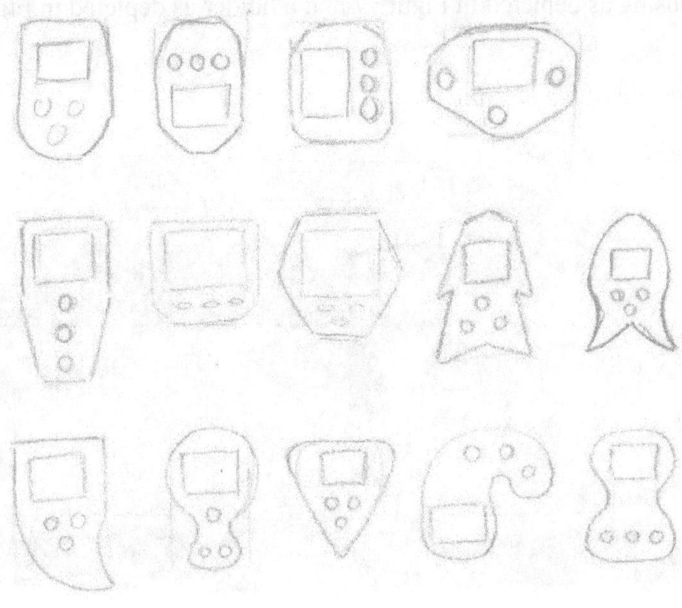

The design of the housing was made using a solid modeling computer aided design program, Solid Works. For developing the model, the size of the electronics and other parts that should fit into the housing were estimated. The result was the structure shown in Figure 6.

The rectifier and the overvoltage protection (Figure 3) are not included in this housing. These parts are included in a separate housing, which is fixed permanently to the bicycle frame to dissipate the heat that may occur at high speed with the bicycle light turned off. In normal operation with the light on, the overvoltage protection does not produce heat.

Figure 6. Simplified structure of the content of the housing. Red: Microcontroller, green: PCB, blue: LCD, transparent: Acrylic plate

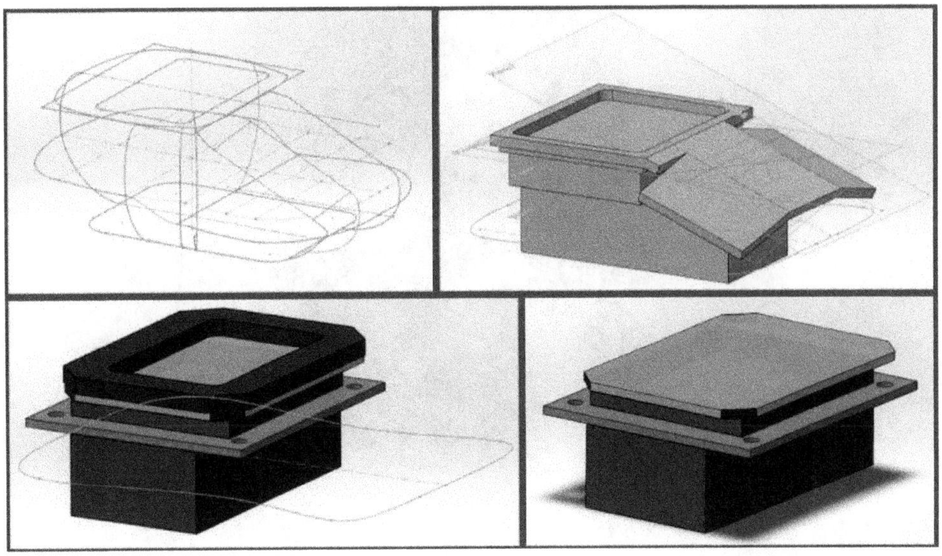

Designing the housing took many iterations until the final model satisfied all requirements. The final model consists of a housing as depicted in Figure 7 and a holder as depicted in Figure 8.

Figure 7. Sectional drawing of the housing

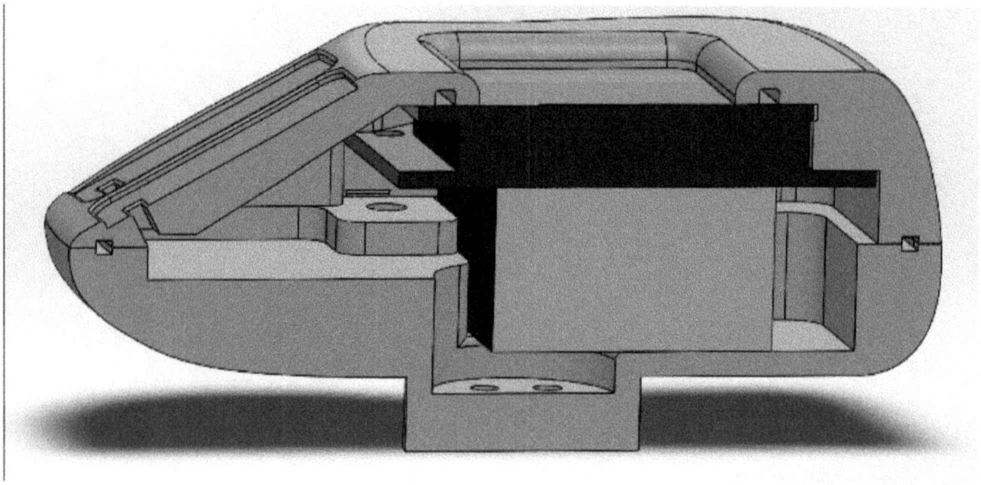

With a quarter turn, the housing is fixed to or removed from the holder as theft protection.

The housing consists of two parts held together by four screws. With this design, the electronics can be mounted in the housing. On the bottom part, where the housing is connected to the holder, four holes are foreseen as being needed for the contacts to connect the housing to the hub dynamo and the brake light. On the front side of the housing, three waterproof and dust-resistant membrane switch buttons can be glued.

Figure 8. Handlebar-mounted holder

The holder can be fixed to the handlebar of the bicycle. The holder contains the contact probes which connect to the electronics within the housing. The contact probes have nominal travel of only 1.5 mm, which ensures a reliable electrical connection after mounting the housing. The housing is fixed to the holder via magnets.

The housing and the holder were printed with a 3-D printer and tests were made to verify the assembly. As the electronics was not created as a PCB but only tested on a breadboard, no final function test with housing and electronics could be performed.

CONCLUSION

This project was performed by a multidisciplinary team (industrial design, computer science, quality management) formed of international students. Each team member was able to contribute their individual experience and strengths to solve the challenges resulting from the requirements. The technical solution was implemented in a prototype, and this device demonstrated the proof of concept. The original requirements included the features of Bluetooth connection to smart devices, which could not be realized due to a lack of electrical engineering skills in the team. The international team created innovative ideas and new aspects to solve the classical engineering task. The European Project Semester is an excellent framework for broadening the perspectives of local project participants and international students in a variety of ways.

ACKNOWLEDGMENT

The authors wish to thank Renate de Kok for her contribution a member of the EPS team that designed and developed the Bicycle Speedometer Plus.

REFERENCES

Arduino. (2019). *Arduino Uno*. https://www.arduino.cc/

Barr, M., & Massa, A. (2006). *Programming Embedded Systems*. O'Reilly.

Beeline. (2022). *Bike Computer Beeline Velo | Beeline EU*. https://eu.beeline.co/pages/beeline-velo

Breymann, U. (2016). *C++ Eine Einführung*. Hanser Verlag. doi:10.3139/9783446449121

Cateye. (2022). *Velo 9 | Products | CatEye*. https://www.cateye.com/intl/products/computers/CC-VL820/

City of Münster Presse- und Informationsamt. (2018). *Bicycles*. https://www.stadt-muenster.de/english/transport/bicycles.html

Conway, B. E. (1999). *Electrochemical Supercapacitors: Scientific Fundamentals and Technological Applications*. Springer. doi:10.1007/978-1-4757-3058-6

Coward, C. (2015). *3D Printing*. DK Publishing.

Fahrradbeleuchtung. (2012). *Shimano-Nabendynamos: Typen und ihre Unterschiede*. https://fahrradbeleuchtung-info.de/shimano-nabendynamos

FAZIT Communication GmbH. (2018, April 1). *Germany – land of bicycles*. https://www.deutschland.de/en/topic/life/sports-leisure/germany-land-of-bicycles

Garmin. (2022a). *Garmin Edge® 1030 | Bike GPS Computer*. https://www.garmin.com/en-US/p/567991/pn/010-01758-00

Garmin. (2022b). *Edge® 25 | Bike Computer | Garmin*. https://www.garmin.com/en-US/p/508489

Hemsing, R. (2018). *SmartHalo Review – Bring Your Bike to Life | Momentum Mag*. https://momentummag.com/smarthalo-product-review-bring-bike-life/

NXP. (2016). *LPC176x/5x User manual*. https://www.nxp.com/docs/en/user-guide/UM10360.pdf

Omata. (2022). *Omata | Smart GPS Bicycle Speedometer*. https://omata.com/

Smith, C. (2016). *Lezyne Super GPS Enhanced Review | GearLab*. https://www.outdoorgearlab.com/reviews/biking/bike-computer/lezyne-super-gps-enhanced

Wahoo Fitness. (2022). *Instructions | Elemnt Bolt GPS Bike Computer | Wahoo Fitness EU*. https://eu.wahoofitness.com/instructions/elemnt-bolt

Yiu, J. (2010). *The Definitive Guide to the ARM Cortex-M3*. Newnes.

KEY TERMS AND DEFINITIONS

Arduino: Single-board microcontroller and the accompanying software, based on Arduino technology.
Bluetooth LE: Technique for connecting computer devices wireless, LE means Low Energy.
Data Retention: Storage of digital data for later use.
GNU: Free software, mass collaboration project founded by Richard Stallman.
Hub Dynamo: Device for generating electric current from mechanical power on a bicycle's hub.
Speedometer: Device for measuring the relative speed between a vehicle and its ground or medium.
Supercapacitor: High-capacity capacitor with a capacitance value much higher than other capacitors, proper as a temporary power supply for devices with low energy consumption.

Chapter 14
Current Inclusive Design Projects for Social Innovation

Marina Puyuelo
Universitat Politècnica de València, Spain

ABSTRACT

The issue of design for social innovation is closely linked to citizen participation and the inclusion of the different groups that make up society, particularly the most disadvantaged or at risk of being marginalized. The new product and service design paradigm requires a specific sensitivity in which education and research have to be combined to produce sustainable innovation. A key aspect in the future of design will be cross-disciplinary knowledge and disciplines more adapted to local needs and specific user groups. Hence, as a matter of constructing the future, it seems that the business model has to be re-invented in collaboration with all stakeholders. This chapter gives a sample of the innovation processes experienced in the EPS program, which reflect the interrelationship between open and participatory design methods linked to the issues of accessibility and inclusion in different contexts.

INTRODUCTION

Immersed in an unquestionable crisis of resources of all kinds, we have analyzed the critique of design, which requires more integrative approaches that focus on achieving real improvements. The issue of design for social innovation is closely linked to citizen participation and the inclusion of the different groups that make up society, particularly the most disadvantaged or those at risk of being marginalized. The new product and service design paradigm requires a specific sensitivity in which education and research must be combined to produce sustainable innovation.

From this perspective, design now faces a dilemma along with an undeniable crisis of resources, which seeks new and more integrative approaches and focuses on achieving real improvements. As Manzini (2015) points out about sustainability and design, in most cases, design is more a part of the problems than the solutions, and this situation must change. It is necessary to activate awareness in the designer to initiate a new dialogue between needs and users, available techniques and technologies, and the necessary balance between what is natural and artificial. Since the nineties, numerous professionals and authors

DOI: 10.4018/978-1-6684-2309-7.ch014

Copyright © 2022, IGI Global. Copying or distributing in print or electronic forms without written permission of IGI Global is prohibited.

from different sectors (Dannoritzer, 2010; Puyuelo et al., 2011; Thackara, 2015) directly denounced the unconscious design that led to the system of programmed obsolescence and its dramatic consequences.

However, design is much more complex than objects and their materiality; it is a fascinating construct where emotional needs encounter practical daily experiences. Accordingly, design triggers constant relations with everyone's life by proposing new habits, beliefs and commitments. In a changing context in which demography, the environment and technology are transforming our way of life and our concept of well-being, the design sector must respond to the new needs of people and society. As we stated earlier, the new design paradigm must necessarily interrelate education and research in an integrative way to achieve effective solutions that must be innovative and sustainable.

Based on this perspective, the objective of this chapter is to show how students may feel like agents of the change in society through design. Three design projects are featured, which have been developed at the University, fostering research and participation methods in authentic, local places. These experiences foster creative synergies between different cultures, communication practices and potential businesses and provide disruptive solutions for issues that respond to social diversity, new technologies, existing production systems and current needs.

The current chapter includes five additional sections providing the background, comprehensive design education, illustrative design proposals and solutions, future research directions and the conclusion.

BACKGROUND

A new participatory culture has emerged from the digital and immaterial space, since it is a public platform that enables people to work together and collectively classify, organize and create new information (de Couvreur & Goossens, 2010), facilitating the development of collective intelligence. In the design field, trends related to the concepts of open design, collaborative design, and innovation communities (von Hippel, 2006) are calling for plural research that can be implemented by different approaches in order to respond to the challenge of innovation for the contemporary world; where their central axis resides in the change of production methods, the re-creation of the social habitat (Romero et al., 2012) and the design of the commons. The book "Teams of Teams" (McChrystal et al., 2015) describes a new structure model based on Networking which comprises strong communication and power teams. These global changes and the complexity of the systems that have led to network thinking have given rise to the need for a transformation in the way of addressing problems by moving from a static, rigid and categorized linear structure to a dynamic reticular structure, which is open to changes (Weinberg, 2015).

Accordingly, the voice of users as individuals has become more powerful and their decision-making is essential. Users and customers are non-conformists and are increasingly demanding in their expectations and requirements; consequently, they have more self-confidence in their own premises (Sánchez de la Guía, 2019). Today, the complexity and diversity of social and environmental demands are positioning innovation as the only resource for future growth and development at both local and global levels. Users, clients, customers, and producers can share their experiences and work as experts in some contexts. Providing tools for expression and participation means achieving a more active role in the design process (Kristensson et al., 2004).

Accordingly, there are new challenges, demands and lifestyles that require tools for the innovation and development of strategies and products that should reveal another way of enhancing value, which is more focused on people, rather than on consumers (Sánchez de la Guía & Puyuelo, 2019).

Participation is shown as a concept or relational space where different actors converge to contribute, co-manage and assume responsibility for the development of projects. In other words, participation involves interaction between parties and there is a need for these parties to share a willingness to discuss, listen and respond; they must be competent, prepared and have access to the required information and skills. The subtle local tactics of resistance and private practices available to the common person have been proven, as described in the nineties by de Certeau & Mayol (1998), reclaiming their own autonomy from the forces of commerce, politics and culture and claiming their capacity as producers of valuable solutions. We are currently witnessing an empowerment of the "learn or do it yourself" approach inspired by communication technologies. Likewise, users and citizens demand responses to everyday problems by requesting design to advocate participation methods in the projects, intervention tactics for the public space and open sources (Imbesi, 2016).

The international federation of benchmark Living Labs in Europe and worldwide (ENoLL) provides co-creation, user engagement, testing and experimentation facilities that target innovation in many different domains such as energy, media, mobility, healthcare or agrifood in order to produce open innovation ecosystems (ENoLL, 2013). The goals of these urban living labs include innovation, knowledge development for replication and increased urban sustainability emphasizing the need for supported, local solutions (Steen & van Bueren, 2017). The activities to reach these goals which always involve real-life use contexts are: co-creation with the actors (users, private and public actors, institutes, etc.) to shape the innovation process, and the insistence on feedback gathered from use and evaluation, empowering them for decision making.
Figure 1 shows the cards developed as tools to communicate several relevant statements.

Figure 1. Tips & Tricks for Building a Sustainable Living Lab
(ENoLL, 2021)

By exploring the connections between research, design education and practices, this paper features the potential of educational projects to improve inclusion and social innovation in the city of Valencia, including users, citizens and startups, in order to spread awareness and design culture in society. These projects, developed for international students, are based on the hypothesis of the need to share Design as a central part of any innovation process, especially, when thinking about collective places. This perspective clearly illustrates the required interrelation between design methods, businesses, local crafts, creative and communicative design practices. It is a fact that the comprehension of these methods arises from the specific aspects and intersections of the means and procedures of local experiences and new procedures available at the present time.

In short, social innovation represents an evolution in the way social problems are solved, by seeking to create value for the community. In this case, this goal can be achieved through design. Since it is necessary to articulate the ability to combine individual and collective interests, the instruments applied to the design project for social innovation represent a discontinuity of traditional project methods because they are posed in a challenging way.

COMPREHENSIVE DESIGN EDUCATION TO ASSEMBLE STAKEHOLDERS

To encourage design projects with this co-philosophy perspective, students must be aware of the designer's position as a mediator who requires permanent contact with end-users. This awareness leads to the emergence of the local as a main proximity context, where projects may focus on developing the design culture with the experience of real situations.

This approach was presented to European Project Semester students in a controlled context, which proposed ethnography and fieldwork methods to them. They had to observe and obtain data from people, groups, spaces, and objects of use in daily life to make them personally understand how all these things define a system of life.

Thus, it is necessary to mention the essentials of public space as a valuable scenario for life outdoors and co-design:

- Features such as public domain, anonymous, zero cost, and unrestricted use.
- The need for full accessibility of all citizens regardless of their physical or sensory characteristics.
- Multi-functionality concerning the diversity of users.
- Everything that can trigger the complete development of its social mean.

Based on these parameters, it is possible to assess the quality of public space according to the intensity and quality of the social relations it facilitates, as well as the ability to stimulate and promote symbolic identification, expression and cultural integration (Merino & Puyuelo, 2021). Hence, the design of products for public use becomes essential because they are the elements that determine the habitability of public space, encourage citizens' participation in the common environment and, consequently, highlight urban space as a living space, thus emphasizing the city's image and playing a representative role within the community. They are active supports of public life, which enable new forms of participation and civic commitment (Puyuelo & Merino, 2021).

With this approach, the aim of the projects in every course is to obtain the knowledge required to define a framework in order to improve the living conditions of specific groups of people as well as society as a whole.

One of the main strengths of the public space as a work context is that everybody has their own experience as a citizen somewhere and here in Valencia, they have to build a new experience in this city. Furthermore, urban elements provide the opportunity to use knowledge from very different study fields. Taking advantage of these features and the multidisciplinary EPS teams, which make it possible to analyze the situation from different perspectives and backgrounds, the project was developed through the Design Thinking process (Brown, 2009), by using public places as scenarios for life and collective interests. In this context, students can play the role of agents of change and innovation and figure out the project's possibilities according to their profiles and interests. Figure 2 shows a visual representation of the Design Thinking process.

Figure 2. Diagram of Design Thinking phases
(Harris, 2021)

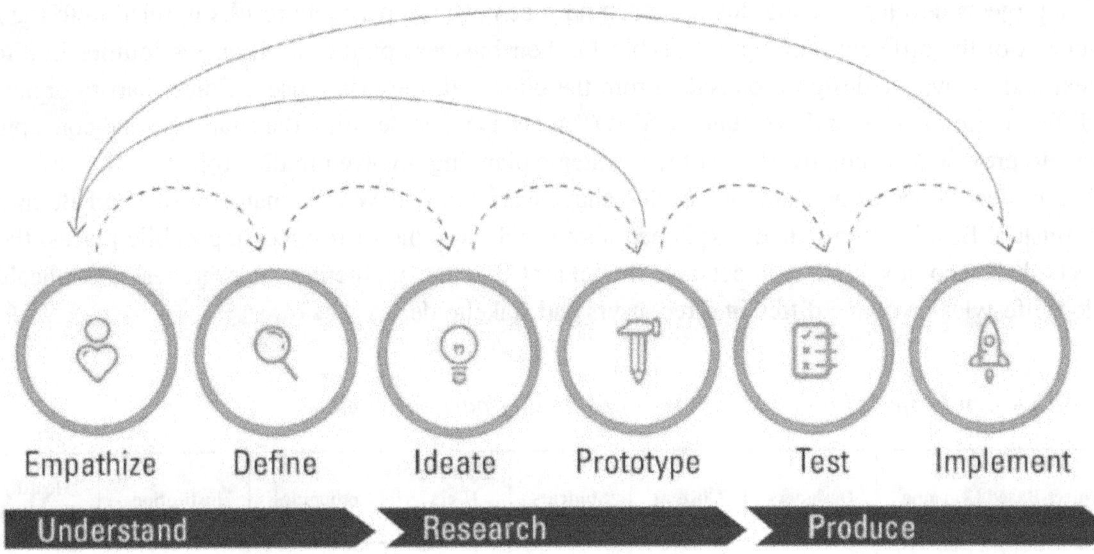

Materials and Methods

It has been a common educational practice for design educators to propose the topic or the object that students should work on in their projects. This made it easier for the students to directly start their research by searching for the characteristics and requirements needed to set up the design problem, usually in relation to market demands.

That approach actually gave a secondary role to the student in terms of decision making because the motto for the design project was, on most occasions, a premise defined as an opportunity already detected in the market or for a specific company. Furthermore, that approach has critical aspects when discussing collective goods and urban design characterized by continuous changes as environments for a diversity of users and interests. In this sense, the inspiration and experiences of the Urban labs implemented in northern Europe provide a new way to work and understand public spaces. In these projects, we experience appropriate strategies of project-based learning and hands-on skills to empathize with the users and the scenario. In this phase, the initial research begins by defining the stakeholders map for a given situation to find out what the needs are in an open way. Students gather knowledge from these groups of people to understand and transform the groundswell of interest in that area into tangible, sustainable initiatives that can co-produce positive change at a grass roots level (Fathers, 2017). They have to formulate a hypothesis and define their challenge, which is eventually related to the composition of the EPS team.

In this environment, students work as a professional design team, with plural research and the contribution of aesthetic values, practical representation, knowledge, color, and digital skills to co-develop proposals. They practice design as an iterative process with the different stakeholders, and this process concludes when they test their projects at different levels. They also test their hypothesis by creating experiences of the solution with users and other people who may be involved. As stated earlier, the target group members for these projects are the citizens and park users of all ages; hence it is necessary to be flexible to split into other groups based on their activities or interests.

The projects developed using this approach have achieved remarkable results in relation to the involvement of the different participants (Table 1), their business plan or scaling possibilities in a real context and innovative design proposals. From the outset, all the teams use a Gantt chart to organize the different group tasks, and eventually a SWOT analysis when defining the challenge and conceptual design, to provide a general overview of the strategic planning involved in the project.

The projects selected as examples in this chapter are Hortus, WAW what a waste! and Be aware OpenBustop. Besides the methods explained above and the topic of re-inventing public places, these projects share a common will to understand the Smart City concept, increasing awareness about healthy outdoor lifestyles involving different direct users and stakeholders.

Table 1. Design Projects for collective use – stakeholder participation

Project-Based Learning	Designers	Citizens	Students	Specific Users Groups	Companies	Institutions	Healthy Practices
SmarFactory	*	*	*			*	*
FIT	*	*	*	*	*		
Origame	*	*			*	*	
Hortus	*	*		*	*	*	**
WAW what a waste!	*	*	*		**	*	
Be aware OpenBustop	*	**	**	*	*		*

(Own elaboration, 2021)

As mentioned above, to set up the project, the first question students have to answer is what challenge they want to tackle (EOLA & Limitless, 2021). And they select it after research and fieldwork conducted in different urban areas, where most people spend their time outside. The students usually have to spend time taking strolls and visiting the city to gather information about what already exists and the lack of products or services which can be improved. In Valencia, these studies led to the specific scenario of Turia Park, a riverbed transformed into an urban park in the nineties. The participants were grouped into five to six student teams from different backgrounds and countries. In the three cases presented, they carried out a joint analysis of each different context or situation interrelating the aspects involved in the design, with an intentional commitment to improve the system and have a widespread impact on society itself. Visual system mapping tools, surveys, and fieldwork were used to obtain a plural and consensual view of each case.

SOLUTIONS AND DESIGN PROPOSALS

All these experiences stimulate an active learning environment where the student learns by developing and applying his/her knowledge for a real purpose in collaboration with others. Progressively, the student gets in touch with different stakeholders and elaborates a co-working concept. From an educational perspective, they are giving shape and materials to ideas that will result in a relevant experience (Schön, 1987). Furthermore, the entire process is situated in real contexts, with a practical significance that generates greater motivation in the student (Gerritsen, 1999).

Turia Park is the public place for the research and design scenario of the projects: WAW what a waste and Hortus. This urban park embodies some of the priorities of the Smart city concept because it coincides with environmental trends for a healthy City, open gyms, third youth living, and priority on "soft" transport services such as cycling (Casado et al., 2015).

In the Empathize phase, the teams need to collect information from users. Every year, a short public survey is conducted about which existing facilities are highly valued and what the interviewees found to be lacking in that urban environment.

It is usually conducted three times per week at different hours of the day during a month. People are chosen randomly for every course, covering different user profiles such as residents and visitors as well as people from different groups: citizens, young athletes, older people, and tourists. The result has changed over the years. It depends on the new facilities, the influx of public in different years, or the change in outdoor space perception in the previous year, after the lockdown due to Covid19.

Based on the fieldwork and survey study, the teams have to find out their work area and define their challenge. In order to provide tools for expression and participation of the stakeholders and potential public, students need to prepare different documents to communicate with them in order to achieve a more active role in the design process (Kristensson et al., 2004). People participate in the different problem definition, ideation, prototyping and testing phases. On the one hand, they point out different needs, such as the lack of signs or toilets some years ago, or specific problems in relation to uses or lighting. On the other hand, they express their opinions about the students' proposals and participate in the design definition.

WAW Project

This proposal has a somewhat "adaptive-flexible" design; it is an entrepreneurial profile concerning the development of a recycled material linked to local possibilities and needs. The research was based on the statement: "It is possible to design a product for public spaces made from recycled material that can be inspiring, inviting and give a better understanding of how important it is to use recycle materials". The product must communicate a straightforward, sustainable process to the final user (Figure 3).

The idea is focused on using the public space as a powerful medium to share ecological problem with a global audience while at the same time, producing innovative solutions that are adaptable to local needs. The designed products line starts with a modularly shaped bench, the triangle symbol of recycling plastic, and a system to park and lock the bicycle. It also provides several possibilities for other urban elements made with this concept by combining recycled materials that make each element unique.

Figure 3. Modular pieces of multiple recycled waste materials
(Vento et al., 2011)

The combination of analyzed and proposed materials is easy to recognize such as plastics melted together in different compositions, tectan and steel.

To provide the public with better know-how of which materials can be recycled and how important it is to do it, a webpage has been set up with the explanation and details concerning the idea.

This open-concept project develops its business plan with the involvement of interested users, raw materials (waste materials), companies and agencies.

Hortus Project

The Hortus project has an alternative focus on a particular group of users as the center of a social innovation strategy. After in-depth research on the users of this park, the elderly group appears to have fewer facilities and opportunities to enjoy this park. The fieldwork showed that few areas were sufficiently adapted for the elderly and only plain benches and elements for practicing some physical exercises were available for them.

Several studies of concepts, potential locations, size and shape of available space, the abilities and limitations of the elderly, physical and mental constraints, interests and age appropriateness, led to many proposals, which were evaluated and discarded. Finally, a series of three products were designed, especially adapted to them so they could enjoy the park and practice some activities in the open air (Figure

4). Table games, a telescope for bird watching, and planning meetings and dates to play music and dance were some of the results of this proposal.

Figure 4. Hortus project: students explore the qualities and accessibility of the place to design outdoor products for the elderly
(Stanicka et al., 2016)

All these elements have a flexible use: the chair and table have adjustable seats to help people sit and stand up with armchairs and spin in the chair to use the table. The binoculars have a double grip for easier and safe support while observing trees or birds, also allowing different heights as well as their use by a wheelchair user. The music board is a touch screen with a very simple interface, which works with solar panels that display notices, music, dance or gym classes, information, or advertising.

Be Aware: a 'Green Public Mindset'

The last example, Be aware, was proposed in the framework of upgrading a transfer point and raising awareness for eco-friendly living. Its aim was to promote research, education and excellence in design applied to the city's daily commuting system (Figure 5).

It is important to note that in this case, the team members were studying architecture, mechanical engineering and chemical engineering which led them to agree on the idea of working on a "message", rather than an object as a product. They perceived an opportunity to reach a broad target group, inviting participation by bringing their own seat, using solar energy, creating a relaxed atmosphere in a small place where people are bored, hence they can be open to new things and experiences. The installation not only reaches bus users, but also passersby, potentially stressed or bad-tempered people, etc.

The bus users are mainly commuters, people who use the bus as a way of transportation from work to home. The main motivation for using the bus is that it is cheap, fast and easy compared to other transportation methods.

The study focused on creating a product and a service that the users could interact with and consider themselves as participants instead of passive consumers. The study also focused on how it is possible to raise awareness and change people's mindset to make a lasting change by using design.

In the preliminary studies and survey, a big group of the respondents, 50%, used the bus as their main daily transportation. Others used it at least a few times a week. When asking the respondents how they wanted to upgrade the bus stop, half of them chose free Wi-Fi. When asking more specifically about eco-friendly wishes, most choose that they would appreciate protection against emission gasses and the use of solar energy. It is an excellent site to test the use of touch screens with other people and see how they are used outdoors in a shared context; the possibility to share a recipe of the day with healthy ingredients, showing how to make it sustainable.

Figure 5. Be aware OpenBustop – Take your chair and let's talk
(Claw et al., 2014)

The aim was to involve the users in the product, like the "guerrilla garden project" in England. They created something with a big contrast compared to the surroundings, which is actually a good way to get attention. They performed an activity that was playful, made people smile, and inspired people to get involved. The following aspects were considered: the idea had to inspire people to get involved in the product. Furthermore, putting a smile on the citizen's face was considered to be a very good thing. So even if the message provided to the users had to be very serious and important, it was clear that it should be communicated in a fun and clever way.

As they wait, the installation invites people to look at different materials and find the re-used elements, small, simple things in a new and creative way that people generally do not associate with a bus shelter. These elements were able to attract people's attention and, at the same time, deliver new messages concerning eco-awareness. Furthermore, new or old seats could be included on this Bus Stop.

The project can be replicated and used at many places and it can have a major influence on the surroundings by encouraging people to take the bus, which is an eco-friendly way of travel.

FUTURE RESEARCH DIRECTIONS

This academic approach seeks to contribute to open design concepts that provide insights and practices about emerging future and multidisciplinary trends which can improve the new scenarios of life. Engineering and design can make these projects feasible and successful.

All these academic proposals, concepts, and numerous topics should be explored further, to make environments more inclusive for people of all ages, as well as safe and valuable for as many people as possible. It would be suitable to work on final prototypes to test some of these concepts in real life.

One ideal goal to be achieved in the current context of design practice is to develop a model that expands design training beyond science and technology, with the aim to train visual intelligence, ethical sensitivity, and aesthetic intuition (Findeli, 2001).

CONCLUSION

This chapter has provided a perspective on the breadth of design practices for social innovation led by research, participation and co-design at different public places in Valencia. The new product and service design paradigm requires a specific sensitivity in which education and research must be combined to produce sustainable innovation.

The article explains how the concept of participation has spread due to different factors, such as technologies, culture, and how current design projects are understood from an overall perspective, where the different stakeholders must play an active role. Based on this perspective, the EPS outlines an internationally leading platform, where different kinds of skills and knowledge are available to face the projects: design skills, knowledge about materials, mechanical production methods, sociology, business and what is more relevant, international backgrounds and cultures. In these three projects for instance, individual talent and teamwork are promoted; engineers, designers, and other scientists jointly develop interdisciplinary metropolitan solutions. The team supervisor works as a promoter to initiate projects with local situations involving both private and public partners and, above all, citizens and users.

The reviewed literature facilitated the selection of the approach for the urban labs, which have been the basis to launch projects in our educational context by sharing design with different stakeholders in each case, in this local context. We strongly believe that these types of projects enhance the sensitivity towards diversity of users and relevant collective issues, hence preparing professionals for the inclusion of non-experts in the design process by giving tools to them so they can face complex demands and an open attitude towards new initiatives.

The results reveal the potential of collaboration with users, citizens, and companies in different formats to generate more adapted innovative solutions and expand the culture of inclusive and solidarity design, as part of the improvement of society as a whole. If we seek more public and social awareness about design, we should demonstrate the designer's capability to be a mediator in the transformation of ideas into solutions. In this aspect, design education is the means to establish the new role of this profession. As expressed by Brown & Wyatt (2010):

The notion that there is nothing wrong with experimentation or failure, as long as they happen early and act as a source of learning, can be difficult to accept. But a vibrant design thinking culture will encourage prototyping—quick, cheap, and dirty—as part of the creative process and not just as a way of validating finished ideas.

This is what we have initiated with these design projects for public spaces, which foster a free lifestyle for inclusion and social innovative practices.

ACKNOWLEDGMENT

This research received no specific grant from any funding agency in the public, commercial, or non-profit sectors, however the School of Design Engineering of Universitat Politècnica de València always supported the activities required by the students.

REFERENCES

Brown, T. (2009). *Change by Design: How Design Thinking Transforms Organizations and Inspires Innovation*. HarperBusiness.

Brown, T., & Wyatt, J. (2010). Design Thinking for Social Innovation. *Stanford Social Innovation Review*, *8*(1), 31–35. doi:10.48558/58Z7-3J85

Casado, M. G., Revert, C., Sales, V., & Veral, S. (2015). Smart Cities Trends: Tendencias en las Ciudades Inteligentes y oportunidades para los sectores del hábitat [Smart Cities Trends: Trends in smart cities and opportunities for sectors habitat]. ITC. AIDIMA y ASCER.

Claw, S., Folkertsma, L., Oliveira, H., Stoel, M., & Thor, L. (2014). *Be aware Mindset OpenBustop* (EPS Report). Universitat Politècnica de València.

Dannoritzer, C. (2010). *Comprar, tirar, comprar* [Buy, throw away, buy] (Documentary 28/12/2017). http://rtve.es/v/1382261

de Certeau, M., & Mayol, P. (1998). *The Practice of Everyday Life: Living and Cooking* (Vol. 2). University of Minnesota Press.

de Couvreur, L., & Goossens, R. (2010). Design for (every) one: Co-creation as a bridge between universal design and rehabilitation engineering. *Proceedings of the Seventh International Conference on Design and Emotion.*

ENoLL. (2013). *Tips & Tricks for Building a Sustainable Living Lab.* https://enoll.org/wp-content/uploads/2017/08/tips-tripcs-final-upload.pdf

ENoLL. (2021). *European Network of Living Labs.* https://www.enoll.org/

EOLA & Limitless. (2020). *SOCIA: Social Innovation Initiative Logbook.* https://www.socialenterprisebsr.net/wp-content/uploads/2020/10/SOCIA-Self-Assessment-Logbook-vFIN-pdf.pdf

Fathers, J. W. R. (2017). Engaged Design. *The International Journal of Design in Society, 11*(2), 23–29. doi:10.18848/2325-1328/CGP/v11i02/23-29

Findeli, A. (2001). Rethinking Design Education for the 21st Century: Theoretical, Methodological, and Ethical Discussion. *Design Issues, 17*(1), 1–13. doi:10.1162/07479360152103796

Harris, P. (2021). *Interpersonal analyses of bulimia: Normal-weight and obese* [Unpublished doctoral dissertation]. Universitat Politècnica de València.

Hoffman, K. (2015). *Co-Design, Not Redesign* (Event Apart Video). https://aneventapart.com/news/post/co-design-not-redesign-by-kevin-m.-hoffman-an-event-apart-video

Imbesi, L. (2016). *Design for Social Innovation in Canada.* Common Ground Publishing. doi:10.18848/978-1-61229-811-5/CGP

Kristensson, P., Gusyafsson, A., & Archer, T. (2004). Harnessing the creative Potential among Users. *The Journal of Innovation Management, 21*(1), 4–14. doi:10.1111/j.0737-6782.2004.00050.x

Manzini, E. (2015). *Design, When Everybody Designs: An Introduction to Design for Social Innovation.* The MIT Press. doi:10.7551/mitpress/9873.001.0001

McChrystal, S. A., Silverman, D., Fussell, C., & Collins, T. (2015). *Team of Teams: New Rules of Engagement for a Complex World.* Portfolio/Penguin.

Merino, L., & Puyuelo, M. (2021). El espacio público como el contexto del diseño y la innovación [Public space as a context of design and innovation]. In M. Puyuelo (Ed.), *Todo sobre Diseño* [All about Design] (pp. 47–60). Editorial Universitat Politècnica de València. https://riunet.upv.es/handle/10251/166499

Puyuelo, M., & Merino, L. (2021). The public space as a design observatory. The SmartTuria Hub project in Valencia (Spain). In *Proceedings International Conference on Built Environment, Science and Technology* (pp. 52-43). SRM Institute of Science and Technology.

Puyuelo, M., Merino, M. D., & Fernández, N. (2011). Design Challenges Today: An Overview of Its Sustainable Principles. *Design Principles and Practices an International Journal, 5*(4), 63–78. doi:10.18848/1833-1874/CGP/v05i04/57896

Romero, G., Mesías, R., Enet, M., Oliveras, R., García, L., Coipel, M., & Osorio, D. (2012). *La participación en el diseño urbano y arquitectónico en la producción social del hábitat*. CYTED. http://cdam.unsis.edu.mx/files/Desarrollo%20Urbano%20y%20Ordenamiento%20Territorial/Otras%20disposiciones/Participaci%C3%B3n_dise%C3%B1o_urbano.pdf

Sánchez de la Guía, L. (2019). *Modelos y Métodos de Participación del Usuario en el Diseño de Productos y Servicios* [Models and Methods of User Participation in the Design of Products and Services] [Unpublished doctoral dissertation]. Universitat Politècnica de València.

Sánchez De la Guía, L., & Puyuelo, M. (2019). Dealing with Users Participation in Design of Products and Services. *The International Journal of Design Management and Professional Practice*, *13*(1), 21–29. doi:10.18848/2325-162X/CGP/v13i01/21-28

Schön, D. A. (1987). *Educating the reflective practitioner: Towards a new design for teaching and learning in the professions*. Jossey-Bass.

Stanicka, N., Zwinkels, J., Darmody, C., Larke, C., Ball, E., & van Nerum, P. (2016). *Hortus (EPS Final Report)*. Universitat Politècnica de València.

Steen, K., & van Bueren, E. (2017). *Urban Living Labs: A Living Lab Way of Working*. AMS Institute for Advanced Metropolitan Solutions. Delft University of Technology. https://www.ams-institute.org/news/urban-living-labs-living-lab-way-working/

Thackara, J. (2015). *How to Thrive in the Next Economy: Designing Tomorrow's World Today*. Thames & Hudson.

Vento, E., Kurz, M., Sarv, M., Mutlu, E., Evans, R., & Christensen, N. (2011). *WAW what a waste* (EPS Final Report). Universitat Politècnica de València.

von Hippel, E. (2006). *Democratizing Innovation*. The MIT Press.

Weinberg, U. (2015). *Network Thinking: Beyond Brockhaus Thinking*. Murmann Publishers.

ADDITIONAL READING

Aparici, R., & Osuna, S. (2013). La cultura de la Participación. *Revista Mediterránea de Comunicación*, *4*(2), 137–148. doi:10.14198/MEDCOM2013.4.2.07

Chasanidou, D., Gasparini, A. A., & Lee, E. (2014). Design Thinking Methods and Tools for Innovation in Multidisciplinary Teams. *Proceedings of the NordiCHI'14 Workshop Innovation in HCI: What can we Learn from Design Thinking?* https://chasanidou.com/assets/articles/2014-multidisciplinary.pdf

Chesbrough, H. (2011). *Open services innovation: Rethinking your business to grow and compete in a new era*. John Wiley & Sons.

Curedale, R. (2013). *Mapping Methods: for Design and Strategy*. Design Community College.

Curedale, R. (2013). *Design Thinking Process and Methods*. Design Community College.

Ordenes, F. V., Theodoulidis, B., Burton, J., Gruber, T., & Zaki, M. (2014). Analyzing Customer Experience Feedback Using Text Mining: A Linguistics-Based Approach. *Journal of Service Research, 17*(3), 278–295. doi:10.1177/1094670514524625

UK Design Council. (2010). *Multi-disciplinary design education in the UK*. https://www.designcouncil.org.uk/sites/default/les/asset/document/multi-disciplinary-design-education.pdf

KEY TERMS AND DEFINITIONS

Co-Design: A design process where the work evolves with different people in a workflow, which increases the quality of the result through collaboration.

Community: A wide range of people who can support or be beneficiaries of the project. Multidisciplinary users who can provide feedback, different impressions, knowledge, or assistance for different steps of specific projects.

Innovation Communities: Research carried out by different groups involved in particular cases or situations to respond to current innovation challenges.

Living Lab: User-centered and open innovation ecosystems using systematic co-creation approaches with public-private-people partnerships, integrating research and innovation processes in real-life communities and settings.

Open Design: A form of co-creation where the users may design and develop products or services by using and sharing design information.

Social innovation: An evolution in the way social problems are faced and solved to create added-value for a community.

Urban Lab: A strategy to seek proposals and solutions which integrate participatory methods for urban planning.

Chapter 15
The Individual Survival Unit: Transporting Patients in a Protective and Stable Environment in Disaster Scenarios – a Case Study

Philip Breedon
Nottingham Trent University, UK

Martin Higginson
Nottingham Trent University, UK

Francesco Luke Siena
https://orcid.org/0000-0002-6908-7365
Nottingham Trent University, UK

Michael H. J. Vloeberghs
Nottingham University Hospital Trust, UK

ABSTRACT

The Individual Survival Unit project was delivered under the European Project Semester exchange program. Four students from varying multidisciplinary education backgrounds and nationalities undertook the four-month project. The project investigated and addressed the design and feasibility of a unit that can be used to protect and increase the survival time of an injured individual within a disaster zone. The designed solution provides a stable and protective environment for an individual whilst avoiding further injury during transportation. To enable the design of an optimal solution, disaster areas were analyzed and identified, providing a focused design problem requiring exploration and resolution. Essential basic life support equipment was researched, listed, and then ranked by the importance according to the literature review based upon interviews with experts. The final concept was presented through technical drawings, computer-aided visualization and renders, animations, interface simulation, and a real-life situation presentation.

DOI: 10.4018/978-1-6684-2309-7.ch015

Copyright © 2022, IGI Global. Copying or distributing in print or electronic forms without written permission of IGI Global is prohibited.

INTRODUCTION

Every academic year, as part of the European Project Semester (EPS) exchange program hosted at Nottingham Trent University (NTU), the Medical Engineering Design Research Group (MEDRG) provides a "Live" project brief focused in the medical design sector in collaboration with external partners. The aim is to provide students with a unique learning experience of engaging with medical engineering design specialists, medical experts, and product designers thus enriching the student learning experience.

The Individual Survival Unit (ISU) project in collaboration with NTU, MEDRG and Nottingham University Hospitals Trust (NUH) offered a unique experience to help students develop design and engineering skills whilst engaging in a competitive and fast-changing business environment. The semester-long activity was designed to educate and prepare students to work in an industrial/research capacity within a multi-national, multi-disciplinary team whilst being supported by academics and industrial supervisors. The ISU project was also delivered alongside a study program comprising of subjects that supports the undertaking of an industrial or research-related group project thus helping build a fundamental skill and a specialist skillset associated with a given discipline.

This case study demonstrates how the ISU project group developed a grounded knowledge base in a specialism (medical device/system design) whilst also learning about economics, management, communication, and languages. This set of skills is important when working within the medical device design industry and with a wide variety of stakeholders in multiple countries and continents. The ISU project demonstrates how it is possible to complete an integrated engineering, design, or business project whilst developing a range of interpersonal skills when working within a cross-cultural group. This chapter presents an overview of the organizational partners involved in the project delivery, a detailed case study of the Individual Survival Unit (ISU) project, student supervision arrangements, and finally, a review of several technological, organizational, and management arrangements leading to recommendations for future project and EPS program delivery.

ORGANIZATION BACKGROUND & PROJECT PARTNERS

The ISU project was delivered collaboratively by three key partners, NTU Product Design Department, MEDRG and NUH. The accessibility to these three key resources enabled students to gain access to a wide range of expertise and opportunities, including student/project feedback, opportunities to engage with project partners and access to supporting resources.

Nottingham Trent University

NTU is a public research university based in the city of Nottingham, England. NTU has a long track record of academic excellence, attracting top academic staff across numerous disciplines, and has been driving innovation since 1843 establishing itself as one of the UK's most renowned creative universities.

Since 2003, NTU has invested heavily (over 450 million GBP) in estates and facilities, revamping historic buildings, and building state-of-the-art new facilities. NTU's commitment to further improving both the student experience nationally and internationally, as well as its research capacity, can be recognized through recent estate developments. This includes the award-winning Interdisciplinary Science and Technology Centre (ISTeC), the construction of a brand-new Global Lounge building and the development of a brand-new University Hall and Music Centre. In addition, NTU has committed to building a new 23 million GBP Medical Technologies Innovation Facility (MTIF), a new 23 million GBP dedicated School of Engineering used as a teaching and research facility, and finally, a 9 million GBP new Enterprise Innovation Centre. This investment drive was NTU's commitment to providing the best possible education establishment, allowing it to achieve a number of significant accolades.

Nottingham Trent University – Product Design Department

The Product Design subject discipline is part of NTU's School of Architecture, Design & the Built Environment (ADBE). ADBE academic and technical staff have a wide range of expertise to help students develop as creative and analytical thinkers, with the freedom to discover and progress their own individual design style, visions, and philosophy. This is further fostered through a vast network of contacts with commercial organizations and industry professionals, in addition to active research groups of national and international standing. The product design department offers several courses at both undergraduate and postgraduate levels, including BSc (Hons) Product Design, BA (Hons) Product Design, BA (Hons) Furniture & Product Design, MSc Design: Products & Technology and MA Design: Products & Furniture.

The suite of BA/MA product design courses at NTU concentrates on the design of innovative products, systems, and services, including consumer goods, electronic gadgets, sustainable products, sports and healthcare equipment, packaging, and transportation. In addition, the courses offer an opportunity to develop innovative designs for the consumer market in areas of commercial and domestic furniture, interiors, lighting, and homewares. However, the BSc (Hons) Product Design course, which supports and contributes to the EPS program, concentrates on providing a technology-focused design education.

BSc (Hons) Product Design focusses on the underlying technology of product design, with projects that combine both form and function. The course concentrates on the design of products for manufacturing, digital, industrial, electronic, and medical applications, placing technology and scientific advancement at the core of the design process and developing knowledge and skills in materials and new or emerging technologies, including mechanics and electronics. Students reflect, model, test, modify, develop, and improve their ideas through commercial projects with industry.

Medical Engineering Design Research Group

Based at Nottingham Trent University, the MEDRG is an interdisciplinary group with a vast network. The MEDRG brings together a wide range of academics, clinicians, surgeons, healthcare companies and other medical professionals across a wide range of research specialisms. The MEDRG focuses its research on six themes, including software-enabled surgery, medical devices and systems development, robotic technologies for surgical applications, medical technologies and interaction design, mixed reality and immersive systems for medical applications and additive manufacturing. The MEDRG utilizes dedicated resources to enable product development with expertise and equipment in 3D Printing and additive manufacturing, injection molding, silicone casting, electronic prototyping/manufacturing, virtual and augmented reality, computer-aided design and simulation and numerous testing facilities, etc.

The MEDRG seeks to address the changing needs of medical research both within the UK and globally by conducting research in the following sectors:

- Medical device development and additive manufacturing processes.

- Health management and rehabilitation - including mobile applications.
- Biomaterials – including dental, cardiovascular, ophthalmics, and orthopedics.
- Surgical robotics.
- Emergency medical care, systems, and procedures.

The MEDRG has extensive knowledge in developing medical devices and technologies with links to UK and international industry and healthcare providers. In addition, the MEDRG offers a range of core services to students, staff, and external partners. These services include innovative product development for the medical industry, medical and surgical robotics, rapid prototyping, multi-material printing, flexible electronics printing, translation of medical technologies for product manufacture, fundamental and applied research for novel medical technologies, virtual and augmented reality development, hardware interface design/development and user experience/user interface design and development.

Nottingham University Hospitals NHS Trust

Based in the heart of Nottingham, NUH provides services to over 2.5 million residents of Nottingham and its surrounding communities. In addition, specialist services are provided for a further 3 to 4 million people from across the region (Nottingham University Hospitals NHS Trust, 2019). NUH employs around 15 000 across three sites, Queens Medical Centre (QMC), Nottingham City Hospital and Ropewalk House, to provide services such as an Emergency Department (A&E), Major Trauma Centre and the Nottingham Children's Hospital. NUH has a national and international reputation for several specialist services, which include services designed around stroke, renal, neurosciences, cancer, and trauma. In addition, NUH acts as a teaching trust which has strong relationships with numerous universities within the East Midlands. NUH plays a vital role in the education and training of doctors, nurses, and other healthcare professionals within the UK and internationally.

ISU PROJECT BRIEF

Diverging from standard clinical practice, Professor Vloeberghs was asked to prepare and present a paper on medicine in space for the Leicester Space Day in 2005. Reviewing available literature showed that space medicine is based on injury avoidance, comparable to the industrial Hazard and Operability Study (HAZOP) protocols for event avoidance. One of the weaknesses in any situation is protecting an injured party and vice versa, the protection of peers in case of a serious event, i.e., nuclear, or chemical contamination.

A well-known issue in emergency medicine focuses on the delivery of basic life support to a severely injured individual. Common immediate threats follow the ABC rule, i.e., Airway, Breathing, and Circulation. A severely injured individual also quickly loses body heat, creating extra physiological problems and worsening their condition. The consideration of these core elements lead to the original Individual Survival Unit (ISU) concept for outer space medical emergencies (Figure 1).

The ISU was planned to be a modular unit made for specific conditions but with the common goal of being an individual unit that protects the individual from the harsh environment, keeps physiological parameters (temperature and oxygen intake) constant and is intuitive to use with minimal training. As a collaborative project, this concept concentrates on medical, engineering, materials, and ergonomic knowledge to improve the safety of transport of injured individuals in a harsh environment and contain the threat posed by potential contamination.

Although inspired by a hypothetical demand for emergency care in space, the project was selected by a group of EPS students with various educational backgrounds to offer a solution to emergency care/transport in harsh environments. Two potential space scenarios are presented in Figure 2, where an ISU is utilized for patient care both inside and outside of a space station.

The fundamental problem to be addressed is to design an ISU that can extend the survival time of an injured person in disaster areas. The key function is to deliver medical care in a stable environment in which a seriously injured or ill person can be protected, or an injured or uninjured person can be safely transported and avoid further injury during transport. This could be in the form of an incubator similar to the system used in Neonatology. Currently, there is no product solution like this for adults.

Figure 1. Original Individual Survival Unit Concept
(Credit: Prof. Michael Vloeberghs)

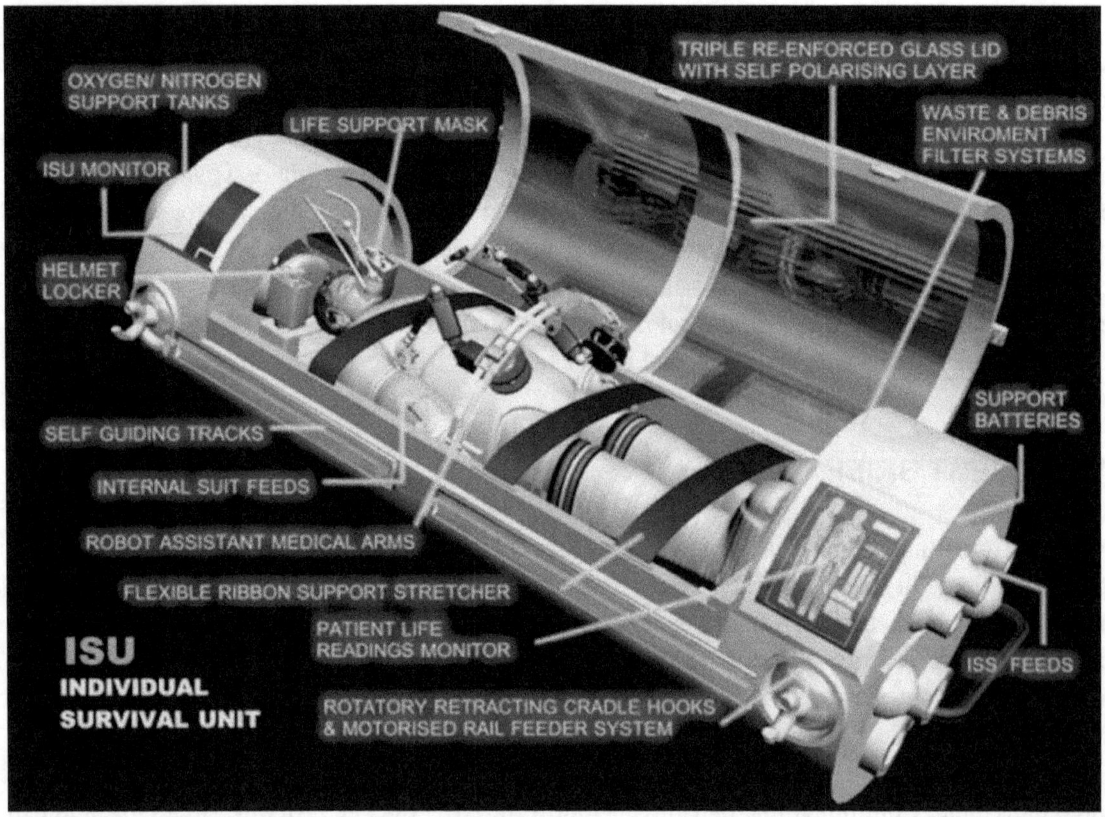

Figure 2. Individual Survival Unit Space Scenarios
(Credit: Prof. Michael Vloeberghs)

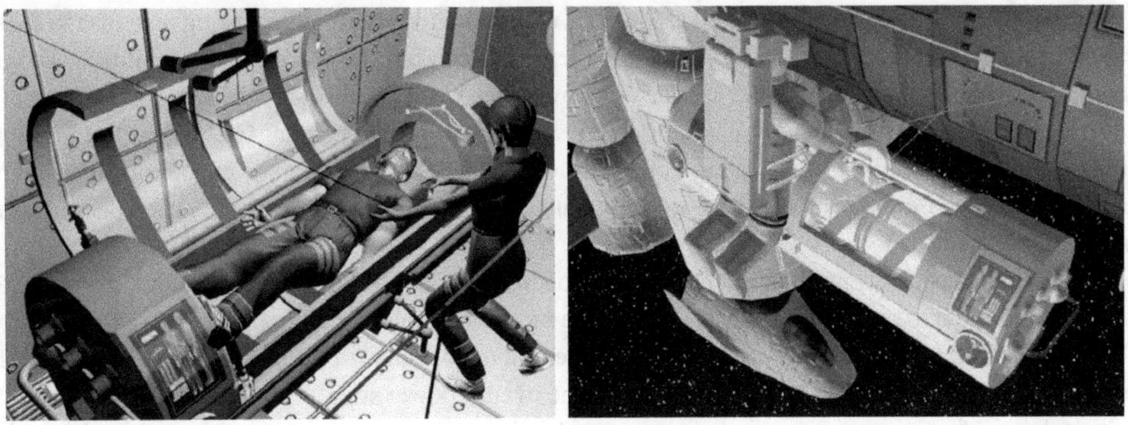

The controlled environment could be used to keep a person warm or cold and protect them from a hazardous environment, i.e., radioactivity, chemical, biological or bacteriological. Also, patients may need to be removed from a contaminated area without being a hazard to those accepting them i.e., move to a separate location, inspect, hose down, decontaminate. The system can be used to transport the wounded to safety or remove able-bodied people from catastrophic conditions. Potential commercial use includes marine environments, board room security, in-vehicle model for senior politicians, battlefields, disaster areas and home use.
Key product considerations include:

- Stackable, easy access, room for power supply, oxygen, ventilator, infusion pumps; each individual package can be developed singularly as there is a need for innovation for all of these.
- Materials, light, easily transported, resistant, shock resistant, resistant to corrosion, chemicals, positive pressure to keep contaminated air out.
- Adaptation to current civilian and military transport and power supplies.

CASE DESCRIPTION

Project Design Specifications

The aim of the project was to create a stable setting in which to protect a person exposed to a severe environment and to avoid further injury during transportation. Potential commercial use of such a product includes marine environments, battlefields, and disaster areas. The general purpose is to allow first aid to be delivered in addition to the stabilization of the patient by on-site teams, enabling them to then transfer the individual through the hazardous environment to a 'safe-haven' for further management.

Alternatively, in the proposed application, i.e., in disaster or rescue situations, the ISU can be used to evacuate subjects from locations through a hazardous environment to prevent exposure to an external factor, i.e., dirty bombs, biological threat, marine rescue and as an ultimate potential rescue scenario in suborbital flights or commercial orbital structures.

An injured person may need to be taken out of a contaminated area without being a hazard to those accepting them, such as when moving them to a separate location, inspecting their condition, hosing down, decontaminating, etc. The device needs its own power supply, heater/cooler, infusion pump (if required), a system of monitoring vital parameters, a communication system, Global Positioning System (GPS) receiver, and a beacon in case one gets lost (marine environment or chaotic location). The device is envisaged to be used to transport the wounded to safety or remove able-bodied people from catastrophic conditions.

Key design considerations focused on the inclusion of essential life support equipment. This included but was not limited to respiration and blood pressure monitoring and the appropriate filtered ventilation of the unit, including temperature control. It was essential to consider life support equipment that would be required for the ISU as this helped to determine the overall power supply requirements to realistically support the unit for the required period of transport. A "plug and play" system was adopted for the equipment inside the ISU, allowing for potential reconfigurations depending on the needs of patients and a particular disaster type. The design of the unit also considered the use of 'standard' emergency response equipment, particularly spinal boards for transporting patients. This was to ensure that the patient could be placed easily and safely within the ISU. Finally, an external user interface was designed to provide doctors/medics with appropriate vital patient data during patient transport.

To provide a focus for the specific design features for the ISU, a "disaster type" was chosen, namely earthquakes. Although earthquakes are not the most common natural disaster (flooding is the most common disaster with earthquakes being second most common), earthquakes have a far greater negative effect on a community due to the impact on the population infrastructure and the number of directly linked fatalities. Key objectives were identified for the project and included the exploration of similar transportation concepts, disaster impact investigation, and identifying the basic life support equipment requirements for the ISU needed to stabilize the injured person during transportation. Several academic staff at Nottingham Trent University and a visiting professor (Consultant Pediatric Neurosurgeon, Nottingham University Hospitals Trust) provided the ISU project supervision to ensure consistent project feedback.

Designing The Individual Survival Unit

The design of the ISU was formulated from the research conducted by the project group who utilized a range of primary and secondary research methods to assess qualitative and quantitative data. Primary data collection was based on expert surgical and clinical opinion and feedback, focusing on the needs of the patient, and identifying the essential life support systems that would be a prerequisite in the final design solution. Secondary data collection focused primarily on patient transport systems and identifying specific life support equipment and the relevant technical specifications. Initially a detailed review of patient transportation systems was undertaken and evaluated to gain an understanding of the wide range of existing systems and solutions that could be adapted for use in the ISU concept. A key element to securing a patient in any medical emergency scenario is using a stretcher or backboard. Firstly, stretchers were reviewed as they were considered one of the most important aspects of the design. It is a primary patient transportation method utilized for medical services on battlefields (Crome, 1939). Fundamental to a stretcher's performance are its lightweight properties and comfortability requirements; the modern stretcher can also be attached to intravenous poles, monitors, oxygen tanks and be adapted for the attachment of various other equipment solutions.

As part of the qualitative analysis several stretchers were identified and reviewed to assess their capabilities. The Survival Air Stretcher typically used in the United States of America is an inflatable stretcher with wireless monitoring self-supplied with lithium-ion batteries and a solar array (AirShip Technologies Group, 2019). The Micus Stretcher is a compact detachable spine board capable of being connected to medical equipment; this contains a power supply and is capable of being adapted for different kinds of transportation (Gustavsson & Lagerquist, 2003).

Inspiration was also derived from current medical equipment utilized to maintain a patient's medical status. Infant incubators are commonly used in hospitals and act as a unit that creates a stable and suitable environment for premature children who require life support. Incubators act as self-contained units for babies and can regulate body temperatures and ensure the ideal environmental conditions can be created, whether this is by adjusting temperature, providing an environment free from infections, allergens, excessive noise, or light that can cause harm. Air humidity can be regulated to ensure the integrity of the skin. Mittal (2015) suggests that temperature, humidity, light, and oxygenation need to be measured and regulated as the smallest of changes can result in significant harm to newborns with intensive care needs. These are all parameters that the ISU will require.

However, the requirements identified from a stretcher and infant incubator are not enough to create a stable environment worthy of an ISU. As such, a review of equipment located in an ambulance was completed. Equipment necessary for medical professionals and their clinical tasks, including heating and cooling systems and hygienic design considerations, were identified as key design criteria (Snook, 1972; Hignett et al., 2009). According to Ferreira & Hignett (2005), the four main clinical tasks that frequently occur in an ambulance include checking for a pulse, identifying blood oxygen saturation, oxygen administration, and monitoring the heart through cardiac monitors and blood pressure. All these tasks require specific equipment which requires integration into the ISU.

To identify where an ISU would have the greatest positive impact, statistics were analyzed for the Risk of Disasters in Consequences of Extreme Events for 171 Countries – based on the WorldRiskIndex developed by the United Nations University Institute for Environment and Human Security – using the World Risk Report of 2018 (Heintze et al., 2018). This report identified that the most exposed continents in the world are Asia and Africa. The most common disasters prevalent in these areas are floods, earthquakes, and storms. Meanwhile, most deaths caused by natural disasters are from earthquakes and extreme temperature events.

Basic life support (BLS) is a level of medical care that is used to maintain airway patency in addition to supporting breathing and circulation without the use of equipment (Handley et al., 2005).
The key points for basic life support are a) airway, b) breathing, and c) circulation:

a) The protection and maintenance of a clear passageway for gases (principally oxygen and carbon dioxide) to pass between the lungs and the atmosphere.
b) Inflation and deflation of the lungs (respiration) via the airway.
c) Providing an adequate blood supply to tissue, especially critical organs, to deliver oxygen to all cells and remove metabolic waste via blood perfusion throughout the body.

These basic life support features helped generate the list of necessary emergency medical equipment that was to be included in the ISU. This equipment list ensures that basic life support and maintaining a patient's life can be achieved during an emergency; this information is also implemented into the design parameters and product design specifications (PDS).

Problem definition, identification tasks, task exploration, creative phases, and detail design stages are fundamental to the design of any product. However, one of the essential activities to be undertaken is the creation of a product design specification. By identifying specific design and technical requirements, the overall design requirements for the ISU can be ranked in order of importance and appropriately designed. The criteria that needed to be considered was also ranked against the project objectives.

The developed product design specification focused on several key areas: disaster types, functional requirements, aesthetics, performance requirements, dimensions, medical equipment requirements and standards, communication equipment requirements, and material selection. Once the PDS was completed, the overall design requirements were listed and weighted as presented in Table 1 after consultation with stakeholders.

Table 1. Individual Survival Unit Summarized Design Requirements

Criteria	Weighting (%)	Description/Key Questions
Accessibility	15.0	Rescuer's operation requirements, ease of use, and user operative times.
Patient Visibility	15.0	Materials used for the ISU lid/doors must be transparent to allow patient visualization during transportation.
Usability For Emergency Workers	15.0	Interaction considerations for individual equipment located within the ISU.
Weight	10.0	The unit's weight must be minimal to allow for ease of maneuverability.
Water & Airtight	10.0	Airtight seals are required to ensure the unit is airtight and watertight.
Space Utilization	10.0	Space requirements identified based upon patient ergonomic data.
Hygiene & Sterilization Capabilities	7.5	How can the ISU be designed to prevent the unit from being contaminated? The ISU must be easily cleaned and sterilized.
Strength	7.5	What stresses will the ISU be placed under? (Impact, Hardness, etc.)
Production Complexity	5.0	What manufacturing processes can be utilized? (Minimal & Simple)
Storage	2.5	How can the units be stacked and stored?
Transportation	2.5	How can the handles, trolleys, and wheels be attached to the main unit?

Preliminary Designs

The design process for the ISU was based on the findings derived from the primary and secondary research conducted. The concepts generated (Figure 3) during the design process were carefully considered to ensure the required quality for the final design and project solution could be achieved. Numerous concept generation and sketching activities were conducted to quickly communicate and exchange key ideas to inform several concept solutions. This allowed the integration and implementation of the conducted research in the design ideas leading to the formulation of development concepts and final solutions. To elaborate on this further, concepts of the ISU were broken down into four areas:

1. **Transportation**: Studied during the research and concept development processes undertaken. The ideation of transportation systems suitable for the ISU was considered in parallel to the disaster type chosen. The choice of transportation methods impacts the general weight of a concept and the interaction between the rescuers and the victim.
2. **Form:** The form/shape of the ISU is informed by the dimensions of an adult stretcher and the morphology/anatomy of the human body. The focus was placed on ergonomic data from the average male and female heights; this allowed the basic definition of the ISU dimensions to be set. This then allowed design work to be conducted on the shape of the ISU to address critical criteria identified in the PDS, i.e., the weight of the system, the comfort for the victim, sterilization, hygiene design considerations, etc.
3. **Accessibility:** After defining the ideal scenarios for using the ISU, a key design consideration was the accessibility and insertion of a patient into the ISU. The main criteria used was to find the easiest and quickest solution by which entry into the ISU can be achieved, whilst also considering mechanisms that are not complex and are convenient for the rescuers to use.
4. **Interaction:** When a patient is secured and protected within the ISU, it is essential that the rescuers/medics can provide medical care where necessary and intervene again when further patient stabilization is required. The solution is determined based upon the logistical requirements set out by the rescue organizations to ensure that the designed ISU can facilitate patient manipulation, decrease intervention time, and increase the likelihood of patient survival. Inspiration was drawn from existing emergency service vehicles.

Figure 3. Concept Generation/Ideation Process
(Hengsadeekul et al., 2019)

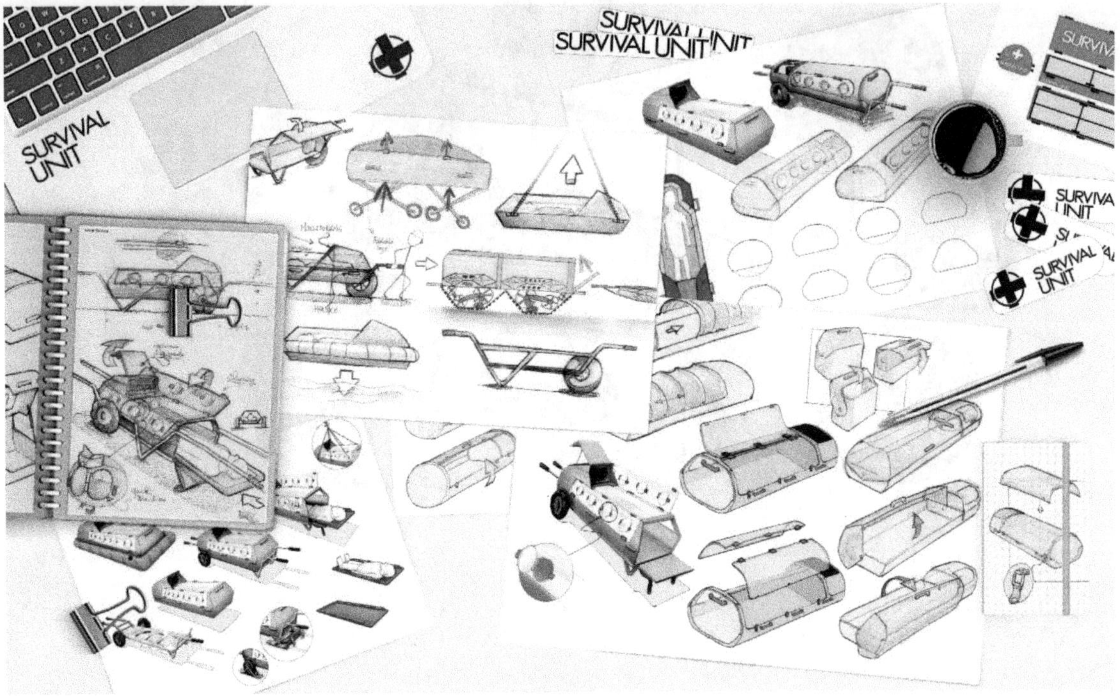

The student group considered three ISU concept designs. The project originally aimed to assess the design feasibility of a unit that could increase the survival rates of an injured person who would not normally have access to such life support equipment immediately following a disaster.

The Final Design

The final design of the ISU presented in Figure 4 and Figure 5 has been produced to be as simple and effective as possible to ensure a patient's survival and ensure that it is easy to maintain, operate, and be adaptable for various scenarios. The design of the ISU utilizes the minimum number of parts required to make this a functional system. This has been completed to ensure a reduced price for the ISU to maximize potential impact. The shape and material studies completed allowed the design to be lightweight without reducing the comfort for the patient whilst also ensuring maximum usability for the rescue teams/emergency care staff.

An outer glass casing has been integrated into the ISU to always allow visualization of the patient during transportation. This solution also allows the patient to see the exterior environment to reduce the likelihood of claustrophobia or anxiety linked to confinement. A control screen has been integrated into the top of the ISU to act as a display for monitoring patient vital signs. This will need to be connected to all the medical equipment located within the ISU in addition to any remote monitoring sensors.

Figure 4. Component & Layout Design of the Individual Survival Unit
(Hengsadeekul et al., 2019)

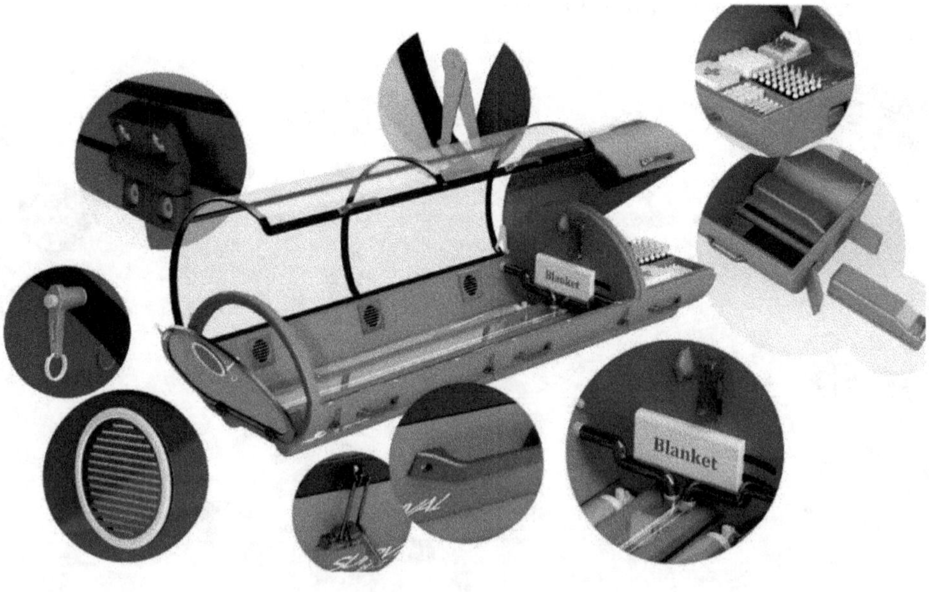

Figure 5. Individual Survival Unit Final Concept Visualization
(Hengsadeekul et al., 2019)

Mechanism design considerations have been reviewed and implemented to ensure the glass windows/ doors open utilizing a pivot action. This allows rescuers to intervene and provide emergency medical care quickly when required. The opening of the ISU ensures total patient access; this element is to be manufactured as one part. The integrated pivot mechanisms maintain and assist movement of the doors/ windows during their rotation; the door placed on the backside of the ISU opens outwards, ensuring patient access. Lock systems are placed on the ISU; however, these are released quickly to allow the ISU to be opened and closed quickly. However, the relevant seals have been integrated to ensure compartmental integratory from outside elements. Finally, a security key allows the necessary compartments to be locked to reduce the likelihood of theft, thus only granting access to the rescuers or medical teams.

The mobility and transportation requirements for the system focus on how the ISU can be maneuvered and transported to different environments depending on the present elements. Moreover, the ISU has been designed to ensure that the patient can be treated irrespective of the elements even in the most challenging terrains, thus ensuring that the system can be adapted accordingly. In its current state, the transportation opportunities considered include seven scenarios, including air transportation, water transportation, etc.

To allow the ISU to function on land, it can be moved using handles located on the unit as per normal usability considerations with standard medical equipment. The frame can extend underneath the ISU to increase the length of handles for difficult terrains. The extra length of handles located at the front and rear of the ISU provides multiple locations where the unit can be held from. Meanwhile, the carrier handles can also be adapted for wheelbarrow and ski sled scenarios, thus allowing the ISU to be used in boggy land, ice, and snow environments ensuring the unit's capability to be maneuvered with limited physical support.

Alternative attachments designed into the system consider other transportation methods such as water mobility or air transportation attachments. An inflatable raft can be added/attached underneath to ensure the ISU can float and be driven/pulled on the water to transport the unit. The handles at the side of the unit were also designed to allow connection to ropes/wires retracted from air-based vehicles such as helicopters. Lastly, for mass transportation and storage, the unit can be stacked upon each other to transport many units simultaneously. However, additional framing around the outside of the units will be required to allow mass shipping/stacking.

Further research and design work was conducted based on a user interface to communicate medical information and GPS data (Figure 6).

Figure 6. User Interface Design of the Digital Communication Screen
(Hengsadeekul et al., 2019)

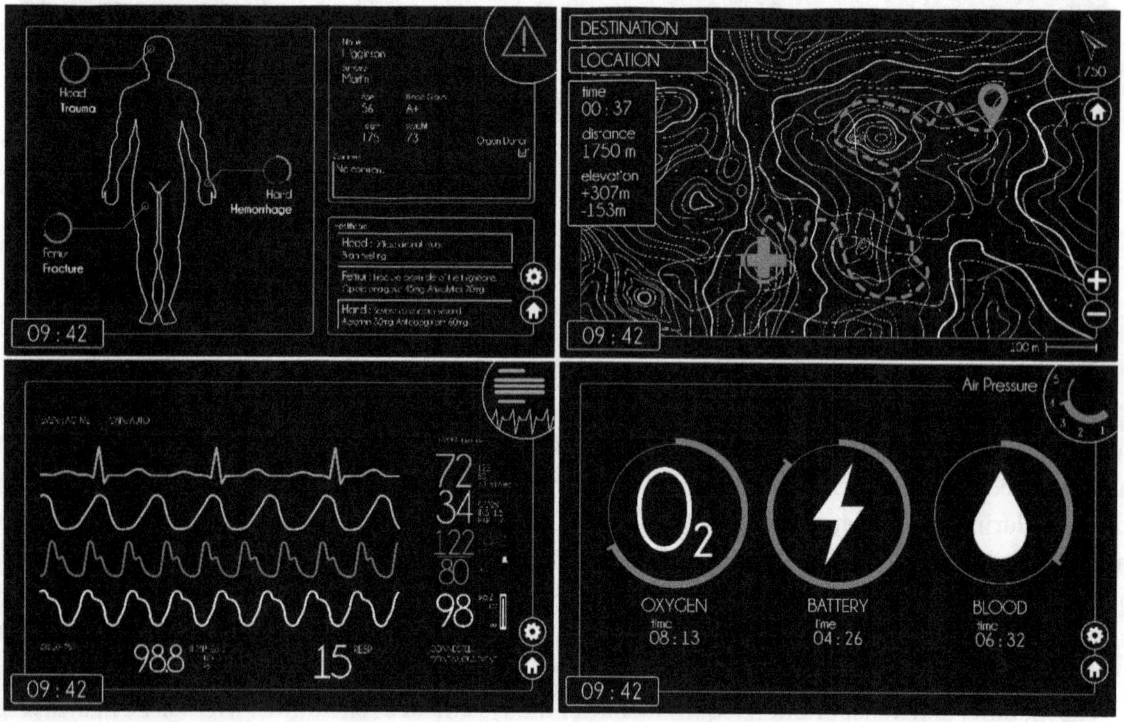

This interface was created with the idea of simply sharing the necessary patient information to the medical teams allowing rescuers to make informed decisions on the health status of the patient but also monitor the relevant medical equipment usage and capacity. An essential function intended for implementation is the ability to display the necessary diagnostic information of all components of the ISU, i.e., structural integrity. The integrated computer will record all patient data, which can be easily streamed to the doctor at the emergency care unit/hospital, who will assume responsibility for the patient once they arrive at the designated facility.

Emergency Requirements & Power Supply

The main concern in terms of the technologies specified relates to the power supply requirements for the systems within the ISU. Whilst estimates can be made in relation to the overall power requirements and individual equipment loads, clearly practical testing would ultimately substantiate the actual requirements with tests based on maximum load conditions. It is essential to determine the feasibility of the proposed power supply and its run time until battery exhaustion.

From a student perspective, calculating the appropriate power requirements was challenging initially but clearly pivotal in providing a realistic solution for the ISU and associated equipment. Once again, the multidisciplinary nature of this project was highlighted by the need for the student group to address and calculate these power requirement figures. Figure 7 shows the estimated power requirements for the ISU. BLS and monitoring are key considerations when considering the appropriate technology for the ISU. The design considered the use of existing BLS technologies that could be used within the ISU based on a "plug and play" system. A key consideration when specifying the BLS equipment was related to the power requirements of individual items and the potential total power load. Table 2 highlights the main specification requirements for the ISU.

Figure 7. ISU Estimated power requirements
(Hengsadeekul et al., 2019)

Power Autonomy	Operation	Time (min)
	Medical equipment operating at full power with the heater/cooling element functioning at 100 % capacity.	102
	Medical equipment operating at full power with the heater/cooling element functioning at 75 % capacity.	136
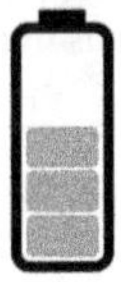	Medical equipment operating at full power with the heater/cooling element functioning at 50 % capacity.	204
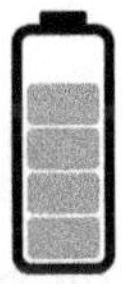	Medical equipment operating at full power with the heater/cooling element functioning at 25 % capacity.	409
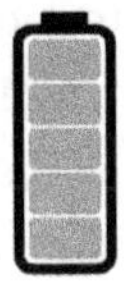	Medical equipment operating at full power without using the heater/cooling element.	686

Clearly, the form factor and overall size of existing medical equipment to provide BLS had to be considered to ensure the individual equipment/units can fit within the ISU. In addition, the design necessitated the need for a heating/cooling system which potentially could provide the highest power drain on the system. Once again, taking a multidisciplinary approach, the student group had to investigate and recommend the appropriate power supply to meet the potential needs and demands of the system.

Careful equipment selection and determining the appropriate and realistic power requirements ensure that basic life support and maintaining a patient's life can be achieved during emergency situations.

Table 2. Key technology components and considerations

Requirements	Purpose
Air Control System	Control the air temperature inside the ISU to ensure regulation/maintenance of the patient's body temperature.
Charging System	Charging the ISU in preparation for its subsequent use.
Connection Ports	Connection of all the necessary medical equipment with the monitoring and display systems.
Emergency Light	Presentation of the ISU to the emergency responders and injured patients.
Fire/Heat Resistance	Prevention of further patient injury caused by external fire and heat.
GPS Device	Identification and notification of the location of the ISU.
Heating/Cooling System	Control the temperature inside the ISU to ensure regulation/maintenance of the patient's body temperature.
Hygienic Design	Prevention of contaminants coming into contact with the patient or ISU user.
Internal Light Unit	Improving the ambient lighting environment to ensure patient visibility is maintained and the patient is also aware of the environment and not left in a dark enclosed environment.
Monitoring Systems	Presentation of the patient and ISU data to the rescue and emergency response teams to ensure patient maintenance.
Power Supply	Must be capable of generating enough power for the ISU systems and be self-sufficient during immediate use.
Sterilization Design	Access to all sections of the system, no cavities that are not accessible, no sharp edges, rounded interior; all these requirements allow the device to be sterilized.
Two Way Communication	Two-way communication systems ensure communication can be maintained between the emergency response team and the patient.
Water/Gas Leak Prevention System	Prevention of O_2 leakage. Watertight seal for the ISU to prevent external fluids from leaking into the system.

REFLECTIONS ON EPS DELIVERY AND MANAGEMENT

Project Supervision & Support

Enhanced project supervision was provided by academic staff and a visiting professor throughout the course of this project. The project director provided weekly project supervision to monitor the progress of the student work. This was supported by a second supervisor who assisted during the weekly project meetings when required and was also available for additional project meetings. This dual supervision process ensured constant project feedback could be provided whilst also improving knowledge transfer due to improved accessibility to a wide range of expertise, thus contributing to the successful project outcome. Including healthcare professionals and discipline specialists as project supervisors added more significant value to the project supervision and thus significantly influenced the quality and viability of the project deliverables.

In addition to the project supervision, access to a dynamic multidisciplinary research group (MEDRG) consisting of specialists such as product design engineers, human factors and human computer interface specialists, software developers, robotics, simulation, and automation developers ensured that additional resources relevant to specialized sectors are available and applied to the research work accordingly. This is imperative when considering the design of any medical device or equipment. The importance of using a multidisciplinary design team during the design process cannot be overstated and often influences the success or failure of a project.

Alongside the specific supervision requirements pertinent to the project brief, students also received general guidance and tuition regarding key components that underpin the EPS project delivery. Teamwork and communication skills are honed through discrete activities along with Project Management and Research Methods teaching to ensure the group has the correct tools, methodologies, and underpinning knowledge in the subject outside of that provided by the supervisory team. Other support subjects are similarly designed to create a good group dynamic and foster good working relationships.

Students are also given access to several design studios, high-quality engineering laboratories, extensive modern studios and workshop facilities, information technology and computer aided design suites. All NTU students, including EPS students, undertake an induction in workshop safety and practice to access and utilize workshop equipment as part of their project outcomes. All these facilities are operated by a dedicated technical team who are on hand to provide students advice at any time. Access to design studios utilized by both the undergraduate and postgraduate cohorts of students promotes the studio-based culture that encourages collaborative learning, ensuring that students experience a range of creative perspectives, allowing the students to gain an understanding of the product market, social needs, and end-users.

Organizational Challenges

As part of the EPS application process, students are asked to choose their preferred project options and second preferences; this information, alongside EPS rules, is utilized to determine group structures. This is further supported by a skills audit undertaken when the students arrive at NTU during the University's welcome week. The ISU project brief proved extremely popular, and the demand outweighed the number of student places available. The student group took up the project quickly after the briefing held at the beginning of the project. Initially, the limited understanding of resuscitation and emergency care had to be overcome. Only basic 'mechanical' understanding was required for the ISU project as the group rapidly focused on the engineering challenges involved. The initial idea of space rescue gave the group extra impetus, particularly since the device was directly applicable to "live" situations. An added attraction is that it is an entirely new concept with a host of applications and a contemporary feel, mainly because of the advent of autonomous vehicles, where a direct link with the ISU can be made.

Projects offered through the EPS scheme have a significant impact and can provide additional value when shown as "live" project briefs. Other benefits are linked to the provision of project support as these briefs are often linked to external organizations, specialists, and professionals. It can become increasingly difficult for active researchers to allocate time to manage these projects. However, connecting these projects to current research activities and existing live project briefs can be extremely rewarding for both students and their supervisors.

To allow the practical completion of EPS briefs, allocation of resources is critical. However, there are obvious resource implications currently met from the Product Design department's budget unless an industrial partner is prepared to support the costs. With the current cohort sizes (approximately 20 to 25 students per year), this has not been too problematic, but if the number grows, this may become more of an issue in the future.

The EPS project groups have partial use of a defined workspace. Still, as is the case for Product Design's Undergraduate population, dedicated rooms do not exist due to a centralized resource timetabling system and growing student numbers across the sector. There are ample resource rooms around the University where small groups can work, though it would be ideal to have EPS working spaces, similar to some partner institutions.

Student Experience

The group successfully achieved the project's aim, and the concept of the survival unit was successfully established. The project's feasibility study has been created, which will lead to further research and development. By undertaking the Individual Survival Unit (ISU) project, the students have developed and demonstrated the acquisition of new knowledge in various areas by engaging in the dynamic area of medical device/equipment research and design.

Working on a live brief alongside external professional collaborators often promotes student interaction and engagement and consequently has many synergies to the EPS ethos. Student feedback highlighted that:

The team were able to proudly state that they really enjoyed working on this project as well as feeling grateful for the opportunity to explore and gain new experiences as part of the program.

Offering students projects that have a real-life impact to which they can relate, whilst enabling them to work alongside external professional collaborators, continues to be an attractive and desirable option for students.

CONCLUSION

Future development work required for the ISU project must be split into several different areas including the construction of a prototype that visually represents the ISU and the development of individual working prototypes of the mechanisms/mechanics of the ISU; this should then be followed by the development of a fully functional scaled prototype suitable for bench testing. If successful, a full-scale basic working prototype for a simple trial and feedback in standard situations, i.e., military, mountain rescue, marine, etc., should be developed. Based on the student work developed, it is evident that the most powerful parts of ISU are the innovation, the immediate practical uses, and how it demonstrates an unmet demand for safer transportation of injured and non-injured individuals.

The ISU project has helped the student group develop broader skills and attributes essential for graduate engineers and designers to cope with the ever-increasing demands of the global market. By completing a "Live" project, the student group has developed an understanding of a broad range of disciplines such as economics, management, communication, and languages whilst tackling a complex medical product/system design project. Upon completing the project, the student's reflections referred to the value of the support subjects that underpinned the project and how they utilized their learning to solve a problem brief. The cohort in general and the ISU group were very complimentary about their experience at NTU. One of the ISU group members cited the EPS experience as a contributory factor in his decision to apply for a further semester at NTU.

In conclusion, it is imperative that a multidisciplinary approach is adopted when completing "Live" student projects. Design, engineering, and technology-based projects can help provide and develop student awareness of the impacts of design, engineering, and technology on today's world and society. The diversity of multidisciplinary research projects requires designers to integrate sound design principles and create an awareness of emerging technologies. The methodology and practical approach adopted by EPS student projects is particularly relevant considering the expanding technology market, as such it is now essential that designers have a wider perspective on appropriate technologies and their integration into products.

ACKNOWLEDGEMENTS

The authors would like to acknowledge the Erasmus student group who delivered this project brief: Thawanrat Hengsadeekul, The Hague University of Applied Sciences, Netherlands, Douwe Groot Zevert, Saxion University of Applied Sciences, Netherlands, Morten Dau, University of Southern Denmark and Adrien Feron, Strate School of Design, France.

REFERENCES

AirShip Technologies Group. (2019). *Survival Air Stretcher*. https://www.airshiptg.com/air-stretcher

Crome, L. (1939). The Stretcher Problem. *British Medical Journal*, 2(4111), 819–820. doi:10.1136/bmj.2.4111.819 PMID:20782725

Ferreira, J., & Hignett, S. (2005). Reviewing ambulance design for clinical efficiency and paramedic safety. *Applied Ergonomics*, *36*(1), 97–105. doi:10.1016/j.apergo.2004.07.003 PMID:15627427

Gustavsson, J., & Lagerquist, M. (2003). *MICUS Stretcher*. The Index Project. https://theindexproject.org/stories/micus-stretcher

Handley, A. J., Koster, R., Monsieurs, K., Perkins, G. D., Davies, S., & Bossaert, L. (2005). European Resuscitation Council Guidelines for Resuscitation 2005: Section 2. Adult basic life support and use of automated external defibrillators. *Resuscitation*, *67*(Suppl. 1), S7–S23. doi:10.1016/j.resuscitation.2005.10.007 PMID:16321717

Heintze, H.-J., Kirch, L., Küppers, B., Mann, H., Mischo, F., Mucke, P., Pazdzierny, T., Prütz, R., Radtke, K., Strube, F., & Weller, D. (2018). *The WorldRiskReport 2018: Child Protection and Children's Rights*. Bündnis Entwicklung Hilft and Institute for International Law of Peace and Armed Conflict. https://www.WorldRiskReport.org

Hengsadeekul, T., Zevert, D., Dau, M., & Feron, A. (2019). Survival Unit (EPS Final Report). Nottingham Trent University.

Hignett, S., Crumpton, E., & Coleman, R. (2009). Designing emergency ambulances for the 21st century. *Emergency Medicine Journal*, *26*(2), 135–140. doi:10.1136/emj.2007.056580 PMID:19164631

Mittal, H., Mathew, L., & Gupta, A. (2015). Design and development of an infant incubator for controlling multiple parameters. *International Journal of Emerging Trends in Electrical and Electronics*, *11*(5), 65–72.

Nottingham University Hospitals NHS Trust. (2019). *Our organization | NUH*. https://www.nuh.nhs.uk/our-organisation

Snook, R. (1972). Medical aspects of ambulance design. *British Medical Journal*, *3*(5826), 574–578. doi:10.1136/bmj.3.5826.574 PMID:5069643

KEY TERMS AND DEFINITIONS

ABC First Aid (ABC): The method of preserving life, preventing harm, and promoting recovery by the means of providing basic life support through essential airway, breathing and circulation techniques.

Basic Life Support (BLS): The level of medical care that healthcare providers and professionals administer to people with life-threatening illnesses or injuries up until they can be given full medical care by advanced life support workers.

European Project Semester (EPS): A program of study offered to students who have completed at least two years of study by several European universities as an exchange program.

Individual Survival Unit (ISU): A modular medical unit manufactured for specific disaster conditions capable of suitably protecting individuals from harsh environments whilst maintaining an individual's physiological parameters (temperature and oxygen intake).

Medical Device Development: The development process by which medical devices and systems are designed and realized based upon a medical engineering or medical design challenge.

Multi-Disciplinary Design Team: An approach whereby a single team of individuals from a wide range of different disciplines and expertise collaborate on a design/engineering problem.

Product Design Specification (PDS): A document which details the essential design requirements which a product is required to provide as defined by the key stakeholders.

Chapter 16
The City as a Laboratory:
Concepts, Designs, and Experiments in Higher Education

Isabel Gasch
https://orcid.org/0000-0001-7036-4481
Universitat Politècnica de València, Spain

Aurora Martínez-Corral
https://orcid.org/0000-0001-8222-0864
Universitat Politècnica de València, Spain

Nuria Matarredona-Desantes
Universitat Politècnica de València, Spain

ABSTRACT

Treating cities as laboratories and as learning and experimental objects has many advantages playing an important role in fostering and supporting the circular transition. A city is an accessible and inexhaustible resource for considering motivating challenges to those learning any subject, and its nearness and complexity make it ideal for university students. This chapter explains this concept and describes the method and results obtained by two different examples and a related workshop with groups of students of the European Project Semester. The first was a learning experiment on the inclusive city carried out during the academic year 2018/19, and the second was on the redesign of the Valencia bus station as a challenge during the 2020/2021 academic year, including the workshop as preparatory activity. In this chapter, the authors explain a reflexive-type activity that contributes to exploring the city as a complex environment collaborating through university activity in the necessary transition towards more sustainable cities.

DOI: 10.4018/978-1-6684-2309-7.ch016

Copyright © 2022, IGI Global. Copying or distributing in print or electronic forms without written permission of IGI Global is prohibited.

INTRODUCTION

The European Project Semester (EPS) at Universitat Politècnica de València (UPV) provides a set of Authentic Learning opportunities. Very often, one of the options includes the use of the city as a lab because cities are an accessible and inexhaustible resource. For this purpose, professors from different UPV departments collaborate to define, supervise, and evaluate the projects based on their complementary expertise. These projects also allow learning from stakeholders like neighborhood associations, technical staff, and private and public institutions managers.

The main goal of this chapter is to explain why a city could be a fantastic learning and experimental object. Obviously, students have a background on the subject, and it is easy to find the appropriate level of complexity for creating exciting scenarios and tasks.

To achieve this goal, the authors explain the rationale and describe methods and results from three examples:

1. An EPS project called Cities, also for kids. It was a learning experiment on the inclusive city carried out during the academic year 2018/19 (Steerman et al., 2019).
2. A workshop about the key aspects of critical complex city installations, with special emphasis on a bus station. The activity is based on Kolb's reflexive learning theory, understanding learning as the creation of knowledge through experience, but which needs reflection to be produced. Thus, learning is a conversation between thinking and doing, and it involves blending theory and practice.
3. The redesign of the Valencia bus station as an EPS challenge proposed for the 2020/21 academic year (Berger et al., 2020).

Thus, the chapter is structured in the three parts mentioned above, each one detailed from the contextualisation to its own conclusions or, in the case of the workshop, from the ideation and objectives to the contents to be worked on, which are complemented by a general introduction and final conclusions.

BACKGROUND

To consider any activity connected with a city is a constant challenge involving all the subjects. It also means entering the dynamic, ever-changing space in which we live, work, and relate to others, especially where everybody has his own experiences and perceptions of the existing collective image. This is a constant challenge and responsibility that can be interpreted in an infinite number of ways.

Therefore, as a learning experience, it identifies, includes, and involves emotions and gives rise to curiosity and debate. As it is carried out within the EPS framework, it provides enriched feedback from the visions and contributions of other cultures and ways of city living, besides being an added value to the interchange experience. Working on an aspect of the city in the creative EPS framework with teams of different nationalities, using the heterogeneity of the diverse backgrounds, is a fantastic opportunity to learn not only through mutual enrichment but also by meeting the actual environment and its dynamics, experiencing the object of the analysis in person. This sensorial enrichment is a motivating contribution to the final result, as it considers the proposed objective as approachable and relevant, and this is a guarantee of motivation and success. Teamwork, leadership, practical thinking, innovation, and effective communication are some of the transversal competences that reinforce the students during the course of this activity.

Carrying out a specific activity in the city also means both practical and economic or normative problems besides considering diversity, one of its intrinsic components: functional, gender, and social diversity, as well as considering the different collectives. Greater sensitivity, empathy, respect and humanity are other values that indirectly enrich the activity.

In fact, actions related to activating the city for normally under-considered population segments, such as children, are becoming more and more frequent in town planning. For example, the creation of pedestrian zones and 15-minute cities are cutting-edge subjects currently being considered by the European Union (EU) authorities in relation to children's welfare. The results are inclusive and accessible spaces that favor equality and communal life. Trying to show the relevance of the context of the case under study, in this occasion the city, its collectives and problems and the possibility of acquiring "soft skills" or indirect values through it is one of the considered premises.

A valuable activity is to evaluate the experience of each group of students concerning the problem being studied. Assessing the problems under study, what is to be debated, how they are related, how they experience the place and how they solve the different questions that arise. Reference is made to the use of a style of reasoning in which an attempt is made to formulate questions through understanding the phenomena, i.e., questions are asked that must be answered utilizing the information collected by observation of the context impregnating the problem. The solution, therefore, is not derived from the problem but is modeled on it.

The student must conceptualize, apply, analyze, synthesize and actively and skillfully evaluate the information obtained or generated through observation, experience, reflection, reasoning, and communication to guide his actions. The ideal is to use the universal intellectual values that transcend the divisions between subjects: clarity, exactitude, precision, coherence, relevance, solid proof, good reasoning, profundity, amplitude and equity. One of the EPS' significant advantages is that it combines strictly scholarly contents, terms and timetables (to which most of the studies are subjected) with creativity, flexibility and various ways of obtaining results, i.e., a balanced acquisition of soft and hard skills.

As mentioned in the introduction, it is not the first time the city is used as a lab within the EPS. During the last decade, there have been some projects related to the city of Valencia (Gasch et al., 2012; Gasch et al., 2014). The students have worked designing the "Parque Central", creating an urban plan of the area of Mestalla stadium (which will remain free when Valencia C.F. move to the new stadium being built in the Avenue of the Valencian Courts) or redesigning some areas of the Cabanyal district.

CASE STUDY 1 – CITIES, ALSO FOR KIDS?

Basis of the Activity

Town planning is the result of a set of requirements that have given priority to units of production over those of reproduction or those associated with informal work, especially since the industrial revolution.

In recent decades, cities have been consolidated to facilitate motor transport to take commuters to work, most of whom are male breadwinners. Activities carried out in other contexts are thus given less importance, such as home deliveries, journeys to school, or visits to medical centers, thus under-rating population segments such as children (Figueras, 2007).

In this regard, Tonucci (2004, 2009) proposed that "Kids should have the same space to play as adults have for parking". As a city that is better for kids is better for everyone. Figure 1 is an example of the illustrations created to disseminate this concept.

Figure 1. A place to play
(La Città dei Bambini, 2017)

Various studies have dealt with what is known as the geography of childhood, which studies the gradual disconnection of kids from public places. The natural appropriation of constructed spaces has declined in favor of ever-more normalized spaces. Growing up in a city thus means surviving, in a certain way, an environment perceived as hostile in which only certain areas permit collective games and vital open-air activities away from schools. Therefore, the interaction of kids with the city is limited to certain spaces generally designed by adults considering their vulnerability and directives regarding quality and safety.

However, even if it is true that there is growing awareness of questions such as safe school routes, cities are still far from allowing kids' free and creative development (IAEC, 2020).

Aims

In the 2018/19 academic year, a group of EPS students found themselves considering whether there were any innovative solutions to be able to hack urban hardware by means of apps or software to provide kids with new ways of interacting with their immediate constructed environment, enabling questions such as identifying with a place or healthy physical and mental evolutionary growth.

This project was created to improve city space, make children feel safe, not excluded from the activities around them, and make the urban area a place where they could reach their maximum potential while raising awareness and questioning the existing solutions for making big cities suitable for everyone.

Method

For this, the students evaluated, individually and collectively, their own experiences and compared them with the reality of a city like Valencia. The international group confirmed the tendency to make children invisible in the city and the globalized response to the need to have public play spaces. Figure 2 shows one example of mapping and ideation activities.

Figure 2. Example of team activity for mapping and ideation
(Credit: N. Matarredona)

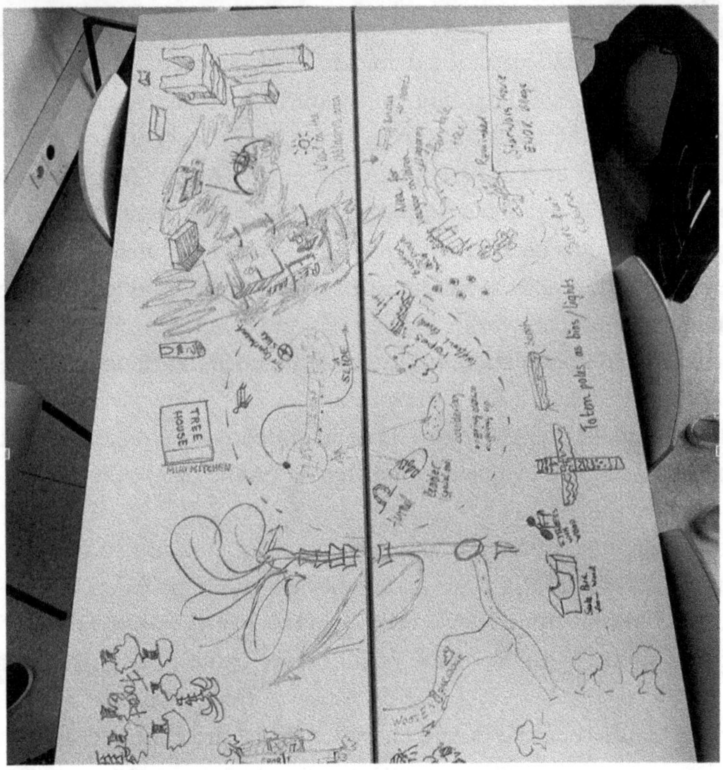

Their conclusions highlighted that the basic fact recognized initially was the difference between children living in cities and those living in the smaller towns or the countryside. Even if children in other countries have different opportunities or social levels, the basic facts match each other.

Every member of the group loved to play outside in the open air in places like forests, gardens, streets, or fields. Gadgets were not that important because the environment provided basic elements to build toys and invent games.

After reflection, the group started an investigation to contextualize their own experiences by analyzing case studies.

Research and information are needed to learn about what has been done in the past and to learn about successful and less successful ways of generating change. Analyzing these facts, they could determine why and how these attempts have been successful and trigger new and innovative ideas to achieve similar results.

Data were collected through the analysis of successful case studies and then analyzing them to discover existing problems that became the basis for generating ideas to create a solution to achieve the project's aim.

The key questions were transformed into a 'Top 10 criteria' that summarized the emotion, thinking, and learning that this project aimed to fulfill.

1. 4. Explorative: part of growing up, learning.
2. 5. Challenging: acceleration, danger, action, adrenalin, different from everyday movement.
3. 6. Innovative: something new, fulfill the demands, entertainment, get people outside.
4. 7. Nature: in contrast to city life, learning from nature, not busy (relaxing).
5. 8. Rest area: safe place (withdraw from activity), parents, intergeneration, comforting, sun/shadow.
6. 9. Colorful: make it look friendly, mark happy places, learn from colors (warnings, etc.).
7. 10. Mobility of playground: available for the whole city and different social levels, flexibility.
8. 11. Safety: important parents and children, if not they will not come again.
9. 12. Team games: interaction, activity, learning the meaning of winning and losing, becoming social, learning the importance of teamwork.
10. 13. Creative: learn to use imagination, improve intelligence, learn new things, skills.

In the subsequent ideas phase, the group actively and collectively conceptualized ephemeral proposals to be tested in the city itself and evaluated their impact on the daily activities of children that come across these activities.

In the process of finding a solution, the team shifted from creating something physical to a more important factor of the atmosphere of a place and awareness of the problem.

Giant objects are used in playgrounds everywhere in the world. A playground full of big everyday stuff is truly an exciting sight. The idea of the Giant theme came from the case studies found in the research phase.

The team then visited Gulliver Park in Valencia to interact and experience this unique playground. Essentially, this is a large-scale, 70-meter-long playground, where Gulliver is lying on the ground with his clothes and his hair forming ramps, slides, and stairs where kids can play. The question is, why are giant objects so interesting?

In order to better understand this issue, the team performed several experiments in the city of Valencia and concluded that not much effort is needed in terms of changing a place's atmosphere.

Three different experiments aimed to prove that firstly it is possible to change the atmosphere of urban areas to be more child-friendly, and secondly that not many resources are needed to entertain children. According to the research, huge elements of everyday stuff are fascinating for children, and it was decided to create and use the objects listed below.

- Dices
- Bubbles
- Chalks

Managing a proper experiment also includes thinking about expectations and, in the end, drawing a conclusion from the experiment: Which expectations were right or wrong or how the children reacted?

Small and Large Dice

Children are expected to carry small dice in one or two hands. They may try to open and shake them, and it could be that they try to throw them like standard cubes. They could also start building a small tower by putting one cube on another.

Dice were made out of cardboard boxes, wrapping paper, and some tape to strengthen the edges. Dots were marked by spray paint and round stencils. These materials were ecological, low-price lightweight cardboard and this made it possible for the children to carry and throw them. One huge dice, a few medium-size, and hand-size dice were created. The experiment took place on the promenade of Malvarrosa Beach.

After dropping the dice on the ground and leaving the nearby area, a group of children instantly started running to play with them. Surprisingly, they reacted quickly and with enthusiasm. In fact, passers-by were interested, including older people (see Figure 3).

The children figured out lots of ways to interact with the dice. The most common way was just throwing the dice into the air and watching it hit the ground. Some of them built dice towers and then pushed and kicked them down. They also tried juggling with the smaller dice, rolling the biggest one and hiding behind it like Hide 'n' Seek.

This experiment discovered an easy way to entertain children. Excluding the box-building process, it was all done in 10 minutes with little effort. The children did not care about the quality and details of the dice, but instantly started running towards them to play. Most expectations were fulfilled. Perhaps it is not usual to play with a big dice but they just started doing it spontaneously. The dice were something familiar which children used differently.

Figure 3. Picture of spontaneous reaction of children enjoying the dices
(Credit: N. Matarredona)

Blowing Bubbles

Blowing bubbles seems to be magic for most people, especially kids.

The experiment took place in the Plaza de la Virgen, which is a hot spot for tourists. There were often many families around, including people from different countries.

The children instantly started watching, smiling, and playing and tried to catch and burst the bubbles. This involved jumping and running. Parents stopped what they were doing and started watching their children playing. Most of them smiled the whole time and took pictures and videos of their children. Even passers-by stopped, watched, took pictures, and smiled.

Most of the expectations matched the children's behavior. Putting bubbles in children's areas does not need to cost a lot, and the children instantly started interacting. The atmosphere in this area was transformed into a fun place with smiling people and happy children (see Figure 4).

Figure 4. Picture of children interacting and enjoying the bubbles
(Steerman et al., 2019)

Chalk in The Streets

This experiment took place in a children's play area, but it does not have any playground attractions. It is just an open space.

The kids first just watched what was happening and hesitated to join the project team drawing things on the ground, but after a few minutes, the space became crowded with children playing, jumping, running and drawing.

The new elements, such as chalk, brightened up the atmosphere where the children got together and learned from each other.

These three experiments showed how easy it is to get children to entertain themselves and learn from each other with minimal effort. Not much money or effort was necessary to change their behavior for the better and the experiments proved that there are multiple ways of catalyzing change in our cities for children to develop autonomously and feel integrated.

It is only about providing opportunities to make cities more child-friendly. Without exception, the city areas are designed only for adults and children are often separated from daily life into small controlled playgrounds.

During the experiments, the children started conversations, discussions and being creative and showed how easy it is to change city atmospheres so that kids can participate (see Figure 5).

Figure 5. Picture of natural engagement of children enjoying chalk art
(Steerman et al., 2019)

Results and Conclusions

The issue that had to be faced at the beginning of the semester was the situation of children in cities. The problem stated that places prepared for children are often closed areas, are not challenging enough for them and do not improve their creativity and possibilities of self-development. The mission was to make cities aware of the need to give more attention to children and create a solution to help them reach their maximum potential and have fun.

The goal of the project was to shed some light on how minor actions can catalyze significant changes for kids to be part of the cities and make it evident that it is possible to easily hack the software to make the hardware, the city configuration, more livable.

The project started by researching children's behavior and many places in cities were visited to gain an insight into the situation. After that, many methods of developing the project were used, defining the problem, suggesting a solution, and achieving it.

After planning and carrying out the experiments, it was concluded that children react very quickly, even to small changes from adults. Experiments were prepared using cheap, easily available and affordable materials that had a big impact on the children in the city environment. The children have much fun and the adults as well. Also, the activities had a positive impact on their creativity, intelligence, and visual-motor coordination.

The results achieved were surprising. Not only did the project group realize how little effort is needed to make children interact more with the city and feel more like a part of it, but the group also brought people's attention to the current situation children are in. This awareness is a promising sign that, since the project is now on the right path, it can interest other people and could be continued by EPS students in future semesters to make a considerable change.

Future "Cities also for Children" project groups could work on:

- First designs (based on the concept idea).

- Market analysis (find cities and spaces where a project could and should be realized).
- Making people aware of the problem.
- Calculating the cost of the solution.
- Creating information events.
- Finding interested parties for the project (i.e., city councils, governments or UNICEF).

The project must be continued because the mission of "creating cities also for children" is not over yet. As overhead mentioned, cities that are better for kids are better for everyone. Moreover, there is still a big room for action to be taken.

WORKSHOP RETHINKING OF THE VALENCIA BUS STATION

EPS includes diverse workshops, seminars and other activities that help the development of the participating teams and achieve good results. Most of these activities are non-specific, such as communications or project management, but there are others designed to boost the work of specific projects. Below we describe in detail a workshop on essential aspects of critical complex city installations, with special emphasis on, for example, a bus station.

The activity is based on Kolb's reflexive learning theory (Kolb, 1984), which understands learning as the creation of knowledge through experience, but which needs reflection to be produced. It is thus considered as a dialogue between thinking and doing and involves combining theory and practice, thought and action (Schön, 2017; Osterman & Kottkamp, 1993).

The aim is to show how to work with a complex element in a city. In this case, the element is a bus station, and the work is redesigning a bus station. After an introductory presentation, several groups of students worked on the essential aspects that the complex system had been divided into. The aim was to make them aware of the station as a whole and capable of figuring out the complexity and the large number of variables that designing a bus station involves. They had to be able to think outside their own discipline, widen their vision and accept the contributions of their companions from other disciplines.

For the workshop, the students were divided into five groups to deal with the five essential aspects of the study of the bus station: functions, construction, aesthetic, urban and abstract. These were based on their personal experiences in the field of architectural research on rail transport infrastructures. The groups were meant to work as "cooperating research groups", i.e., the group tasks were shared out among the members according to their skills or interests, organizing and taking the necessary decisions to carry out their different tasks. For the final session of decisions to be taken by all the groups together, a document was drawn up with a summary of all the main conclusions and observations reached.

The introduction contained the specific theoretical foundation the activity was based on, with a short presentation of the different activities (functional, constructional, aesthetic, etc.). Suggestive images related to the topic dealt with (Figure 6) were introduced to reinforce each of the main aspects. Reference was made to the questions meant to guide their reflections as they worked on each of the different proposed activities and the need to combine theory and practice, thought and action, as they had been told at the beginning.

Figure 6. Slide from the aesthetic aspects group
(Credit: A. Martínez-Corral)

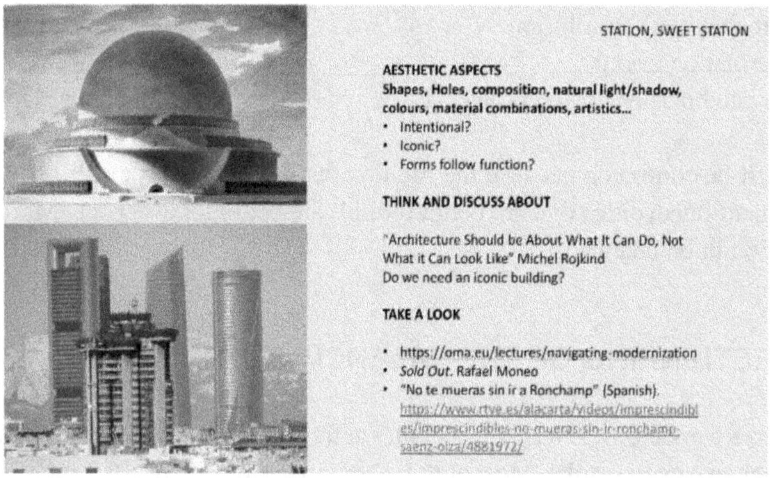

The presentation ended with a "to be continued", meaning an open and dynamic activity and method. According to the results, sensations and the necessary feedback from the students, changes could be made to the contents or new activities, or scenarios could be introduced in order to achieve a dynamic, fun, integrating and effective learning activity.

This initial presentation was of necessity given in a classroom or workshop with a projector, but the rest of the activities were carried out in different places: from the station itself to iconic places in the city to provide inspiration for some of the scheduled debates.

Since EPS students were involved in the activity, English was used in the workshop and its estimated duration was two sessions of two-and-a-half-hour each with a break so that all the groups could work, and a final joint one-and-a-half-hour session (30 minutes per group) to agree on the conclusions of all the aspects dealt with.

Basic Aims of the Activity

The first aim, implicit from the title of the workshop's introductory presentation, was the reflection and assessment of the public space as an essential concept ("Station, Sweet Station", instead of "Home Sweet Home"), or a public space like everybody's home, a place of free transit, use and expression (Delgado, 2011). The redesign of the bus station initially made this reflection possible to undertake the design of the urban space and the building itself. Another of the implicit initial aims, taking the Alsa bus station in Gijón as an example (included in the Gijón General Plan Catalogue), was to show that bus stations are and can also be considered as part of the national or industrial heritage (see Figure 7).

Figure 7. Front of the Gijón bus station
(Credit: A. Martínez-Corral)

Intentionally showing the Gijón bus station, in use since 1941, was an invitation to reflect and debate the need for austerity so that the new station could be built without wasting resources. The students had to consider sustainability, the use of renewable energy, sustainable materials, and tight budgets in their work. In short, to contribute to the understanding that more efficiency, comfort, and beauty is possible with less consumption of non-renewable resources and to think about the long-term impact of their project on the city.

To sum up, four concepts are managed whose reflection and debate provide the necessary foundations for the final specific design: public space, heritage, sustainability, and town planning.

Contents and Strategies

The proposed workshop is a preparatory task for the case study 2 "Station, bus station".

Working on each one of the aspects and giving equal importance to all, the activities were broken up to encourage reflexive analysis from personal experience and combined with other visual narrative activities.

An open, dynamic, and changing methodology is proposed so that, depending on the results, content is updated and/or replaced. Feedback with the student is essential. The structure of the development of the activity proposes:

- Explanation of the task where each student is clear about the task to be performed and its relevance.
- Team composition (skills, backgrounds, diverse knowledge).
- Activities of interdependence, interaction but also individual responsibility. Readings/viewings are presented individually and then, using the round-robin technique (Kagan's structures), which is versatile and straightforward, both are put into practice. The round-robin technique is based on posing a relevant question/topic about what has been read/viewed, allowing time for reflection after which the students, in turn, and with the same amount of time, present their point of view

on the question posed and, finally, a sharing of ideas. Timing is considered essential for shorter or more detailed answers. In this case, 5 minutes for each group member and 5 minutes for the group response, which gives 30 minutes for the development of the activity. The workshop is designed so that each of the activities in each group can be carried out in 30-45 minutes.

The tables below (1-5) show the design of the activities for the five aspects. The sources to work from, included in the reference list, are intended to be open access, preferably accessible online and in English.

Table 1. Details of the activities proposed for Group 1

Group 1. Functional Aspects	
Type of activity/Source	**Example**
• Analysis of specific documentation on bus station design (The National Association of City Transportation Officials, 2022) • Graphs and sketches.	• Reading design guide. Selection of concepts (functions, routes, flows, hierarchies, access, acoustics, etc.) • Sketches and graphs
• View of bus station projects (Architizer, 2022; Archdaily, 2022)	• Learning different bus station solutions. Comparing urban, formal functional solutions. Programs, capacity, materials, budgets, norms, etc.
• Fieldwork. Valencia bus station. Graphs and sketches.	• Testing dimensions and variables for design (turning radii, size and shape of bus stops, leisure zones, access, emergency exits, etc.
• Planimetric analysis/aerial photos (Generalitat Valenciana, 2022) • Graphs and sketches.	• Maps with urban milestones/accesses/infrastructures.
• Personal reflection and group discussions.	• The preponderance of air conditioning in buildings. Need versus design and sustainability. Climatic considerations for Valencia (round-robin).
• Reading and reflection of Junkspace (Koolhaas, 2001)	• Summary and personal interpretation of proposed Reading. Common Decisions.

(Credit: A. Martinez-Corral)

Table 2. Details of the activities proposed for Group 2

Group 2. Construction aspects	
Type of activity/Source	**Example**
• Analysis of specific documentation on sustainability in architecture (materials, resources, crafts and local products, green decks, organic architecture, etc.) (Archdaily, 2022; Dezeen, 2022)	• Individual analysis of the aspects. Summary of main characteristics. Things in common.
• Selection of materials/techniques/installations (Tectónica, 2021)	• Acquire knowledge of materials, construction, installations, etc.
• Make a scheme of structural functioning.	• Make a structural scheme of the bus station.
• Set of construction	• Make model (basic structural elements of the bus station).
• Personal reflection and group discussions.	• Importance of construction in building design. Construction conditions shape and space (round-robin)
• View documentary “Material Matters” (Harboe, 2015).	• Analyze, reflect, summarize construction aspects. Extrapolate station design.
• Read article “Five Most Renowned Green Architects” (Architecture Art Designs, 2019).	• Read, study, know, investigate green architecture.

(Credit: A. Martínez-Corral)

Table 3. Details of Group 3 activities

Group 3. Aesthetic aspects	
Type of activity/Source	**Example**
• Observation and analysis. View photos. • Sketches.	• Choose a representative building (universal, from your own city/country), analyze volume, shapes, empty spaces, unique compositional elements. • Make a scheme.
• Fieldwork. • Analysis of maps and aerial photos.	• Identify iconic buildings in Valencia. • Place them on the map. • Make a scheme of volume/proportions/essential compositional elements.
• Question of personal reflection and group discussion. Read "Architecture should be about what it can do, not what it can look like: In conversation with Michel Rojkind" (Belogolovsky, 2019) Watch documentary "Is iconicity good for architecture?" (de Graaf, 2015)	• Do we need iconic buildings? • Group discussion (round-robin)
• View documentary "Do not die without going to Ronchamp" (Saénz de Oiza, 2014)	• Group discussion.
• Watch talk on "Navigating Modernization" (Koolhaas, 2010)	• Listen and discuss in group
• Read book "Escritos de arquitectura" (Pérez Arroyo, 1993), composed of short independent chapters.	• Each student reads one brief chapter. Reflection and group discussion.

(Credit: A. Martinez-Corral)

Table 4. Activities for Group 4

Group 4. Urban aspects	
Type of activity/Source	**Example**
• Observation and analysis. • Sketches (Keskeys, 2017) • Free graphic tool: City Roads (Ovacen, 2021)	• Analysis of city districts. • Synthetic schemes of the district.
• Analysis of "Master planning and design" (Gehl people, 2022)	• Each group selects one of the projects, analyzing and commenting for the others.
• Analysis of maps/aerial photos of Valencia (Generalitat Valenciana, 2022)	• Analyze transport infrastructures in Valencia and principal connections/accesses. • Make scheme.
• Watch lecture on "Cities for people" (Gehl, 2011)	• Listen and discuss in the group.
• Read "Arquitectura: Lo común" (Fernández-Galiano, 2019)	• Groups select text included and comment on its conclusions with the others.
• View documentary "The human scale" (Dalsgaard, 2020)	• View and group discussion

(Credit: A. Martinez-Corral)

Table 5. Activities for Group 5

Group 5. Abstracts aspects	
Type of activity/Source	**Example**
• Observation of sensations. Touch and vision of architectures. Take photographs.	• Each student selects a building in the city or part of it for its texture, color, the play of light and shade, etc. Show and comment briefly using the Pechakucha format (Klein & Dytham, 2003)
• Read "The eyes of the skin" (Pallasmaa, 2016)	• Read individually, reflect and comment in the group
• Personal reflection and group discussion: Architecture must engage with people. "Narrative architecture is a disaster" – Eduardo Souto de Moura (Igloo Magazine, 2012)	• Comment and debate together (round-robin)
• View the short film "5 cities, 5 Places, One Day" (Robinson, 2011)	• View together. Exchange impressions.

(Credit: A. Martinez-Corral)

CASE STUDY 2 – STATION, BUS STATION

Introduction

The Valencia Bus Station was opened to the public in 1970 on the 9th of February. The bus station was originally called the "Menéndez Pidal building" but became known as the Northern Bus Station due to there already being a bus station in the south since 1966. Since its opening, the bus station has not had any improvements made to it.

The bus station is located on the northern border of the Turia River Park, very close to the city center of Valencia. It is surrounded by small shops and a big hotel. To the south, it borders on Turia Park and to the north on a small park, but there are no direct footpaths from the station to either of those parks. To the west of the station is the big shopping mall Nuevo Centro, which also doesn't have a direct connection to the bus station but is easily reachable by road nonetheless.

Concerning public transport into the city of Valencia, the bus station offers a space for taxis in front of the main building as well as a bus stop for the local EMT buses. There is also a metro station close by – Turia – but it is not directly connected to the bus station.

Basis of the Activity

The students were exposed to various methods of implanting infrastructures, understood as opportunity places, after the analysis of obsolescence and the possible rehabilitation options, bearing the technical and political options in mind. Students should actively and skillfully conceptualize, apply, analyze, synthesize and evaluate information gathered from, or generated by, observation, experience, reflection, reasoning, or communication, as a guide to belief and action. Ideally, they used the universal intellectual values that transcend subject matter divisions: clarity, accuracy, precision, consistency, relevance, sound evidence, good reasons, depth, breadth, and fairness.

A group of 2020/21 EPS students was given the job of redesigning the bus station, a key enclave for the city of Valencia, to put the above into practice. There is a tremendous complexity involved in this task, but the possibilities of the EPS creative framework are huge, with no more conditions than the limits of their imagination.

The Design Thinking approach (Razzouk & Shute, 2012; Panke, 2019) was used to do this project. The first phase was the investigation/search for information, for which they had to carry out a bibliographical search both on the Internet and in libraries or files, combined with fieldwork. They also visited the bus station, its surroundings, notable buildings, infrastructures, and places. These places' location was found by examining maps and aerial photographs and acquiring information like the position and scale of the site within the city. It was interesting to combine this analysis with a study of similar interventions in other cities to understand their comparative size. The final task in this phase was to make a questionnaire to be answered mainly by users, workers, residents, and tourists, to get an idea of different points of view and the problems to be solved.

The next stage was the ideas stage, in which the information obtained in the first phase was used as the basis for debating ideas/concepts using techniques such as brainstorming, brainwriting and sketches for making subsequent decisions. The decisions were taken through SWOT analysis to understand better the strengths and weaknesses, opportunities and threats. This kind of analysis is a tool used to study the situation of a project through a square matrix that includes internal factors (Strengths and Weaknesses) and external factors (Threats and Opportunities). The objective pursued by the SWOT analysis is to analyze what are the competitive advantages of the case studied, studying its own characteristics and the external characteristics of the market in which it is located.

A decision matrix was created to evaluate a series of characteristics that were listed according to the importance/relevance they could have in the project (Martínez-López & González, 2020; Kans, 2021). Not only were its requirements considered, but also the project's viability. In our roles as active observers, we contributed to establishing different milestones in the process to help and motivate the students. Different activities were carried out to keep the classroom/street connection alive. After deciding the final concept, it was refined to add important details and even a physical model was created.

Results and Conclusions

The first step in the bus station project was to define its objective, for which the students made their first visit to the bus station. To start off, their first impression was that the bus station was in very bad condition and did not adequately reflect the image of Valencia. It was dirty, some parts of the walls and ceiling were run down, it was not clear where to go, there was a lot of wasted space, and the bus timetables were hardly readable. In terms of language, there were also difficulties, as the staff did not speak English and the signs were only in Spanish, while a website of the bus station was lacking and no renewable energy sources were used.

After this, they defined the project's aim as follows:

Our goal is to renovate the Valencia Bus station so it will fit in with Valencia's image. We want to improve the user experience by adding a website or mobile application. We will also apply renewable energies and a benchmark to set this station apart from others.

Before starting on the project, the students used the Work Breakdown Structure (WBS) to divide it up into work packages, make further plans and work more efficiently. The WBS is a hierarchical tree structure that describes a project and breaks it down into smaller and more manageable parts. The goal of a WBS is to make a large project more manageable. Breaking it down into smaller parts means that different members of the project team can get work done simultaneously, which translates to higher team productivity and easier project management overall. (Brotherton et al., 2008).

The students also formed working groups to take over different roles according to their academic profiles.

After the first phase of searching for information, they passed to the ideas stage. Figure 8 shows the three concepts of the bus station building they selected from the sketched ideas from which the final version was to be chosen.

Figure 8. Concepts of the bus station building obtained during the Ideas phase
(Berger et al., 2020)

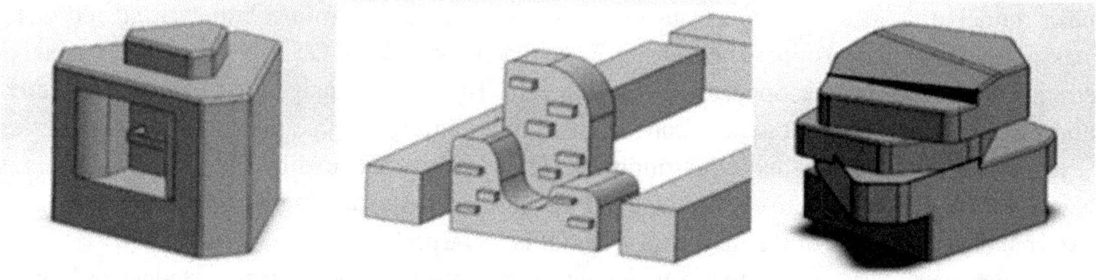

After assessing the three models, they used the above-described techniques to decide to go on working with Concept 3 and started the process of producing a more elaborate model.

The new model had a total height of 28 m and fitted well into its surroundings. It was designed in such a way that the floors were rotated in relation to each other, creating terraces and gardens on the roofs of the floors below. This rotation created, from all perspectives, a futuristic and innovative design that met the requirements and wishes identified on the survey. The roof structure was modified by a curved line at a certain angle to add elements of organic shapes to the building. Figure 9 shows the principal building's final design.

Figure 9. Final Design – Main Building
(Berger et al., 2020)

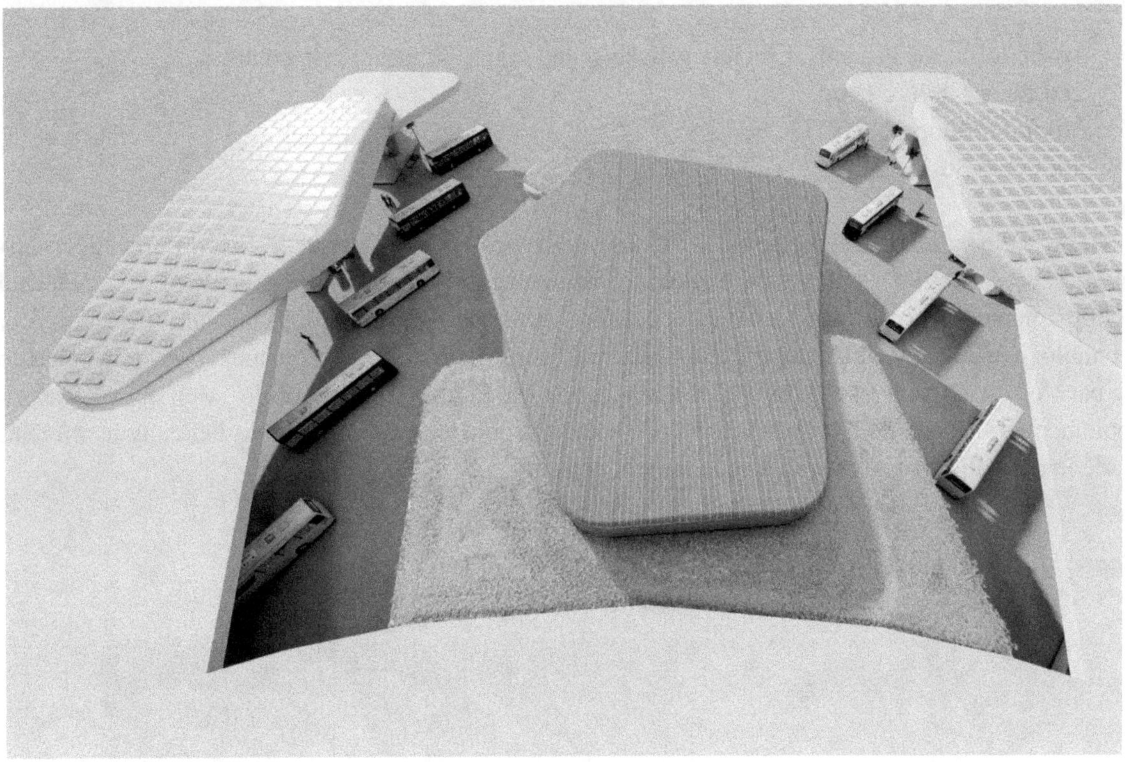

Like the main building, the roof structure of the platforms also consists of elements lying on top of each other with different alignments and geometry, thus creating a futuristic impression. Here, as well, curved forms are used for some elements. In the lower structure of the platforms, the elements are kept in an organic style. The triangular cut-outs are intended to intensify the image further and create the impression of a bionic design. Figure 10 shows the platform zones.

Figure 10. Platforms of the bus station
(Berger et al., 2020)

In the interior, the individual floors of the bus station were assigned different themes. as can be seen in the following list:

- 1st floor (on the ground): Efficiency, tickets, and other bus travel-related areas.
- 2nd floor: Chill & Food.
- 3rd floor: Shops.
- 4th floor: Business.

Besides the building design, they also worked on the possibility of incorporating different alternative energy sources. In the first phase, the students considered options to be implemented and how they could be implemented. It was finally decided to opt for solar panels – SunPower Maxeon 3 (Maxeon Solar Technologies, 2021) – which are the most efficient panels available. These panels would be placed on the back of the station, on the roofs of the platforms (see Figure 10). The brille-soleil design was also implemented. As can be seen in Figure 11, the design improves the station aesthetically as it creates shade in the building and provides access to the terrace.

Figure 11. Brille-soleil in the bus station
(Berger et al., 2020)

Through the project work fulfilled, the team succeeded in achieving the goals of the task and successfully completed the project. As a result of the project, a virtual model of the bus station was created, which is very close to reality thanks to the use of the latest software and provides an exact picture of which elements will be used. The building offers excellent future potential through the use of current components with regard to renewable energies, fits very well into the image of Valencia through its extraordinary design and could contribute to the expansion of the city's tourism industry in the long term.

CONCLUSION

An interdisciplinary approach to the city is fundamental to understand its complexity and propose, after due reflection, creative actions that improve the quality of life of residents without compromising that of future generations. In such a scenario, it is vital to create integral and integrating ecosystems solutions that offer a range of innovative responses.

A learning environment such as EPS can generate this indispensable international interdisciplinary dialogue to obtain a successful response to change the future of its citizens.

Promoting talent and interaction between the different representatives from both inside and outside the university is key to obtaining successful responses to the successive challenges that continually appear in this context, and that also affect those outside the university.

The activities described in this document show that it is precisely this multidisciplinary and international character that produces efficient and creative proposals, the type required to meet future challenges.

Besides the good academic results, the students involved appreciated the personal and professional growth fostered by the EPS framework. This is attested by the growing interest in participating in the program.

EPS is continuously evolving and improving thanks to the creativity and innovation of the participants and the collaboration of an extensive and international framework. It constitutes a singular attraction to exchange students, strengthens bilateral relations with prestigious universities, and creates an appropriate environment for the development and evaluation of transversal competencies such as applying practical reasoning, innovation and effective communication.

ACKNOWLEDGMENT

The authors wish to convey their sincere gratitude to the diverse institutions, companies, professionals, and citizens that collaborated in the work described in this chapter, especially the Escuela Tècnica Superior de Ingeniería del Diseño (UPV).

They also wish to thank the EPS students and the Higher Education Institutions of the EPS network for their active participation in generating the necessary framework and making it possible over the years.

REFERENCES

Archdaily. (2022). *Bus station projects*. https://www.archdaily.com/

Archdaily. (2022). *Sustainable architecture*. http://www.archdaily.com

Architecture Art Designs. (2019). *Five Most Renowned Green Architects*. https://www.architectureart-designs.com/worlds-five-most-renowned-green-architects/

Architizer. (2022) *Bus station projects*. http://architizer.com

Association of Educating Cities. (2012). *Educating Cities*. http://www.bcn.cat/edcities/aice/estatiques/espanyol/sec_educating.html

Belogolovsky, V. (2019). *Architecture should be about what it can do, not what it can look like: In Conversation with Michel Rojkind.* https://www.archdaily.com/910732/architecture-should-be-about-what-it-can-do-not-what-it-can-look-like-in-conversation-with-michel-rojkind

Berger, T., Clarke, K., Koning, N., Kravchenko, D., & Runowski, B. (2020). Processing and development of ideas to solve existing problems in the local bus station of the city of Valencia (EPS Report). Valencia: Universitat Politècnica de València.

Brotherton, S. A., Fried, R. T., & Norman, E. S. (2008). Applying the work breakdown structure to the project management lifecycle. *2008 PMI Global Congress Proceedings.* https://www.pmi.org/learning/library/applying-work-breakdown-structure-project-lifecycle-6979

Dalsgaard, A. M. (2020). *The human scale.* https://www.youtube.com/watch?v=tZcqJoaFIN0

de Graaf, R. (2016). *Is iconicity good for architecture?* https://www.oma.com/lectures/is-iconicity-good-for-architecture

Delgado, M. (2011). *El espacio público como ideología.* Los libros de la catarata.

Dezeen. (2022). *Sustainable architecture.* https://www.dezeen.com/

Fernández-Galiano, L. (Ed.). (2012). *Arquitectura: lo común. Fundación arquitectura y sociedad.* Fundación Arquitectura y Sociedad. http://arquitecturaysociedad.com/wp-content/uploads/2019/07/ARQUITECTURA_LO-COM%C3%9AN-comprimido.pdf

Figueras, P. (2007). *La Ciudad Educadora.* Ajuntament de Barcelona.

Gasch, I., Fuentes-Durá, P., & Perry, D. (2012). European Project Semester. In Jornadas de Innovación Educativa 2012 (pp. 132-136). Academic Press.

Gasch, I., Fuentes-Durá, P., & Puyuelo, M. (2014). EPS, un proyecto multidisciplinar internacional. In *Jornadas de Innovación Educativa y Docencia en Red* (pp. 866–877). IN-RED.

Gehl, J. (2011). *Cities for people – A lecture by Jan Gehl.* https://www.youtube.com/watch?v=KL_RYm8zs28

Gehl people. (2022). *Masterplanning and Design projects.* https://gehlpeople.com/work/projects/

Generalitat Valenciana. (2022). *Visor cartogràfic de la Generalitat.* https://visor.gva.es/visor/

Harboe, G. (2015). *Material matters.* https://www.youtube.com/watch?v=JqIsZsBLkSkInternational

IAEC. (2020). *Charter of Educating Cities.* https://www.edcities.org/wp-content/uploads/2020/11/ENG_Carta.pdf

Igloo Magazine. (2012). Interview with Eduardo Souto de Moura. *Igloo Habitat & Arhitectură, 126.* https://www.igloo.ro/interviu/

Kans, M. (2021). Engineering education development – a business modelling approach. *Higher Education Evaluation and Development, 15*(1), 53–77. doi:10.1108/HEED-02-2020-0003

Keskeys, P. (2017). *Epic Overview: Benjamin Grant on the Sky-High Art of Aerial photography*. Architizer. https://architizer.com/blog/practice/materials/epic-overview/#.YTnIGa6JnAM.link

Klein, A., & Dytham, M. (2003). *PechaKucha — Klein Dytham architecture*. http://www.klein-dytham.com/pechakucha

Kolb, D. (1984). *Experiential Learning: Experience as the Source of Learning and Development*. Prentice Hall.

KoolhaasR. (2001). *Junkspace*. https://www.readingdesign.org/junkspace

Koolhaas, R. (2010). *Navigating modernization*. https://www.oma.com/lectures/navigating-modernization

La Città dei Bambini. (2022). https://www.lacittadeibambini.org

Martínez-López, J. I., & González, C. M. A. (2020). Evaluation of Decision-making Quality using Multi-attribute Decision Matrix and Design Thinking. In *2020 IEEE Global Engineering Education Conference (EDUCON)* (pp. 622-629). 10.1109/EDUCON45650.2020.9125174

Maxeon Solar Technologies. (2021). *Maxeon Solar Panels*. https://maxeon.com/us/node/1163

Osterman, K. F., & Kottkamp, R. B. (1993). *Reflective practice for educators: Improving schooling through professional development*. Corwin Press Inc.

Ovacen. (2021). *Cómo crear mapas personalizados, sin calles, elegantes y en 1 minuto*. https://ovacen.com/mapas-personalizados/

Pallasmaa, J. (1996). *The eyes of skin*. https://arts.berkeley.edu/wpcontent/uploads/2016/01/Pallasmaa_The-Eyes-of-the-Skin.pdf

Panke, S. (2019). Design Thinking in Education: Perspectives, Opportunities and Challenges. *Open Education Studies*, *1*(1), 281–306. doi:10.1515/edu-2019-0022

Pérez Arroyo, S. (1993). *Escritos de arquitectura*. Pronaos.

Razzouk, R., & Shute, V. (2012). What Is Design Thinking and Why Is It Important? *Review of Educational Research*, *83*(3), 330–348. doi:10.3102/0034654312457429

Robinson, C. (2011). *5 cities, 5 places, one day*. https://fundacion.arquia.com/mediateca/filmoteca/explora-documental/

Saénz de Oiza, F. (n.d.). *No te mueras sin ir a Ronchamp*. https://www.rtve.es/alacarta/videos/imprescindibles/imprescindibles-no-mueras-sin-ir-ronchamp-saenz-oiza/4881972/

Schön, D. A. (2017). *The Reflective Practitioner: How Professionals Think in Action*. Routledge., doi:10.4324/9781315237473

Steerman, S., Tuominen, A., Gregersen, M. S., Mickiewicz, K., Linsen, M., Schuler, L. M., & Grabowska, K. (2019). Cities, also for children. Strategies to rethink where we live (EPS Report). Valencia: Universitat Politècnica de València.

Tectónica. (2022). *Constructive materials and techniques*. https://tectonica.archi/

The National Association of City Transportation Officials. (2022). *Bus station design guide*. https://nacto.org/

Tonucci, F. (2004). *La ciudad de los niños: Un modo nuevo de pensar la ciudad*. Fundación Germán Sánchez Ruipérez.

Tonucci, F. (2009). Ciudades a escala humana: La Ciudad de los Niños. *Revista de Educación, número extraordinario 2009*, 147-168.

KEY TERMS AND DEFINITIONS

Authentic Learning: Refers to a wide variety of educational and instructional techniques focused on connecting what students are taught to real-world issues, problems, and applications.

Circular City: Cities where the use and life of resources, products and assets are extended and optimized, and resource extraction and waste generation are minimized. Cities applying the principles of the circular economy.

Design Thinking: Is an iterative process to understand users, challenge assumptions, redefine problems and create innovative solutions to prototype and test.

Experiential Learning: Engaged learning process whereby students "learn by doing" and by reflecting on the experience.

International Association of Educating Cities (IAEC): Is a non-profit Association constituted as a permanent collaborative structure uniting local governments committed to the Charter of Educating Cities, which is the road map of the cities that comprise it.

Ionic: In the field of Architecture, buildings that have historic or cultural values of their own or are even connected to a significant place or person.

Learning Environment: Refers to the diverse physical locations, contexts, and cultures in which students learn.

Observation: Is the active acquisition of information from a primary source. It employs the senses and can involve the use of instruments.

Town Planning: Is the planning and design of all the new buildings, roads, and parks in a place to make them attractive and convenient for the people who live there.

Chapter 17
Embedding EPS Program in Multi-Party Cooperation at The Hague University of Applied Sciences

Wander Herman Colenbrander
The Hague University of Applied Sciences, The Netherlands

Kim Poldner
The Hague University of Applied Sciences, The Netherlands

ABSTRACT

This case study describes a special edition of the European Project Semester at the course Sustainable Packaging Design and Innovation at the Faculty of Industrial Design Engineering at The Hague University of Applied Sciences (Fall Semester 2017). In this special edition, unique cooperation took place between 12 parties. The parties were three research institutes, six universities, and three companies. Some parties have developed an educational module focused upon sustainable and circular packaging design, including the use of a dedicated tool for life cycle assessment. This module was embedded in the regular EPS. At The Hague University of Applied Sciences, an international class of 16 students worked in four teams on a real-life design assignment. They were offered a wide range of lectures, workshops, pitches, and presentations. The chapter concludes with a review of the followed processes and organizational, managerial, and practical concerns. Although run as a unique edition, all parties discuss to continue this cooperation.

INTRODUCTION

The Hague University of Applied Sciences (THUAS) has around 25 000 thousand students of over 140 nationalities and 1900 staff. THUAS is located on four campuses in The Hague (main campus), Delft, and Zoetermeer. It offers over 65 bachelor programs supplemented by other programs as associate

DOI: 10.4018/978-1-6684-2309-7.ch017

Copyright © 2022, IGI Global. Copying or distributing in print or electronic forms without written permission of IGI Global is prohibited.

degrees, masters, post-graduate, and others. THUAS offers five EPS programs at two faculties (www.thehagueuniversity.com). THUAS has a strategic focus on world citizenship, internationalization, network university, and quality. This is to be practiced both by students and staff.

THUAS started European Project Semester (EPS) in 2014 because the current fast-changing world requires engineers, IT specialists, and designers who are prepared to look at problems from different perspectives and who have learned to collaborate with professionals from other disciplines (Andersen, 2012). EPS network provides expertise and sharing opportunities with higher education institutions focused on design (Puyuelo et al., 2016) and sustainability (Duarte et al., 2020). EPS was also proved to be a good approach for competence-oriented programs (Fuentes-Durá et al., 2014).

This chapter describes a special edition of the EPS on Sustainable Packaging Design & Innovation provided by the Industrial Design Engineering course program of the Faculty of Technology, Innovation, and Society. The special edition was held in the fall semester of 2017.

SETTING THE STAGE

The uniqueness of this edition lies in the fact that it involved twelve parties working closely together. The edition was not only a success in terms of cooperation but also highly beneficial to the participating students. The benefits for the students were achieved by a large series of lectures, presentations, and workshops by experts and scholars in (packaging) sustainability, marketing, design, and research. This makes the described EPS edition stand out as an extraordinary event compared to the usual program. The usual EPS set-up offers a design project focused upon a problem where sustainable packaging should be designed by the participating students. An external party usually gives the design project, typically a company supplying a realistic design assignment. The design project is supported by lectures, both by regular staff as external speakers, workshops, brainstorm sessions, excursions, and other activities as team-pitches and presentations (also by participating students). The main objective of this large variety of educational activities is to give students a broad view of the project, to stimulate teamwork (in the international set-up of EPS, six nationalities involved), and to practice working on complex problems with multiple stakeholders.

The EPS has joined a unique collaboration between three research institutes, six universities, and three companies for the special edition. More specifically, the parties are given here:

- **Research institutes**:
  - Netherlands Institute for Sustainable Packaging (KIDV), an institute advising and inspiring companies on sustainable packaging by offering tailor-made advice, factual knowledge, and practical tools (KIDV, 2022).
  - Top Institute Food & Nutrition (TiFN), a public-private partnership for multi- and interdisciplinary research in food and nutrition (TiFN, 2022).
  - Netherlands organization for applied scientific research (TNO) is an independent research organization connecting people and knowledge to create innovations that boost the competitive strength of industry and the well-being of society in a sustainable way (TNO, 2022).
- **Universities**:
  - University of Twente (UTwente) at Enschede (UT, 2022).
  - University of Groningen (UG) at Groningen (UG, 2022).

  - Wageningen University & Research (WUR) at Wageningen (WUR, 2022).
  - HAS University of Applied Sciences at Den Bosch (HAS, 2022).
  - HvA Amsterdam University of Applied Sciences at Amsterdam (HvA, 2022).
  - THUAS University of Applied Sciences at The Hague, Industrial Design Engineering (IPO) (THUAS, 2022).
- **Companies**:
  - EasyFairs, an international company hosting trade shows, conferences, congresses, concerts, and sporting fixtures (EasyFairs, 2022).
  - Struik Foods Europe, a food company, specialized in soups and broths (Struik, 2022).
  - Wessanen, a food company with a sustainable attitude towards health for both consumers and the planet (Wessanen is now called Ecotone).

All these parties have – at least – one thing in common. They all share a specific interest in packaging, packaging design & innovation, and sustainability. The research institutes study these themes as well as the universities. The universities offer courses involving packaging design and marketing-related matters and do research. The food companies produce packaged food, and the trade fair company organizes specialized trade fairs about packaging.

The collaboration was initiated by KIDV. KIDV launched an initiative in 2015 called Scientific Research Program 2015-2019 – Interdisciplinary approach to sustainable packaging (KIDV, 2019). In brief, this initiative is named WOP, abbreviated from the Dutch name *Wetenschappelijk Onderzoeksprogramma.* The following parties are included in this initiative: KIDV, TiFN, and TNO research institutes and UTwente, UG, and WUR universities.

The main objective of this initiative is to provide answers to urgent and relevant questions which contribute to decreasing the environmental pressure of the product-packaging chain in The Netherlands (KIDV, 2019). Within the scientific research program initiative, eleven research themes are formulated. Three examples of these themes are the *Disposal behavior and packaging design*, *The plastic cycle closed,* and the *Recycled PET in new bottles* (KIDV, 2019).

The connection with parties as the universities HAS, HvA, and THUAS as well as with the companies EasyFairs, Struik Foods Europe, and ecotone is made at the research theme *Education module packaging design – From eco-efficient to eco-effective product packaging designs.* The first action was a pilot carried out at HAS (Sept./Oct. 2017). The company Struik Foods Europe joined the pilot. Evaluated and improved with the learnings of this pilot, the *Education module packaging design* was carried out at THUAS (fall semester 2017). The company Struik Foods Europe was not present at THUAS; company Wessanen took their role here as a provider of the actual design assignment. The WOP at HvA is carried out in spring 2019. In April of 2018, the company EasyFairs provided an exhibition stand for all involved parties at their packaging trade fair Empack. At the exhibition stand the results of the *Education module packaging design*, consisting of posters, digital presentations and prototypes were presented to the public. For this reason, the results of the IPO-EPS teams couldn't be displayed to a larger audience at an earlier stage.

This chapter describes the participation of THUAS in the WOP, the scientific research program. Within this specific participation, Wessanen provided the design assignment: design a sustainable (preferably circular) packaging for on-the-go drinks (fruit-based beverages) of the brand Zonnatura.

To summarize, the research theme at THUAS was carried out with the partners of the WOP (three universities and three research institutes) and two companies (Wessanen and EasyFairs) – see Table 1.

Table 1. Scheme of cooperation of all involved parties

Initiative & organization (WOP partners)	Providers of design assignments	Participating universities	Provider of publicity and exhibition stand
• Universities: UTwente, UG, WUR • Research institutes: KIDV, TiFN, TNO, TN	• Struik Foods Europe • Wessanen	• HAS (Den Bosch) • HvA (Amsterdam) • THUAS (The Hague)	• EasyFairs

Within the EPS at THUAS the participating 16 students worked in four groups of four students. The EPS class consisted of five exchange students (from Bulgaria, China, and Germany), two international students (full course), and nine students of the hosting Industrial Design Engineering (IPO) course, see Figure 1.

Figure 1. The EPS group, staff, and jury members at the final presentation

CASE DESCRIPTION

The set-up of the special edition of EPS fall 2017 is composed by the integration of the WOP research theme *Education module packaging design* in the regular EPS Sustainable Packaging Design & Innovation (in brief *Education Module*). The *Education Module* provided the design assignment of company Wessanen. During the preliminary preparation phase all involved parties carefully tuned their objectives, expectations, preconditions, constraints, and other related matters. This was done in a series of meetings. The first meetings had an exploratory character: *can we cooperate, and if so, how will we do that?* Later meetings were focused upon timetables, topics, working procedures, grading, evaluation, and other practical matters. The special edition of EPS was located at the THUAS main campus in The Hague. English is chosen as a language throughout the course for all activities, presentations, and written communication.

Once full cooperation was agreed upon, a detailed timetable was made, covering all ten weeks of the EPS. It was decided to replace predominantly a set of regular EPS lectures with activities of the *Education Module*. The justification to skip these lectures was found easily: most of them were covered by lectures of *Education Module*. IPO formed groups of participating students. They were all given the same design assignment. The *Education Module* has included a design game in the course. Every group member is given dedicated roles during the course within each group. The roles were: manager, engineer, designer, marketeer, and Sustainable Guardian. The last-mentioned role is a newly introduced role, developed by researchers of WOP. In brief, it is a role with special interest and focuses upon sustainability within design processes. The *Educational Module* also offered a newly developed Life Cycle Assessment (LCA) tool, which could be used by the students to do assessments of the to-be-designed packaging throughout the process. Figure 2 shows one of the workshops.

Besides this project work, IPO asks EPS students to prepare student lectures supported by literature research. This was maintained during the course.
The *Education module packaging design* has five perspectives included in the course:

Perspective 1. Integration of sustainability considerations in design processes
Perspective 2. Environmental impact of product-packaging combinations in terms of Life Cycle Assessment
Perspective 3. Consumer behavior at purchase and consumption of product-packaging combinations
Perspective 4. (Sorted) waste collection behavior of consumers and subsequent sorting steps
Perspective 5. The efficiency and effectiveness of (plastic) recycling chains

Figure 2. Workshop LCA tool

These five perspectives are subsequently offered to the students by lectures and workshops, given by guest lecturers invited by WOP.

- **Typical week of the course**:

  – Monday and Tuesday: regular EPS activities.
  – Wednesday and Thursday: *Educational Module* (lectures, design game, and workshops).
  – Friday: regular EPS non-mandatory consults.

The timetable of the reworked EPS is given in Table 2. This table neither includes detailed regular EPS activities nor detailed activities of the *Educational Module.*

Table 2. The timetable of special edition EPS (Nov. 2017 – Jan. 2018)

EPS Week	Activities and Perspectives of Educational Module	Design process phase
1	Introduction, Design game	Introduction, Analysis, Ideation
2	Perspective 1. Integration of sustainability considerations in design processes, Perspective 5. The efficiency and effectiveness of (plastic) recycling chains	Analysis, Ideation
3	Perspective 2. Environmental impact of product-packaging combinations in terms of a Life Cycle Assessment	Ideation Pitch
4	Perspective 2. Environmental impact of product-packaging combinations in terms of a Life Cycle Assessment	Concept
5	Perspective 3. Consumer behavior at purchase and consumption of product-packaging combinations	Concept Pitch
6	Perspective 4. (Sorted) Waste collection behavior of consumers and subsequent sorting steps	Materialization
7	Excursion, Perspective 5. The efficiency and effectiveness of (plastic) recycling chains	Materialization
8	Project workshops, feedback sessions	Materialization, Prototyping
9	Final presentation, wrap-up, evaluations	Final Presentations
10	none	(retakes)

The Educational Module provided one or more trained and dedicated guest lecturers for all perspective-related lectures and workshops. These were researchers, scholars, or other specialists with specific experience and knowledge on the field of the relevant perspective. This led to a total of ten guest lecturers. The manager of the *Educational Module* evaluated and reported each contribution.

The final presentation of the projects was held at THUAS and attended by representatives of all involved parties (EasyFairs, *Educational Module* representing UTwente and TNO, HvA, KIDV, THUAS and Wessanen). Student groups have presented their results through a PowerPoint presentation, supported by posters and prototypes of the on-the-go drinks packaging. Students have also prepared a process report to be included in their grading.

At the start of the course and at the end, students were asked to fill out a questionnaire by the organizing party of the *Educational Module.* The results of these questionnaires are described in a journal paper by Mulder-Nijkamp et al. (2018a). Both achievements and shortcomings of the offered *Educational Module* are discussed here as well. In brief, the objectives of integration were well met and at the same time, the team results paid more attention to LCA aspects at the cost of recycling aspects.

Both THUAS and representatives of the *Educational Module* had prepared a grading form for the students' results (reports, prototypes, presentation). It was decided to merge these grading forms into one dedicated grading form.

Figure 3. The final presentation and best group award

The results of the student groups were evaluated by a four-party jury formed by representatives of EasyFairs, KIDV, THUAS and Wessanen. The jury used five evaluation criteria (equally weighted): Innovativeness (originality, checked on state-of-the-art solutions), relevancy (solution for the problem), potential (feasibility), credibility (well described and motivated), and presentation (narrative, graphics, prototype quality). The best group award was given at the end of the session (Figure 3). Finally, the posters and prototypes were shown at the packaging trade fair Empack on a unique stand, hosted by EasyFairs. The poster is shown in Figure 4 (right). The winning design consists of a set of concentrated fruit juices to be poured in a dedicated bottle and fill up with water. The winning prototype is the packaging on the left and front, as shown in Figure 4 (left).

Figure 4. Some prototypes (left) and the poster on Empack (right)

Figure 5. Brochure KIDV

The jury chose this design as a winner because of its reusability aspects and the fact that it needs less water along the packaging and production chain. Further details are not given here.

Epilogue: the WOP results were presented in an official presentation, attended by all involved parties and many relevant representatives from universities, governments, businesses, and press, including over 150 guests. This final presentation was held on January 17, 2019, at Utrecht. All outcomes are given in a brochure (KIDV, 2019) with QR-codes referring to relevant scientific reports, papers and even dissertations. The brochure is given in Figure 5. Publications that particularly address the issues of the *Education Module* are reported by Maaike Mulder-Nijkamp et al. (2018b).

Technology Concerns

The main technological component in this EPS was the application of a newly developed Life Cycle Analysis/Assessment tool by TNO called Envpack (Ansems & Ligthart, 2018; Ligthart et al., 2019a; Ligthart et al., 2019b). The tool is based on a large database, indicating the environmental impact of a given packaging. The database is the backbone of the LCA tool. TNO has subsequently developed a user-friendly interface, tailor-made for student use. Lastly, TNO has developed the combination of software and hardware. TNO has chosen to use MS Excel as a base for the tool and has done this so that the entire tool, including database and user interface, fits on a standard Universal Serial Bus (USB) memory stick. This has led to a cost-efficient solution that made it easy to make copies and give all four student groups their own stick with the tool. The only practical requirement was to have at least one laptop per group. The concerns here were found at the updates: the students had to hand in the tools to have them updated, which could be done relatively fast and easily.

The tool was ready for use and WOP used the students to improve, update and debug the tool. TNO has frequently reviewed the tool together with the users: the students involved. Figure 6 displays one of the outcome graphs of an LCA calculation.

Other technology used in the EPS were the application of Three-Dimensional Computer Aided Design (3D-CAD) software (primarily SolidWorks) and rapid prototyping equipment, laser cutters and 3D-printers facilitated in the prototyping lab of IPO at THUAS. The prototypes shown in Figure 4 (left) were made this way. Not all participating students master the required software nor the use of the prototyping equipment. By composing the groups, attention was paid to having at least one student per group who was qualified enough to use this technology.

Figure 6. Screenshot of the LCA-tool Envpack in use
(Ansems & Ligthart, 2018)

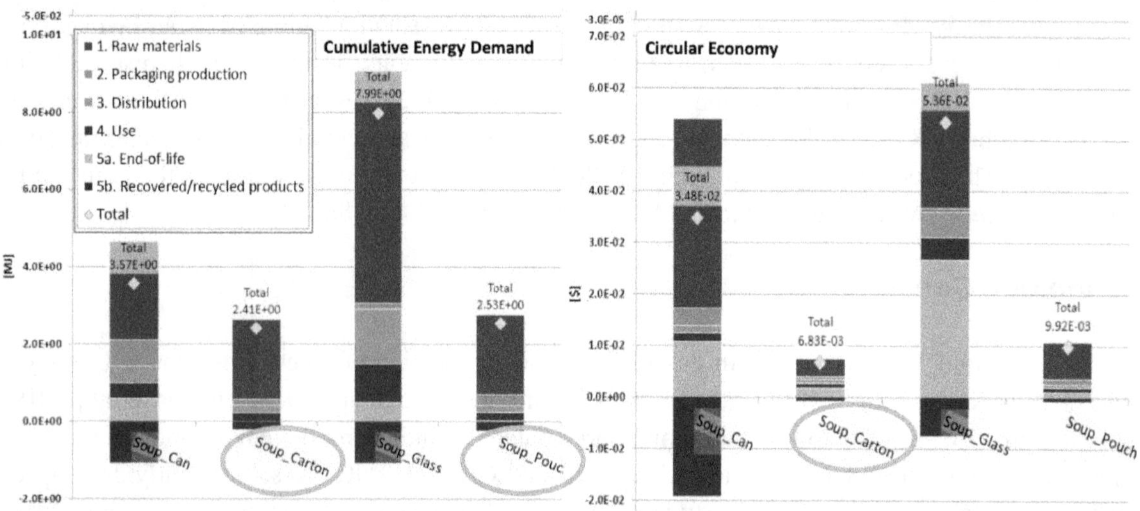

Technology Components

In the regular EPS there are no courses included for the use of an LCA tool, neither for 3D-CAD software and belonging prototyping equipment. For this special edition, the *Educational Module* provided a specialist of research institute TNO to give multiple lectures and workshops to use the LCA tool. Within this edition of EPS, there were no courses for 3D-CAD and prototyping. The students had to make their own choices about how and where to make technical drawings and prototypes. THUAS facilitates only the software and the equipment.

Management and Organizational Concerns

Before the actual start of this particular EPS, the parties involved had several meetings to ensure that the *Educational Module* was well-organized and embedded in the regular EPS. This cooperation had some organizational, educational, and practical implications. Organizational implications were for example: who has the lead? In other words, who is running the EPS and who is responsible, especially towards the participating students? Who will grade the students? THUAS and the board of the *Educational Module* decided to embed the course in the regular EPS. This led to the conclusion that THUAS was still responsible for the whole course module. THUAS had to skip a substantial number of lectures and workshops out of the regular EPS. This was to make room for the *Educational Module*. A special remark is made here about the financial aspects. In fact, there was no finance involved. The *Educational Module* had no fees. It was already financed and the WOP had to find selected student populations/educational courses to test, run and fine-tune their module.

There were some educational concerns to be solved. THUAS/IPO is educating the course of Industrial Design Engineering following a methodological and systematic approach. This had not to be disturbed or alternated too drastically in order to maintain the essential set-up of the EPS. The question was here: how can the educational program of EPS be tuned to the *Educational Module*? Before the start of the

EPS, all involved parties discussed the methods to be used as well as the educational formats (lectures, workshops, etc.). Special attention was paid to the aforementioned five perspectives of the *Educational Module*. THUAS/IPO looked for course components that were double or conflicting with the five perspectives and decided to include these in the to be skipped items.

Since many parties were involved, their representatives often live quite remote from THUAS in The Hague. Some of them had to travel over 250 km to get to THUAS for 2-day lectures and workshops. The question here was: how to solve this? In most cases, the representatives organized overnight stays to make the most out of the day.

CURRENT CHALLENGES FACING THE ORGANIZATION

The most important challenge currently is the continuation of the described *Educational Module*. So far, it has only been run once at THUAS. Both staff and students of THUAS/IPO are very satisfied with the followed process and the achieved results. They would like to do this again. Clear evidence is found in the evaluation questionnaires done by the staff of the *Educational Module* and THUAS/IPO. The WOP has reached its final stage and the researchers are moving on to other activities. Even if the staff could find suitable lecturers to run the program again, another problem occurs: how to finance this? This challenge is not only felt at THUAS but also at the other participating universities of applied sciences HAS and HvA. Currently, the staff of the *Educational Module* is working on a solution. There are lectures ready to be given. There is an LCA tool ready to be used and the entire operation has been thoroughly evaluated. Time is a constraint, too: all the generated material has to be kept updated.

SOLUTIONS AND RECOMMENDATIONS

So far, a solution has not been found to rerun the special edition. Motivation is not the problem here. All involved parties would like to continue the cooperation. After the last debriefing meetings, the parties are still in contact and looking for solutions. One of the possible solutions is a web-based course, to be given and maintained by the staff of the *Educational Module*. There are some concerns at the receiving universities of applied sciences about this solution. They fear that the personal and direct contact of the lecturers and specialists gets lost here, at least to some extent. Furthermore, there might be financial consequences as well. When the *Educational Module* would charge the receiving universities, it will be most unsure if there are resources available to finance this. To summarize: a solution for continuation is welcomed, all parties are still thinking and discussing this, but a final decision has not been taken yet.

REFERENCES

Andersen, A. (2012). The European Project Semester: A Useful Teaching Method. In L. C. Campos, E. A. T. Dirani, A. L. Manrique, & N. Hattum-Janssen (Eds.), *Project Approaches to Learning in Engineering Education*. SensePublishers. doi:10.1007/978-94-6091-958-9_3

Ansems, T., & Ligthart, T. N. (2018). *ENVPACK: LCA-Based Tool for Assessing the Sustainability of Packaging Chains*. https://kidv.nl/media/verdiepingsbijeenkomsten/vd_oogst_wop_verslag.pdf?1.2.1

Duarte, A. J., Malheiro, B., Arno, E., Perrat, I., Silva, M. F., Fuentes, P., Guedes, P., & Ferreira, P. (2020). Engineering Education for Sustainable Development: The European Project Semester Approach. *IEEE Transactions on Education*, *63*(2), 108–117. doi:10.1109/TE.2019.2926944

EasyFairs. (2022). *Visit the future*. https://EasyFairs.com

Fuentes-Durá, P., Puyuelo Cazorla, M., Gasch Molina, M. I., & Perry, D. (2014). European Project Semester: Good practices for competence acquisition. In *Proceedings of the Fourth Valencia Global Conference* (pp.161-172). UPV.

HAS. (2022). *HAS University of Applied Sciences*. https://www.hasuniversity.nl

HvA. (2022). *Amsterdam University of Applied Sciences*. http://www.hva.nl

KIDV. (2019). *Overview Scientific Research Programme 2015-2019 – Interdisciplinary approach to sustainable packaging* [Brochure]. KIDV.

KIDV. (2022). *Getting started with Sustainable Packaging*. https://www.kidv.nl

Ligthart, T. N., & Ansems, T. M. M. (2019b). EnvPack, an LCA-based tool for environmental assessment of packaging chains: Part 2: influence of assessment method on ranking of alternatives. *The International Journal of Life Cycle Assessment*, *24*(5), 915–925. doi:10.100711367-018-1531-z

Ligthart, T. N., Thoden van Velzen, E. U., & Brouwer, M. (2019a). EnvPack an LCA-based tool for environmental assessment of packaging chains. Part 1: Scope, methods and inventory of tool. *The International Journal of Life Cycle Assessment*, *24*(5), 900–914. doi:10.100711367-018-1530-0

Mulder-Nijkamp, M., de Koeijer, B., & Torn, R.-J. (2018a). Synthesizing Sustainability Considerations through Educational Interventions. *Sustainability*, *11*(1), 21. doi:10.3390u11010021

Mulder-Nijkamp, M., de Koeijer, B., & Torn, R.-J. (2018b). Implementing Sustainability Considerations in Packaging Design Curricula. 1-8. In *Proceedings of the 21st IAPRI World Conference on Packaging* (pp. 834-841). Curran Associates, Inc. 10.12783/iapri2018/24465

Puyuelo, M., Fuentes-Dura, P., & Ballester, E. (2016). The opportunity in Industrial Design Engineering. A profile that matches the challenges of contemporary design. *Temes de Disseny*, *32*, 110–117.

Struik. (2022). *Struik Foods Europe*. https://struik.com

THUAS. (2022). *The Hague University of Applied Sciences, Industrial Design Engineering (IPO)*. https://www.dehaagsehogeschool.nl

TiFN. (2022). *Way of working*. https://www.tifn.nl

TNO. (2022). *Innovation for life*. https://www.tno.nl

UG. (2022). *University of Groningen*. https://www.rug.nl

UT. (2022). *University of Twente*. https://www.utwente.nl

WUR. (2022). *Wageningen University & Research*. https://www.wur.nl

KEY TERMS AND DEFINITIONS

Behavior: The way in which one acts or conducts oneself, especially towards others.

Cooperation: The action or process of working together to the same end.

Debriefing Meeting: Reunion about a completed activity or undertaking.

Industrial Design Engineering: Studies that combine the knowledge of industrial engineering with design exploration and research.

Packaging: Materials used to wrap or protect goods.

Parties: Stakeholders interested and involved in a project or activity.

Prototype: A first or preliminary version of a device or a product to show or test some features.

Sustainable Packaging: Development and use of packaging which results in improved sustainability, which reduces the environmental impact and ecological footprint.

Chapter 18
Design and Innovation:
Furniture for Children

Pedro Fuentes-Durá
Universitat Politècnica de València, Spain

Gabriel Songel
https://orcid.org/0000-0002-6546-870X
Universitat Politècnica de València, Spain

ABSTRACT

For more than 10 years at the Escuela Técnica Superior de Ingeniería del Diseño, the European Project Semester experience on design and innovation has been led by the authors due to their experience in researching and developing more than 120 products launched in the market. This expertise is key to driving and supervising a student project until the stage of promotion at a professional trade fair. This fact becomes a milestone in the educational context that requires the full involvement of students, the school, and the university facilities. Applying agile methodologies in educational design projects is more than a learning process. It is an experiential activity nearer to professional skills development.

INTRODUCTION

Every year, one of the assignments for European Project Semester (EPS) students in Escuela Técnica Superior de Ingeniería del Diseño (Universitat Politècnica de València) has been about furniture for children. It starts with an open brief for an interdisciplinary approach so that every team member can contribute to the project's development. It is both a design project and a research project in which students must define the requirements of the product, the target group, the design and manufacturing process, and marketing.

These projects are generally immersed in the furniture sector and mainly focused on furniture and other elements for children. It also implies an immersion in Valencian society and his industrial network. Towards this purpose, shops of equipment for children are visited, helping students to learn about several visual merchandise techniques. At the same time, a leading company in Europe in furniture for children

DOI: 10.4018/978-1-6684-2309-7.ch018

Copyright © 2022, IGI Global. Copying or distributing in print or electronic forms without written permission of IGI Global is prohibited.

as Micuna was visited by the whole group to understand the production process from raw material to packaging of existing products.

The project is developed over four months, and the systematic challenge for all students is to build a 1/1 scale prototype to evaluate all feasibility aspects of the project. The photo shooting session done with children in real photo studios is one of the most amazing moments for students. At that moment, most students understand how vital the prototype is to communicate any idea. In this case, children understand the product and convert this session into a user test session. The pictures from this session demonstrate that the project is ready to be converted into a product.

In fact, in recent years, the results were exhibited in trade fairs like Kind+Jügend in Cologne, Germany, thanks to the collaboration with the international platform for young designers Afilii from Berlin, led by Katja Runge. These opportunities are a special occasion for students to immerse themselves in global markets and receive comments from many different agents in the market. It is the best training to get in contact with real-life projects to develop new products. The objectives of this chapter are to provide a general overview of the innovative design process of EPS students and describe a case study to illustrate the results better.

This chapter aims to provide a general overview of the innovative design process during EPS projects and describe a case study of a particular EPS project in the furniture industry to illustrate the outcomes of this design process.

BACKGROUND

The demand for employability increases consumerism culture and commodification of higher education. It shapes student expectations, and the value of university education in that economic return becomes the main driver (Cheng et al., 2021). The consensus in the literature has been that employability is core to higher education (Mawson & Haworth, 2018). Worldwide studies indicate that teaching-focused University-industry collaborations improve graduates' employability competencies, particularly domain-specific competencies (Borah et al., 2021). As industries and career paths change, graduates are expected to possess some discipline-specific attributes and others transferrable to a broader range of jobs and careers (König & Ribarić, 2019). Whilst most governments' definitions of employability prioritize the development and accreditation of knowledge and vocational skills; many studies reveal that employers are more interested in "softer" skills and attitudes. Employers' emphasis on soft skills also suggests a significant disconnection among government, higher education institutions, and students' groups who focus on practical and vocational skills (Wharton & Horrocks, 2015).

Designers are not only designing but managing beyond the design studio and even deciding upon the activities that need to be done across the business. No single person is likely to have all the skills required for different scenarios, design challenges, and engineering projects. Additionally, the separation of design from manufacturing or implementation often leads to inferior results (Meyer & Norman, 2020). That's why most projects are done by teams, imply deep knowledge and involvement on the subject, and outperform when a design-driven approach is used. Higher education should simulate as much as possible the professional practice and develop the design thinking mindset and UX design beyond design studies (Magistretti et al., 2021).

The speed of change in our society brings designers to understand contemporary conditioning factors, new market situations, and user needs to develop new products and services. Design plays a key role in shaping the world (Puyuelo et al., 2016). That implies many things. On the one hand, solid research must be used to inform design as well as to stimulate ideas about how and when to deploy them effectively (Laurel, 2003). On the other hand, a set of serious design techniques should be chosen and correctly applied to better understand and empathize with potential stakeholders, mainly users, and as a result, build more meaningful products (Martin & Hanington, 2012). All the steps are equally relevant, from the imagination of ideas to the final detailing. Sometimes, deciphering the specialized terminology and moving into unfamiliar areas are difficult but necessary to face a specific problem (Lidwell et al., 2010).

It is essential to consider that design is, naturally, the primary source of innovation (Verganti, 2009). State of the art assists designers with their challenges and reduces the effort required to design across disciplines.

Martin (2009) predicted that Design Thinking would become a crucial tool for individuals and companies to be competitive. Brown & Katz (2019) establish design as a critical factor to inspire innovation in companies. Design Thinking is a robust framework to guide the change of companies when they pursue their goals.

Design should be appropriately managed to contribute to new opportunities for research, product development, and product launch (Borja & Valade-Amland, 2020). Basically, design management is about managing design projects paid for by a client, a business, or an organization, carried out by a designer, a design team, or a design consultancy (Best, 2019).

The combination of diverse perspectives contributes to achieving an innovative culture in the organizations, allowing results to be more systematic and predictable (Morris et al., 2014). Indeed, creativity is the main feature of the people involved, which should balance with business goals (Granet & Hunt, 2021). The creative perspective should be part of the strategy of the organizations to achieve this innovative culture, beyond punctual use of some creative techniques or design outsourcing of determined services (Davis, 2016).

THE DESIGN PROPOSAL

The Pedagogy

The above-cited EPS project assignment, addressed in this chapter, involves a furniture design for children. The subtitle is "innovation in leisure and amusement products and services". Inspiration can be found in specific leisure and amusement activities from the historical and traditional industry approach to cutting-edge technology and services. Each semester, teams decide a specific name for the product, product line, or company they created. This is a very particular approach to brief the project due to experience developed in former years. Usually, design briefs are very closed and limited (for instance, designing a chair or a toy). The choice to formulate an open brief for furniture for children is deliberate. It is based upon practice-based experience developed in former years.

This project aims to identify relevant new business areas for products and services in the leisure and amusement sectors. A practical application will be made as an essential part of the project, including a real scale prototype and a business plan. As a consequence of the open design brief, despite the common brief as a starting point, different types of final results are achieved on each occasion: themed parks layouts, amusement rides, merchandising products, mechanical playgrounds, toys, or furniture for children. In the EPS experience, students come from different backgrounds, countries, and cultures, and consequently, it isn't easy to guarantee that designers will be included in the group. Therefore, not all the group members will be familiarized with design processes. Because of this, an open brief is assigned to let any member contribute to the project development (Fuentes-Durá et al., 2015). Additionally, when the project starts with a general research process, the group is the one to decide what product is going to be developed according to the users and market needs.

Project teams are composed of four to six international students from different areas and backgrounds that used to come from production engineering, information technology, marketing, architecture, and industrial design.

The project is developed twice a year over four months, from September to December and February to June. The group is full time dedicated to project development, with three presentations and some exhibitions in the process. The first presentation is made to introduce the teams and show the basic elements of the project, i.e., the task, the interested parties, the resources, and the environment. Detailed feedback about the team chart, the objectives, and the plan is provided. A midterm presentation is required to allow a solid review of the project's progress and adapt the plan and documents layout. This interim assessment is often the most valuable. A final presentation is made showing all the results and outcomes of the project.

A weekly meeting is arranged with the supervisor to control whether the project is running perfectly, avoid delays, or solve conflicts if they appear. In any case, it is important to underline that the project's management is not the supervisor's responsibility but the whole team.

The Context

The immersion in the furniture sector, specifically those companies dedicated to children's equipment, and the immersion in industrial design through a project on design research and management are essential aspects of this case. It contributes to a change in teaching and learning practice, bringing students into real and professional scenarios like shops, components suppliers, workshop facilities, finishing labs, photo shooting sessions, logistics, and product exhibitions.

When students are challenged with designing a product line of furniture for children, many concepts and methodologies are necessary. The key elements of the project contents go from trend analysis up to business development. In trends research, the primary purpose is to detect new social values or mainstream social behaviors. Reviewing reports and building lifestyle boards are some of the tasks in this stage.

The market research is usually explained through a map of the existing products so the group can detect new values or new opportunities for some innovative products. A list of shops in the city is provided to students at this stage. Students should pay a visit to them and obtain some data. Shopping areas are a way to get in contact with real life, to see real consumers, usually in another cultural context. The learnings obtained in this stage are a part of the field research process where students can compare different merchandise strategies according to different product typologies and customer profiles.

The list is composed of very well-known shops where they can observe customers' behavior in the shop, understanding what kind of information they require. It is a perfect way to immerse in local society and the sector, especially understanding the way of living in the host country. The shop comprises some crucial places: El Corte Inglés (Children care department and Furniture Department as the department store representative of the upper-medium Spanish market), La casita de Penólope (a traditional baby equipment store), Flexa, Habitat, and Lof (as representative design stores with corners for children equipment), Dulce Bebé, Mothercare and Imaginarium (as general suppliers of children equipment),

Tuctuc and Prenatal (as stores connected to children apparel). Figure 1 shows students researching on-site. During these first explorations, learning, discovering, and teambuilding merge together.

Figure 1. Students visiting shopping areas

The user research in this project is focused on two directions. One goes towards the educational world, and the other focuses on particular homes' needs and market offers.

Once the group has a general view of the market, the user, and trends, the students are ready to decide what product will be the proper one. That is the product that will face all requirements obtained in the research stage.

The Creative Process

Then comes the conception of the new product, where different types of visual creativity techniques are applied. All group members must contribute with new concept proposals in this stage. Although not all of them have drawing skills, they all have a certain knowledge of the concepts evaluated during the market research in the first stage. It is a way to let the students discover that everyone has creative skills. At the same time, it offers an opportunity for the team to experience and respect others' limitations in conceptualization by drawing or description.

For conventional industrial projects, the product development stage follows, which includes visualizing the ideas, building first models or mockups that will be useful for the product development, and advancing towards the technical drawings and prototypes. In this stage, the group members must understand the paramount vantage of building a prototype of their idea. On one side, they obtain a constructional test of the feasibility of the product that has to be manufactured. This building process requires the full involvement of the team to discuss every small detail and obtain group commitment to go further. On the other side, they can test an actual prototype with real users. Additionally, they can organize photo shooting sessions with children, which is a perfect opportunity to test the product and realize that they are close to the final product.

The business development stage is another common field of interest for all participants because, thanks to it, they can imagine and see their projects in real scenarios. The calculation of costs according to market pricing, discovering the possibility of short production series, opens a window to entrepreneurship.

The final test is the presentation on real trade fairs. It means the absolute immersion in the sector they started to explore as spectators and in which they have now become active players.

This participation in actual trade fairs is probably the most exciting moment for the group members (and the supervisor). They realize what they have experimented with and learned along with the project. They fully understand the importance of teambuilding, contributing from all fields of knowledge, and being part of the design research process.
The recommended agenda can thus be summarized as follows:

1. Documentation. Trends research focused on social values. Market research based on existing products. User research regarding educational needs and offers.
2. Ideation. New products conception. Visualization of ideas. Evaluation with an interim report and a mid-term presentation.
3. Project communication. Construction of models and prototypes. Pictures with users. Graphic dossier.
4. Business development. Business model and calculations. Market gap. Shops and websites.
5. Final results. Local exhibition. Final presentation. Participation in an actual trade fair.

THE CUB PRODUCT LINE CASE STUDY

The Results

The following design concept was created in the framework of the EPS in Autumn 2017 at the UPV with Professor Gabriel Songel as the supervisor. Following the high number of interested students, two teams were created. Both teams have worked on the same challenge: Innovatively create furniture for children. Participants come from France, Germany, Latvia, Sweden, The Netherlands, the UK, and the USA.

The teams worked together to design, analyze, and build prototypes of the designed furniture to be exhibited in the context of the Kind+Jügend Fair in Cologne, the most important fair in children's equipment in Europe. The teams' mission was to enrich the lives of parents and children through comfortable, safe, multifunctional, and enjoyable furniture. To create high-quality and sustainable products, it was indicated to consider both the children's needs and the parents' requirements.

Three different products were created: Turn & Learn, Rock and Block, and Slide & Hide. In the following paragraphs, a description of the ideas and the characteristics of the products is provided. Some figures help to understand the features and the scale of the designs.

The first product is Turn & Learn. 'Turn & Learn' is a piece of multifunctional creative furniture with a rotatable desk. It can be used as a blackboard in its closed state, and the children can draw on it with white chalk pens. When turned 90 degrees, it is transformed into a desk where children can perform arts and crafts. The desk includes a storage space for drawings, crayons, and other things children may use for their arts and crafts.

'Turn & Learn' is a piece of furniture intended for children aged three to six years old that focuses on the child's creativity. This furniture creates an environment where the child can be creative with white chalk on the blackboard or use the desk to perform other artistic ventures. Through this, children can develop their creative skills, contributing to their personal growth.

'Turn & Learn' has a unique design with innovative multifunctionality. Apart from being an item of creative furniture, 'Turn & Learn' is also educative furniture. It enables the child to perform homework on it and motivates to study. The child can also keep their books and stationery in storage.

The furniture is made of durable and sustainable wood for a long lifespan and high quality to be used for future generations. The furniture design includes simple shapes, which create a clean and attractive appearance. The main design element is the oval-shaped hole pattern. The shaped holes make the appearance stylish, but it also functions to hang things on the back of the creative desk, such as drawings or clothes. By choosing a timeless look with neutral colors and materials, the furniture also appeals to the parents and can effortlessly fit into stylish homes.

'Turn & Learn' is designed with safety as a priority. It is provided with a stopper to avoid the child getting injured by spinning the desk. Also, the lid of the storage function is closed with magnets, so it will not open while turning it. The desk is provided with a blackboard instead of a whiteboard to prevent the child from using toxic markers. White chalk is non-toxic and, therefore, not harmful when eaten. Chalk is also washable from clothes and skin. In addition, it is easily erasable from the walls and other objects. Lastly, the oval hole pattern is designed. It is not repeated on the bottom of the creative desk to prevent children from climbing the furniture.

Figure 2 shows the brochure of the product 'Turn & Learn', including functionalities and purposes. Printed in A1 format, contributed to product explanation in several circumstances.

Figure 2. Brochure of Turn & Learn

The second product is 'Rock & Block'. This idea materializes a rocking horse with modular cushions that enhance the child's creativity and exploration. 'Rock & Block' consists of a bentwood structure with three cushions that can be taken out and combined in various ways. The furniture helps children learn how to hold their balance by rocking on it. The modular cushions can be piled in different manners, and the structure can be turned upside down and transformed into a climbing mountain.

This rocking horse with modular cushions focuses on creative activities intended for children between 3 and 6 years old. The child can play and rock on it alone or use it with a friend as a seesaw. The height of the rocking horse is 50 cm, the length is 80 cm, and the width is 40 cm. 'Rock & Block' helps children learn how to hold their balance by rocking on it. The child can climb on the structure and further improve their balancing and motor skills when turning it upside down.

The ‘Rock & Block’ enables a versatile use for the child through climbing, rocking, resting, and combining the modular cushions in different manners. The structure is made of bent sustainable and durable wood of high quality for a long lifespan so it can be used safely and for future generations. The three modular cushions are soft and light for the child to maneuver.

The furniture is provided with intuitive handles and feet placement indicators so that the child can play securely. The middle cushion follows the design of the wood structure with a cut-out, which creates space for children to place their feet safely while rocking. The open layout of ‘Rock & Block’ makes the furniture light and safe when it rocks. The empty wooden framework can also be used as a place for hiding for small children, which adds a user possibility to the product.

The minimalistic design of the furniture includes simple shapes of cut-out oval holes. The shaped holes make the appearance attractive and function as a means for climbing and as a handle to keep balance whilst rocking on them. By choosing a timeless look with mostly neutral colors and materials, the furniture also appeals to the parents and can effortlessly fit into stylish homes. Figure 3 shows the product ‘Rock & Block’ brochure, including characteristics and examples of use. The pictures with a real kid boost the message.

Figure 3. Brochure of Rock & Block

Finally, the third product is ‘Slide & Hide’. It is an expandable stool that can be easily transformed into a children’s playhouse thanks to the mechanisms built into the stool’s walls. In its closed position, it is a stable stool that can be used to sit on. When it is extended, the colorful windows slide out of the stool’s walls, and the built-in wooden supports lift the fabric rooftop of a playhouse.

‘Slide & Hide’ is a multifunctional furniture piece that gives a personal space and provides a creative environment for children bordering the playing area in interiors. It is intended for children from three to six years and fits about three to four children, depending on the age. When expanded, the playhouse is three times the size of the original stool. The base increases from 70 cm to 150 cm in width, and the fabric tent goes up to 110 cm. This magic box can be used both at home or in public places and kindergartens, giving a great space-saving solution.

'Slide & Hide' playfully brings exploration possibilities to children. The furniture offers children the opportunity to create their private place to hide, rest and feel safe. The playhouse encourages creative imagination and role-plays with its unconventional-formed windows. Children can explore the oval-shaped hole pattern on the walls of the stool and the hidden features, like different colors and forms of the windows.

Thanks to the color scheme chosen, the product can be easily integrated into any interior without disrupting an existing design concept. Moreover, the playhouse helps balance the home interior and hides the mess as soon as kids' toys are hidden inside it. When children are not playing, the multifunctional playhouse can be easily transformed into a stool by pushing everything back together and putting the lid back on.

Wood was chosen as the primary material for the furniture piece due to its ecological, durable, and sustainable features, which provide a long lifespan and high quality. The product is secure, the edges are rounded, and the expand mechanisms are located behind a wood panel to prevent injuries and enable a safe playing environment. 'Slide & Hide' was intended for a second use. It is a stylish and neutral interior furniture piece for both parents and children in its closed position. Parents can then use it for sitting on (or adding it to the couch for additional comfort) or for children to play on it. Some toys can also be kept inside the stool. Moreover, this storage function remains when the children are grown, and the playhouse is not needed anymore. The life cycle is extended by adding this storage function to the 'Slide & Hide'. Figure 4 shows the brochure of the product 'Slide & Hide', including a stuffed toy, to stimulate the imagination as the crucial element for the playful moments.

Figure 4. Brochure of Slide & Hide

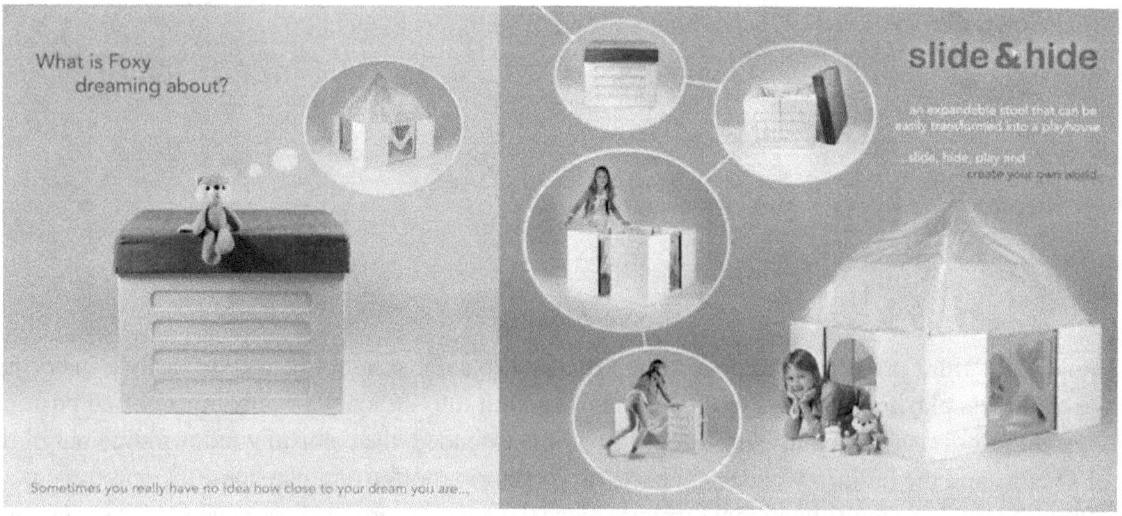

Diffusion and Promotion

The diffusion of the project was done weekly through EPS presentations in front of the students, UPV professors, and international partners. In addition, three singular actions were developed, a city exhibition, a video for the UPV TV, and the participation in a professional fair, as detailed below. Students showed the prototype for three hours during the city exhibition to a vast diversity of potential users and investors. It constitutes an outstanding communication exercise and a fantastic option to get direct feedback. This feedback is a priceless contribution to the detailing phase when the teams finish specifications, calculations, renders, and other technical documents.

To do an even more genuine promotion, the supervisor supports the students to bring the products to real scenarios on trade fairs to test the acceptance of their proposals and develop social and marketing skills because they have to explain their products to professional visitors.

To be present in those trade fair contexts, it is vital to count on specialized promotional platforms that launch and promote new designers. That was the case of Afilii, a well-known German platform focused on design for children, that selected the Cub product line for the exhibition "Play it green" held on Design Post Gallery in Cologne, located in front of the Cologne Fair building. It was the first time for UPV students to be selected on design for children and exhibit in a prestigious design gallery.

To inform a broader audience about the achievements in an international fair of these teams of committed students, a brief release was prepared to be delivered to the media through the Press Office of UPV:

Students of the European Project Semester group from Engineering Design ETSID Universitat Politècnica de València have been selected for the "Play it green" exhibition celebrated at Design Post Gallery in Cologne (Germany) during the international fair Kind+Jügend.

It is the first time that design students achieve to exhibit in this selection, which can be considered a milestone for ETSID. The section was done by the German platform Afilii, which specialized in sound design for children's equipment.

The project developed along one term consisted of working on ideas of pieces of furniture to play with. It is about creating second functions for pieces of furniture for children so that furniture grows with children. Longer life for products is a way to contribute to sustainability from industrial design.

The EPS teams have been formed by ten students from the United States, The Netherlands, Sweden, Latvia, the United Kingdom, Germany, and France, led by Prof. Gabriel Songel.

Figure 5 shows some video captures on UPV TV news.

Figure 5. Video captures of Cub product line from EPS students on UPV TV news

Afilii is an international hub and a platform for meaningful design for kids. Its philosophy is based on the idea that:

Creativity and imagination are necessary conditions for children to live a malleable, meaningful, and content life in our consumer world. If children don't have access to either of these, they will find it difficult to make independent decisions later in life.

Afilii works with designers and manufacturers worldwide to promote children's imagination and creativity across all cultures.

One of the principles of democratic upbringings is the idea that children should "just try it out…". Regarding this, Afilii proposes that children should be given a voice in determining what products and spaces they grow up in and with – ones that they find delight in and that they can use actively and with joy.

All selected projects should answer several questions on which designers should have reflected along with their project:

- What kind of spaces do children need?
- What can products do for them? Are they meaningful and suitable for children? Functional and safe? Of high-quality and durable? Well-thought-out and aesthetic?
- Where and how were they produced?
- Do the products have certifications or did they receive awards?
- Which materials were used in their production?

Afilii is oriented towards both private and commercial clients. The list includes retailers, planners, interior designers, and architects of spaces for children, teachers, caretakers, and children's institutions. Obviously, the platform is also oriented towards open-minded parents of children between 0 to 13 years old.

The exhibition "play it green!" offers a space to create encounters in order to raise awareness for a sustainable and respectful world of design for children. The event had several formats as the exhibition itself, a "play it green!" award, a spotlight, talks on sustainability, and meetings' agenda.

"Play it green!" is initiated and curated by Afilii and took place from September 20-21 at the same time as the Kind+Jugend in the prestigious gallery named Design Post Cologne. It is a showcase for international retailers, interior designers, journalists, bloggers, designers, and children and their families. It exhibits and supports designers and manufacturers worldwide looking for new avenues to explore in designing and producing sustainable, child- and eco-friendly toys, home accessories, and children's furniture. They focus on children's cognitive abilities, agility, creativity, and sensitivity. Figure 6 shows the "play it green!" poster and the main entrance of Design Post Cologne.

Figure 6. Play it green poster and main entrance of Design Post Cologne

The "play it green!" exhibition encourages young creators and gives up-and-coming brands visibility in the global market. Well-established designers and manufacturers are also showing their collections in the Design Post.

The centerpiece of play it green! is an exhibition that will show high-quality and sustainable toys, home accessories, and children's furniture.

The extensive framework program tries to inspire and create awareness for perspective changes. The "play it green!" award will honor a visionary initiative dedicated to social and ecological change.

The lab presents a set of prototypes, student creations, and sustainable materials. Spotlight shows selected publishers for children's books.

There are other highlights for professional and private visitors like workshops for children, subject talks for designers and interested persons, and get-togethers in the evenings. The media partners of the exhibition promote the activities developed in their magazine. Luna Journal was the leading media partner, and they are in charge of the Business Talk "How we can promote sustainable design for kids" with different experts from Korbmayer (DE), Lumokids (FI), Mummy Mag (DE), Plantoys (TH), Senger Naturwelt (DE).

Other exciting issues were discussed: "How can be found the right influencer for a Brand", exposed by Alex Hochstrasser, from the Swiss company MOLUK; "How to foster sustainable education for kids", run by Joris Vanbriel, from the company ecoBirdy; "How to develop a sustainable brand for kids", by Nina Mihovec, from the company ooh noo.

Other relevant institutions like the Institute for Universal Design, the Museum of German Design, and the Karlsruhe Institute of Technology were involved as partners of this kind of event due to its importance in education on sustainable values and its impact on society.

CONCLUSION

Open design briefs for interdisciplinary groups are especially suggestive because every student can contribute to the project's development with their best knowledge and different backgrounds. It is a design project and a research project in which students must define the requirements of the product, target group, development, and manufacturing and marketing. It is a way to get full involvement from the group in the project.

The project is immersed in the sector of furniture in general and particularly focused on children. Moreover, it also means immersion in local society and industrial networks. With this purpose, shops of equipment for children and manufacturers are visited, letting students learn about several visual merchandise techniques.

The project is developed over four months, and the challenge for all students is to cover the whole process, from market analysis to promotion in trade fairs. Due to the available short-term period, necessary agile methodologies are introduced. Extra time was needed for students to organize the fair event nine months after finishing their Valencia experience.

It demonstrates the involvement and commitment of students to the project even beyond the academic period. Construction of 1/1 scale prototypes for feasibility evaluation and photo shooting sessions done with children in real photo studios are opportunities for students to immerse in real professional life. Thanks to that, students understand how important the prototype is to transmit an idea and how the users, children, understand the product and convert this session into a user test session. The output of these sessions proves that the project is ready to be converted into a product. However, the challenge does not finish there. Managing the possibility of exhibiting the prototypes at a trade fair requires the management of funding, logistics, and people. Therefore, these opportunities are exceptional occasions for students to discover international markets and receive comments from many different agents in the marketplace. This case development is an excellent demonstrative example of the best training obtained by getting in contact with real-life projects to develop new products. In this way, the involvement and commitment of students go beyond expectations.

REFERENCES

Best, K. (2019). *Design management: Managing design strategy, process and implementation*. Bloomsbury Visual Arts.

Borah, D., Malik, K., & Massini, S. (2021). Teaching-focused university–industry collaborations: Determinants and impact on graduates' employability competencies. *Research Policy*, *50*(3), 104172. doi:10.1016/j.respol.2020.104172

Borja, M. B., & Valade-Amland, S. (2020). *Design: A business case: thinking, leading, and managing by design*. Business Expert Press.

Brown, T., & Katz, B. (2019). *Change by design: How design thinking transforms organizations and inspires innovation*. Harper Business.

Cheng, M., Adekola, O., Albia, J., & Cai, S. (2021). *Employability in higher education: a review of key stakeholders' perspectives*. Higher Education Evaluation and Development., doi:10.1108/HEED-03-2021-0025

Davis, D. D. (2016). *Creative strategy and the business of design*. HowBooks.

Fuentes-Durá, P., Gaspar, F., & Abarca, J. M. (2015). *Innovación creativa: The NOMAD Travelbox* [Creative Innovation: The NOMAD Travelbox]. In *Proceedings of the Congreso de Innovación Educativa y Docencia en Red de la Universitat Politècnica de València*. Universitat Politécnica de València. 10.4995/INRED2015.2015.1601

Granet, K., & Hunt, H. (2021). *The business of design: Balancing creativity and profitability*. Princeton Architectural Press.

König, L. S., & Ribarić, H. M. (2019). Is there a mismatch between employers' and university teachers' perceptions on graduate employability in Croatia? *Journal of Managerial Issues*, *24*(1), 87–102. https://hrcak.srce.hr/221552

Laurel, B. (2003). *Design research: Methods and perspectives*. MIT Press.

Lidwell, W., Butler, J., & Holden, K. (2010). *Universal principles of design: 125 ways to enhance usability, influence perception, increase appeal, make better design decisions, and teach through design*. Rockport Publishers.

Magistretti, S., Bianchi, M., Calabretta, G., Candi, M., Dell'Era, C., Stigliani, I., & Verganti, R. (2021, October). Framing the multifaceted nature of design thinking in addressing different innovation purposes. *Long Range Planning*, 102163. Advance online publication. doi:10.1016/j.lrp.2021.102163

Martin, B., & Hanington, B. M. (2012). *Universal methods of design: 100 ways to research complex problems, develop innovative ideas, and design effective solutions*. Rockport Publishers.

Martin, R. L. (2009). *The design of business: Why design thinking is the next competitive advantage*. Harvard Business Press.

Mawson, M., & Haworth, A. C. (2018). Supporting the employability agenda in university libraries: A case study from the University of Sheffield. *Information and Learning Science*, *119*(1/2), 101–108. doi:10.1108/ILS-04-2017-0027

Meyer, M. W., & Norman, D. (2020). Changing Design Education for the 21st Century. *She Ji: The Journal of Design, Economics, and Innovation*, *6*(1), 13–49. doi:10.1016/j.sheji.2019.12.002

Morris, L., Ma, M., & Wu, P. C. (2014). *Agile innovation: The revolutionary approach to accelerate success, inspire engagement and ignite creativity*. John Wiley & Sons.

Puyuelo, M., Fuentes-Durá, P., & Ballester, E. (2016). The opportunity in Industrial Design Engineering. A profile that matches the challenges of contemporary design. *Temes de Disseny, 32*, 110-117. https://riunet.upv.es/handle/10251/94457

RungeK. (2013). *Philosophy*. https://afilii.com/en/philosophy/

Verganti, R. (2009). *Design-driven innovation: Changing the rules of competition by radically innovating what things mean*. Harvard Business Press.

Wharton, C. Y., & Horrocks, A. J. (2015). Students' perceptions of employability within their degree programme: highlighting the disparity between what academics believe is included and the student experience. *Proceedings of the 2nd International Conference on Enhancement and Innovation in Higher Education*.

KEY TERMS AND DEFINITIONS

Design for Kids: Focused on kids, visual look or a shape given to a particular object, in order to make it more attractive, make it more comfortable, or to improve another characteristic.

Design Thinking: A user-centered approach to creative problem solving and innovation.

Employability: The skills and abilities that allow people to be employed.

Furniture Design: The way in which an element is planned and made; in this case, things such as chairs, tables, beds, or cupboards, that are put into a house or other area to make it suitable and comfortable for living or working in.

Innovation: A value redistribution thanks to a new or changed product. The term product includes goods, services, or other entities.

Learning: The process of getting an understanding of something by studying it or by experience.

Sustainability: The idea that goods and services should be produced in ways that do not use resources that cannot be replaced and that do not damage the environment.

UX: Encompasses all aspects of the end-user interaction with a product; broadly, with a company, its services, and its products.

Chapter 19
A Fast vs. Slow Fashion Fair on the Global Impact of Local Wardrobe Choices:
Promoting Sustainable Behavior at the University of Antwerp

Sarah Rohaert
University of Antwerp, Belgium

ABSTRACT

Sustainability is a principal research domain and part of the DNA of the European Project Semester at the Faculty of Design Sciences at the University of Antwerp. This chapter describes how a multidisciplinary EPS team collaborated to design an interactive event, exhibit, and campaign to promote sustainable consumer behavior on a fair at the city campus of the university, with a focus on awareness-raising about fast versus slow fashion. Throughout all development stages, the team improved their work-in-progress during weekly coaching sessions with access to the sustainability-related research group and toolkits. This rich trajectory and the resulting fast/slow fashion fair triggered deep reflections on the team regarding fair fashion as well as taught them to plan a major event with many stakeholders, eco-design for the outdoors, and proactively communicate and develop their public relations and negotiation skills. It also raised awareness about the sustainable fashion cause among the students, staff, and the general public who visited the fair.

INTRODUCTION

The European Project Semester (EPS) is a problem- and project-based program of one semester, offered to engineering, product design, multimedia, and business undergraduates by 19 European design and engineering schools. EPS aims to prepare future professionals to collaborate in multidisciplinary teams. Challenging multi-layered projects invite the student teams to think globally and to act locally in

DOI: 10.4018/978-1-6684-2309-7.ch019

Copyright © 2022, IGI Global. Copying or distributing in print or electronic forms without written permission of IGI Global is prohibited.

close interaction with external and internal experts and organizations with profound experience in the project context. Since 2010, the University of Antwerp (UA) has been glad to host third and fourth-year mobility students during the spring semester. The EPS semester at the UA corresponds to 30 European Credit Transfer System (ECTS) units. It includes one Project (20 ECTS) and several project supportive courses (10 ECTS), including, amongst others, Project Management, Teamwork, Intercultural and English Communication, Cross-media Communication Design and Business in the European Union (EU). Sustainability is integrated into the EPS program and is an important criterion for the project definition and the project elaboration, evaluation, and grading. The program provides an opportunity to look across disciplinary boundaries and the competences, necessary to integrate knowledge and skills from different angles in diverse, international teams. The goal of the project (process) herein reported focuses on raising awareness about the impact of the apparel industry on the global south, with special attention to the contrast between fast versus slow fashion, adopting a circular economy approach.

This chapter has four additional sections: the fast fashion problem, the specification of the project requirements, the solution design process and the conclusion.

PROBLEM DEFINITION

First, this section describes the general problems of fast fashion and its environmental and societal costs. Second, an explanation is given about how our EPS project seeks to address these broad and complex problems, namely by raising awareness about the hidden costs of fast fashion.

The Dark Side of Fast Fashion and the Apparel's Industry

High streets showcase the ability of the apparel's industry and marketing strategies to increase the incidence of unplanned, impulsive purchasing by customers, who are seduced and overstimulated by a wide range of clothes, shoes and accessories to choose from at affordable till very low prices.

However, this affordable price comes with severe and secretive hidden social and environmental costs, often not well-known to the general public in the shopping malls. Few consider the long supply chain of the garments' life cycles, starting with the harvesting of cotton or other raw material, to the manufacturing of fabric, the assembly of clothes, to their sales, their repurposing and resale. The transport of apparel around the world creates significant environmental damage because it is mostly done via commercial shipping, one of the highest carbon-emission-intensive modes of transportation. In addition, garment conglomerates have to respond more frequently to charges on issues ranging from exploitative (child) labor practices, to perpetuating global poverty by fighting against wage increases in developing nations, to pursuing an unethical marketing strategy of promoting the overconsumption of cheap, readily disposable clothes, referred to as fast fashion (Bhardwaj & Fairhurst, 2010; Birnbaum, 2008) and thus contributing to high levels of industrial waste (Anguelov, 2016; Diebäcker, 2000; Nimon & Behghin, 1999; Rosenthal, 2007). The process of externalizing the social costs of textiles and clothes and outsourcing the polluting production processes to poorer countries is a result of direct cost-minimization initiatives (Anguelov, 2016). However, it is not only a consequence of direct cost-minimizing initiatives at the firm level, but also of government actions such as industrial recruitment policies that aim to attract businesses in order to help developing nations improve their economies. In addition, the increase in volume of Western second-hand clothes exports to developing nations, which accompanies the increase in fast fashion purchases, threatens the development of local apparel sectors. It also results in a disproportionately large volume of used clothes being finally disposed of in developing nations, where there are few, if any, regulatory standards in waste management (Ahmed & Ali, 2004; Henry et al., 2006; Thomas-Hope, 1998).

Raising Awareness of the Hidden Costs of Fast Fashion and Sustainable Alternatives

A majority of customers do not understand how most products are produced and how environmental damage occurs at each link of the production value chain (Van Doorsselaer & Koopmans, 2020). Even if they have a general idea of what goods are relatively more pollution intensive (oil, petrol, beef meat), they don't often refer to the apparel industry or clothes when they name the most polluting industries and goods. The goal of the EPS team is to reduce this lack of understanding of the downside of fast fashion. By designing and organizing a fair fashion event at the city campus and an online and offline campaign, the team will raise awareness about how responsible clothes consumption behavior can slow down fast fashion (Minney, 2016). The visitors will learn how they have the power to change the way they shop and spend their working lives. The main target audience to address with the interactive event and campaign are the university students, who belong primarily to generation Z (born between 1996-2011) and to generation Y (born between 1981-1995), next to the university staff and the general public, who may also visit the sustainable event.

FROM PROBLEM EXPLORATION TOWARDS PROJECT SPECIFICATIONS

As a departure point, the EPS team members needed to familiarize themselves with the complex topic of the apparel industry, more specifically the fast fashion versus the slow fashion intricacies. Then a strategy and a project plan had to be developed for the design of a campaign and event to translate the obtained insights to the general public in an accessible, engaging and sustainable way.

Exploring the Complexity of the Garment Industry and Value Chain

The team watched "The True Cost" Morgan's (2015) documentary film about the impact of fashion on the planet and the people, which explores and exposes the dark side of fashion. This industry exposé invites the spectators to re-evaluate their shopping habits and ask themselves whether their love of fashion justifies human rights violations and environmental pollution. Other evocative documentaries and reality series were also brought to the team's attention, such as "Stacey Dooley investigates: Fashion's dirty secrets" (Onono, 2018, October 9) and "Blood, sweat and luxury problems" (Endemol Nederland, 2012, April, 2). The latter reports on the experiences of young fashionistas and influencers from the Netherlands, who are put to work in several of the thousands of textile and clothes factories in Dhaka in Bangladesh, one of the largest clothing producers in the world. Also, in-depth comprehensive books exploring fast fashion, its negative impact on the environment and society and explaining eco-design and slow fashion principles, pioneers and best practices were provided to the students. These books included: "The dirty side of the garment industry" (Anguelov, 2016), "This is a good guide" (Eyskoot, 2018) and "Slow fashion, aesthetics meets ethics" (Minney, 2016).

To bring the sheer volume of annually disposed clothes home to the students, and to give an insight into the mechanism involved in repurposing second-hand clothes and the related problematic issues, a guided visit was organized to the headquarters of World Mission Aid. This Non-Governmental Organization (NGO) is known for its red collector bins of used clothes in public spaces. After a first selection at the headquarters' location by volunteers, the clothes are partly sold in the own shop at the distribution center, partly sold to textile traders to gain money to support selected projects in the South and partly shipped to the NGO's charity project partners in developing countries. The students were informed about the evolving role, activities and projects of the NGO, which is critically evaluating its activities and reorienting itself. The students could see the mountains of collected used clothes and accessories, learn about the steadily decrease in quality of the collected clothes at the sorting points and witness the compression of huge piles of clothes into large 60-kg-bales for shipment by volunteers. They learned about the NGO's struggles to maintain its collector points in public spaces due to new legislation and very diverse regulations, rules, policies, taxes and politics by the different municipalities across Belgium. The NGO's communication staff dialogued in-depth about the duplicity and paradoxical nature of exporting Western second-hand clothes to the South. A light was shed on the practices of unscrupulous commercial players involved in second-hand Western clothes export to often illegal, informal markets in the South. The NGO's collaborators also had a fruitful discussion about the challenges of reaching out to new, young volunteers from generations Y and Z and communicating clearly about the NGO's changing objectives and core activities. These activities involve reducing their clothes-on-demand-only export and exploring new ways and strategies to generate the necessary financial means to support development projects in the global south, with a much more pronounced emphasis on project ownership by local collectives.

Another perspective on the global scale of the fast fashion problems was offered by a Belgian representative of the syndicate for the textile sector, who participates in an international network of syndicate representatives who seek to empower textile workers in the South to negotiate better labor conditions and salaries. The syndicate's message is straightforward: only through supporting, empowering and connecting textile workers globally, cheap clothes supply at dumping prices may be reduced. Thus, local European textile workers may safeguard their activities and invest in sustainable business models and shorter supply chains, instead of feeling obliged to outsource their production to poor countries with low salaries to stay competitive at a price level.

Exploring the Target Audience for the Communication and Design Strategy

Even though the largest and fastest-growing rates of environmental activism are among teenagers and young adults, whose global awareness and social consciousness values influence their consumption patterns, their unsustainable apparel-shopping habits are still at odds with their environmental consciousness. Ironically, even those who are sensitive to "ethical consumption" in apparel are skillfully led to apparel products, differentiated with "eco-fashion" marketed attributes, rather than being inspired to buy and use apparel more selectively and carefully (Anguelov, 2016). Therefore, the aim of this project is to raise awareness about the multi-layered complexity of making the industrial garment industry more sustainable globally. To pinpoint contributions everyone can make as a (future) professional and as a consumer with one's own consumption choices and actions seems a feasible, first step. As the participants in the event are quite diverse, different personas have been developed during the multimedia design course. Thus, a better understanding is sought of the different motivations of both sustainability-sensitive and less-sustainability-wise concerned students and staff members, to entice them to attend the event. An interactive expert session with trends agency Trendwolves, specialized in youngster culture and generation Y and Z, sparked new insights on how to communicate with and design for these particular age groups. After diverse conversations with the targeted audience, the team decided to promote the event as an entertainment rather than an educational fair to go to. Consequently, the threshold to attend is kept as low as possible for potential visitors who might be less interested in fair trade and sustainability. Deliberately a positive approach is favored, which showcases best practices and pioneering NGO and entrepreneurs in the sustainable fashion industry, instead of an approach that seeks to shock, shame and blame. Positive examples will inspire the target audience to reflect about business ethics in general and provide them with concrete examples of how they can make a difference with their consumption behavior in particular.

Coaching the Project Management Process

To guide the student team to define early on the project specifications, the to-be-developed items and to make a good planning for the different development stages, supportive courses with assignments and submission deadlines for intermediate results are included in the EPS program. The students start on the first day with a project kick-off session to discuss the project briefing and budget with their project client and supervisors. Hereafter, the project supervisors will have weekly follow-up meetings on the project progress with the student team. The team thus refines the project brief. Based upon their team assessment during the teambuilding course and the project management course during their first two weeks of the semester, they divide the project work in work packages and tasks and attribute roles and responsibilities to each team member. They agree upon Specific, Measurable Achievable, Relevant, Time-Bound (SMART) project objectives and a timeline with intermediate steps to achieve their project goals. At the end of week two, a preliminary project management plan is presented to the responsible professor and a board of all the internal and external EPS supervisors. Based upon the received feedback and a project plan checklist, the plans are refined, resulting in a project plan, which is submitted two weeks later to be approved and signed by all involved parties.

Hereafter, the Business Model Canvas classes challenge the team to develop three different business canvas options. These canvases are presented to and discussed with American Business students and lecturers during the first class of their joint "Business in the EU" course, to obtain an outsider view and feedback from yet another different discipline (Business Economics). Complementary, a trans-Atlantic view is thus added to the already intercultural perspectives inside the EPS group of European, Vietnamese and Mexican students, who are studying in the EU. This course broadens the students' perspective, framing the micro-economics of the own project with the macro-economics of the EU policies.

The week before the Easter break, the students have the opportunity to join an intense 3-day *Masterclass on Sustainability* with sleepovers and catering by a "*no-waste-food-collective*", open to all students and young professionals. Thus, they can live, eat, and breathe sustainability and learn from international guest speakers and workshops in diverse tracks on sustainable living. This Masterclass program is a collaboration between the University Foundation for Development Cooperation (USOS), Act for Change, the Institute for Development Studies (IOB) and other organizations. They empower sustainable change makers with a global mind-set and in-depth insights, skills, and knowledge about opportunities for a more sustainable lifestyle. One of the tracks focused on fashion. It included a session on aesthetics and fluid identities versus ethics, a guided city-walk to sustainable fashion brands, shops and entrepreneurs and a debate. For the debate, participants were asked to take position and argue in favor or against second-hand clothes export to developing countries and import embargos, after testimonials of different stakeholders with different perspectives on the topic. Thanks to diverse participants from the global south and lecturers, a multidisciplinary and multicultural perspective was guaranteed, which often challenged the Western perspective and contributed to enriching discussions. All participants, including the EPS teams, were invited to discuss their "sustainable change" ideas and (EPS) projects during the "open space" session in the evening. Thus, the fair fashion EPS team obtained constructive feedback from a very diverse and highly sustainability-aware audience about their communication plan and the planned activities for their sustainable fashion fair. The interaction after the presentation resulted in a change of name of the event (from "sweatshop" to "fast/fair fashion fair") and in several participants volunteering to promote and to realize the event and to showcase their slow fashion related work.

The week after the Easter break, a thorough formal assessment of the project progress, intermediate results and the project and team process is made by the internal and external supervisors and the students. Hereto the students prepare a midterm project report and presentation, documents which illustrate their project process and teamwork. Each student also submits a self-reflection and a self-and-peer assessment of the contributions of each team member to the team and the project process.

This project assessment procedure is repeated at the end of the semester to access the final results. However, the real proof of the pudding is the Fast/Fair fashion fair itself on the 23rd of May 2019, upon which we report in the next section.

THE DESIGN PROCESS FROM PROJECT SPECIFICATIONS TO REALIZATION

We will now report on the project realization, split in four major parts: (*i*) the communication campaign; (*ii*) the design and production of interactive showpieces to raise awareness about sustainable fashion on the event itself; (*iii*) the selection of the organization who can share interesting context knowledge with the visitors about fair/slow fashion; and (*iv*) the detailed planning of the event.

The Communication Campaign

As a starting point for the communication design, the central concepts of sustainability, fast and slow fashion, had to be accurately understood and defined because "*only by clarifying these topics to ourselves we were able to explain them to others successfully*" (Balsa et al., 2019). As Friedman's speech during the recent Copenhagen Fashion Summit 2019 in Denmark illustrates, the students are not alone in their quest for clear and accurate definitions. The fashion director at The New York Times points out that after ten years of speaking about 'sustainable fashion', "*still no one knows exactly what it means.*" For sustainability, the definition of the Brundtland commission has been adopted: "*sustainable development is a development that meets the needs of the present without compromising the ability of future generations to meet their own needs*" (Brundtland, 1987), which acknowledges the necessary interconnection between the economic, social and ecological dimensions for sustainable development.

Fast Fashion is a term used to describe cheaply made clothes and accessories, often in a style that might be very trend-sensitive. Fast fashion originated because of cheap garment production possibilities in developing countries, combined with an increase in demand for fashionable clothing, supported by an increase in purchasing power of consumers, especially in the Western world. Fast fashion contributes to pollution, often poor-quality clothes and poor working conditions in developing countries. Examples of fast fashion brands are Primark, Zara, H&M, GAP, Forever 21. Slow fashion refers to a deliberate choice for fewer and better-quality items, which are environmentally friendly, ethically conscious and durable "*ever-greens*" rather than trend-driven. Transparency is offered about where clothes are coming from and how they are made. Patagonia is an example of a certified B-Corps brand, meeting the non-profit B Lab's standards of social and environmental performance, accountability and transparency.

The team members indicate that watching "The true cost" (Morgan, 2015) together was a turning point to realize "*we were not part of the solution, but part of the problem, until then... (...) For us "The true cost" was shocking and we truly realized that a change had to be made. After seeing it, we were even more motivated to do our project.*" The documentary also offered them first insights into the current working conditions in the Indian textile industry. A deeper understanding was obtained from Anguelov (2016), who compares the minimum wages in the Indian textile industry of 52 € to the bare minimum cost of living to fulfill one's most basic needs of 195 €. He explains how these kinds of disparities are caused by the enormous economic pressures from the fast fashion industry, amongst other reasons. Discussions with other EPS teams, American business students and interviews with the target audience online and offline taught the team that the ignorance about the dark side of fashion is still large (e.g., several students indicated H&M and Nike as sustainable brands; many had never considered the impact of their clothes' consumption, nor knew about sustainable brands or the meaning of what "sweatshop"). It inspired the team to reflect on how to translate the complex topic into very concrete examples, easy to understand and relate to for the general public.

To reach as many potential visitors as possible, both offline and online communication channels were used. The offline campaign of posters and life interactions, supported with a survey board (see Figure 2), one week before the event, aimed to make students interested and curious about the fair and point them towards more information on the social media channels. A first idea to jump-start the social media campaign was a challenge, asking students to post alternative, more sustainable purchase proposals for 2 € instead of the 2 € T-shirts you can buy from fast fashion brands. The objective was to spark conversation about the dumping prices of fast fashion clothes and generate video content to promote their event. This idea was pitched at a university event which discussed ideas to reach out to generation Y and Z. It was received well by the communication consultant, the present university staff of social services, student administration, the green office and student organizations for development (STUOS). The pitching event allowed the team to establish contacts with the student organizations and obtain their engagement for the event. The team accepted the offer from the communication staff to take over the social media channels of the university on the day of the market event to promote their sustainable message and streamline videos of their event. While exploring the Indian context for inspiration for the visual style for the communication, event and showpiece design, the team became inspired by the Indian "Holi" spring festival, which is known for its vibrant colors. The Holi's festive look & feel inspired the visual style, which was applied to all project-related communication means and interactive showpieces to give a coherent overall look, as Figure 1, Figure 2, Figure 3, and Figure 5 illustrate. A contemporary typeface was selected for its good readability on digital platforms.

Figure 1. Indian-Holi-festival-inspired visual style for communication on social media
(Adapted from Balsa, C., et al., 2019)

Figure 2. Promoting the event & slow fashion also offline and on the regional television
(Adapted from Balsa, C., et al., 2019)

The Interactive Showpieces

A design process involved several iterations, during which the product concepts became progressively more simple, feasible and affordable, within the project timespan and budget boundaries.

This process of simplification was further induced by the very strict safety regulations which apply in the context of the central, protected inner garden Hof Van Liere, where the fair takes place. Several alterations had to be made to the design concepts in order to make them weather-proof and sufficiently safe, solid and sturdy to be used in a central outdoor space of the city campus where the frequent passage of many persons occurs. Also, several accesses have to be kept free to allow emergency exits in case of fire. No heavy objects may be placed on the grass or hung on the trees. The music volume must be reduced, not to interrupt ongoing classes and study in the surrounding historical buildings and university library.

All refined design concepts, the resulting interactive showpieces and their positioning on the inner garden were presented for approval by the staff members in charge of the facility management. The team conducted several functional and stability tests. Risks and usability tests were described and assessed in alignment with the safety regulations during the design phase, a mandatory step in any EPS project. Figure 4 and Figure 5 depict one showpiece's evolution.

After several iterations, the team chose to simplify the manufacturing, assembling, and maintenance of the showpieces to be easily deconstructed, repurposed, or reused upon other occasions, according to the circular economy principles. According to the sustainability philosophy, the second-hand material and left-overs from the workshop were chosen as construction materials.

The students created their own sustainable fashion brand. They developed several fashion related logos and upgraded second-hand T-shirts, received from the "World mission aid" NGO, by treating the tissue and adding their own sustainable label. The first idea involved building a large square structure to hang the T-shirts overhead on the central grass field. As an alternative, a structure in the form of a corridor, compartmentalized in several rooms, was proposed, to be placed in the main passageway, so all passers-by should pass through it and become surrounded by different messages, displayed on second-hand clothes. However, the first concept would be expensive to construct and the supporting pillars were not allowed to be put on the grass. The second concept was not allowed due to security restrictions and stability issues, in case of heavy wind. Finally, the T-shirts were exposed on hired clothes-racks (as happens in the Western world) and piled in a huge heap on a shopping cart, to represent the informal markets in the South and the grabbing madness in the high street during sales days such as Black Fridays. These were disposed on the diagonal paths across the grass fields.

Figure 3. Evolution of the design ideas to present the upcycled clothes line
(Adapted from Balsa, C., et al., 2019)

Two mannequins sided one path, Dolores in black with unicorn flip-flops, representing fast fashion, in dialogue with Alegria in white, representing slow fashion, as shown in Figure 4.

Figure 4. Evolution of the fast/slow fashion mannequin design concept
(Adapted from Balsa, C., et al., 2019)

From the branches of a tree, information panels, designed as XL fashion labels, were hanging on strings. After a first try-out, these were fastened to the ground with tent pegs, so they would no longer swing wildly in the wind. Figure 5 shows the Holi-styled labels with short informative texts and illustrative images to inform the public about the fast fashion related problems and solutions.

Figure 5. Examples of the price tags to inform the public about the true cost of fast fashion (Adapted from Balsa, C., et al., 2019)

One week before the event's "life" promotion was made on the event's location. Hereto students were invited to weave a string across the answers of their choice to questions about their buying and using habits related to clothes and fashion accessories on a Holi-styled survey board, as shown in Figure 2. Thus, students' attitudes towards clothes became visual, string by string.

The Selection of Fair Participants

Several organizations, NGOs, and entrepreneurs, active in the field of sustainable fashion, were screened, selected, and invited to promote their activities and share their stories on the event. It is not evident for them to participate, as many involve only a couple of collaborators. This implied they had to close their shop to participate in an uncertain income event. Many also feared that the prices of their fair fashion products would be considered high for a student's budget, even though university staff and passers-by would also attend. One company specialized in fashionable clothes made of bamboo accepted. Crazy Clage, which repurposes waste material of amongst others beer tins and the furnishing industry in its jewelry design, was also present. The jewelry is produced by prisoners, learning new skills and providing them with pride and some earnings. Nash, an IOB degree student, sold handcrafted sustainable fashion accessories. The proposal of a TU Delft Ph.D. student for sustainable clothing and wardrobe options were showcased to let visitors reflect upon new possibilities to enhance responsible clothing choices. Kringloopwinkel, a social organization, collecting and reselling second-hand items while providing social employment and training to their employees, was present to promote their upcoming collector event with a picture competition in a second-hand décor and a stand promoting their shops and renting services of second-hand furnishing and cutlery for events. Oxfam and WMH, the NGO which collects second-hand clothes, had to cancel last minute. However, WHM provided us with bulks of 80 kg of compressed clothes, packaged in plastic, half the size and weight of the usual size of the ones they export. Thus, students could sit on them and learn about, e.g., the volumes of clothes disposed each year in Belgium and worldwide by reading the attached information labels, which are included in Figure 6.

Figure 6. Seating places made from WMH's compressed clothes bulks with information labels
(Adapted from Balsa, C., et al., 2019)

The event also served as a platform to promote sustainability-related university initiatives. A new course on Community Service Learning (CSL) was promoted. CSL involves close collaboration with social entrepreneurs, whose products were presented at the fair, together with information on how to enroll to this new university-wide optional course. USOS promoted its activities and exposure travels to the South on the fair. The green office coordinator and her environmental service colleague promoted subscriptions for seasonal fruits and vegetables from local farmers and reusable bags.

Antwerp hosts a community of Indians, active in amongst others Antwerp's diamond trade, which the organizers wanted to involve in the fair. An Indian vegetarian buffet restaurant, Aahaar, accepted the invitation to catering for the event with Indian street food snacks. Organic drinks were sold and small Indian sweets, home-cooked by the EPS team, were handed out by the recruited volunteers in exchange for completing a survey about the visitors' experience. Figure 7 shows some of these external stakeholders in action on the day of the market.

Figure 7. Pictures of the expo with the sustainable fashion entrepreneurs and the Indian catering (Adapted from Balsa, C., et al., 2019)

The Detailed Planning of the Event Itself

The planning included the coordination of all the volunteers, the participating organizations and the practical issues, such as the timely reservation and set-up of the different stands and equipment. It involved making an overview map of the different stands and the needed electricity and water supply for each and a "bad weather plan B". It included the timely reservation of the needed facilities, the catering and the final check-up on safety restrictions by the university's facility management staff. Besides, it involved making detailed planning of the day of the event itself, including the arrival times and briefing of the volunteers for the set-up of the hired market stands and the coordination of the organic and vegan welcome breakfast, kindly offered by the CSL organizers for all the volunteers and the participating entrepreneurs. It indicated the working shifts for the volunteers, hosting the stands and selling points during the event, and the ones who would help deconstruct the stands and clean up after the event. The planning also established who of the organizing students would welcome and install the participating organizational representatives and who would support the "Swishing" ladies. The "Swishing" ladies were in charge of hosting clothes swishing, actively promoted on social media to the students, the staff, and the general public, as a closing event during the evening. The ladies had to be shown where they could place their racks and tables to receive and sort the clothes brought by the swishing participants. They also needed access to the room, which would serve as a temporary fitting room for their "customers".

CONCLUSION

This paper reports on the development process and the achieved results of the Fast/Fair fashion fair event and campaign for USOS by a multinational, interdisciplinary EPS team at the University of Antwerp. Within the EPS educational framework, this project process has been steered by supportive and complementary courses, which challenge the students to apply immediately the courses' learning content and their complementary multidisciplinary knowledge and skills to their project work.

Thus, the EPS program seeks to overcome the more traditional educational system, which tends to divide knowledge and skills into study programs and clearly demarcated courses, which are subdivided into different course units or lessons. A side effect of such a traditional system is that students often do not make the transfer of knowledge and skills between subjects, let alone make links with other disciplines and apply theoretical insights immediately to multifaceted, real-life problems in practice. It is precisely this ability to look across disciplinary boundaries and the competences, necessary to integrate knowledge and skills from different angles in diverse, international teams, which the EPS network seeks to enhance. This ability and these competences are essential for design engineers who want to address complex issues such as the sustainability challenges implied by a circular economy approach. The thus acquired skills and competences are key assets for the student's future professional careers.

Sustainable versus fast fashion has proven to be an incredibly grateful topic, close to the students' day-to-day life experience, to bring home the urgency and need to combine insights from different disciplines and areas of expertise to tackle sustainable challenges. The complementary testimonials and feedback of experts from syndicates, human rights organizations, NGOs, sustainable fashion entrepreneurs, innovative technological textile developers, and volunteers have taught the students about the multidisciplinary dimensions to consider to make sense of complex, interrelated systems that shape the fashion industry. This thorough understanding is needed in order to come up with sustainable solutions and to explain this complexity during a public event. The EPS teams, the consulted experts, and the event's visitors, amongst whom the university's rector and the regional television, and the participating organizations confirmed the relevance of this project in terms of raising awareness about the topic and the multicultural, interdisciplinary teamwork approach and sustainable development practices it requires.

The EPS program at the University of Antwerp offered the educational framework to support the practice in and reflection on cross-cultural communication, multidisciplinary teamwork, ethical and deontological concerns, sustainable development practices as well as the development of the technical and scientific competences within the team, composed of design, engineering and communication students. This paper seeks to share our experiences. As true impact can only be achieved through a patchwork of repeated and complementary actions, this paper is a call for action, which seeks to motivate readers to also embark on projects which stimulate responsible consumerism and circular, triple-bottom-line economics, which consider the social and environmental account of a business, as well as its profitability. Thus a way forward to 'restorative economics' (Minney, 2016) may be paved, in which transparency, intercultural dialogue, interdisciplinary collaboration, social innovation and responsible design engineering can thrive.

ACKNOWLEDGMENT

This project would not have been possible without the support of many colleagues from the University of Antwerp, at the faculty of Design Sciences and the department of product development, which hosts the

EPS program and guest lectures from the EPS network. We are grateful for the project budget and input received by USOS, who operated as the external project client. We are also grateful for all the advice and feedback received from the experts from syndicates, fashion designers, the clean clothes campaign collaborator of "world solidarity" and for the second-hand clothes provided by "World Mission Aid" to upcycle and sell during our event to raise money for an NGO, which provides education to children of factory workers in developing countries.

REFERENCES

Ahmed, S. A., & Ali, M. (2004). Partnerships for solid waste management in developing countries: Linking theories to realities. *Habitat International*, *28*(3), 467–479. doi:10.1016/S0197-3975(03)00044-4

Anguelov, N. (2016). *The dirty side of the garment industry*. CRC Press.

Balsa, C., Delrue, V., Hartl, P., Lambertus, A., Rohaert, S., & Scheldeman, M. (2019). *Sustainable market* (EPS Final report) [Unpublished bachelor thesis]. University of Antwerp, Belgium.

Bhardwaj, V., & Fairhurst, A. (2010). Fast fashion: Response to changes in the fashion industry. *International Review of Retail, Distribution and Consumer Research*, *20*(1), 165–173. doi:10.1080/09593960903498300

Birnbaum, D. (2008). *Crisis in the 21st Century Garment Industry and Breakthrough Unified Strategy*. The Fashion Index Inc.

Brundtland, G. (1987). *Report of the World Commission on Environment and Development: Our Common Future*. United Nations General Assembly, document A/42/427. https://sustainabledevelopment.un.org/content/documents/5987our-common-future.pdf

Diebäcker, M. (2000). Environmental and social benchmarking for industrial processes in developing countries: A pilot project for the textile industry in India, Indonesia and Zimbabwe. *Integrated Manufacturing Systems*, *11*(7), 491–500. doi:10.1108/09576060010349794

Onono, E. (Executive Producer). (2018, October 9). *Fashion's dirty secrets* [Stacey Dooley investigates]. BBC.

Endemol Nederland. (2012, April 2). *Blood, sweat and luxury problems* [Reality series]. BNN VARA.

Eyskoot, M. (2018). *This is a Good Guide – for a Sustainable Lifestyle*. Laurence King Publishing.

Henry, R. K., Yongsheng, Z., & Jun, D. (2006). Municipal solid waste management challenges in developing countries—Kenyan case study. *Waste Management (New York, N.Y.)*, *26*(1), 92–100. doi:10.1016/j.wasman.2005.03.007 PMID:16006111

Minney, S. (2016). *Slow fashion: Aesthetics Meets Ethics*. New Internationalist Publications Ltd.

Nimon, W., & Beghin, J. (1999). Are eco-labels valuable? Evidence from the apparel industry. *American Journal of Agricultural Economics*, *81*(4), 801–811. doi:10.2307/1244325

Rosenthal, E. (2007, Jan. 25). Can polyester save the world? *The New York Times*.

Morgan, A. (Director). (2015). *The True Cost* [Documentary]. An Untold Production in association with Life is My Movie Entertainment.

Thomas-Hope, E. (Ed.). (1998). *Solid Waste Management: Critical Issues for Developing Countries.* Canoe Press.

Van Doorsselaer, K., & Koopmans, R. J. (2020). *Ecodesign: A Life Cycle Approach for a Sustainable Future*. Hanser-Fachbuch. doi:10.3139/9781569908624

KEY TERMS AND DEFINITIONS

ECTS: Stands for European Credit Transfer System and is a method of measuring the study program as academic currency.

Event: A planned public or social occasion.

Fast Fashion: Cheaply made clothes and accessories, often in a style that might be very trend-sensitive.

Generation Y: The generation born in the 1980s and 1990s, comprising primarily the children of the baby boomers and typically perceived as increasingly familiar with digital and electronic technology.

Generation Z: The generation reaching adulthood in the second decade of the 21st century, perceived as being familiar with the internet from a very young age.

NGO: A non-profit organization that operates independently of any government, typically one whose purpose is to address a social or political issue.

Slow Fashion: Refers to a deliberate choice for fewer and better-quality items, which are environmentally friendly, ethically conscious, and durable "ever-greens" rather than trend-driven. Transparency is offered about where slow fashion clothes are coming from and how they are made.

Sustainable Development: A development that meets the needs of the present without compromising the ability of future generations to meet their own needs, while acknowledging the necessary interconnection between the economic, social, and ecological dimensions for sustainable development.

Trend-Driven: About isolating and decoding the underlying behaviors—even those that feel irrelevant, extreme, or a little out there—and understanding which has the potential to be used more extensively.

Compilation of References

Abata, D. L., Andersen, A., & Krause, W. B. (2013). Transatlantic Interaction with European Project Semester. In *Proceedings of the 2013 American Society for Engineering Education Annual Conference & Exposition* (pp. 23.1264.1-23.1264.13). 10.18260/1-2--22649

Abe, J., Talbot, D. M., & Geelhoed, R. J. (1998). Effects of a Peer Program on International Student Adjustment. *Journal of College Student Development, 39*(6), 539–547.

Abercrombie, D. (1968). *Elements of General Phonetics*. Edinburgh University Press.

ABET. (2019). *Criteria for Accrediting Engineering Programs 2020-2021*. https://www.abet.org/accreditation/accreditation-criteria/criteria-for-accrediting-engineering-programs-2020-2021/

ABET. (2021). *About ABET.* https://www.abet.org/about-abet/

ABET. (2022). *ABET*. https://www.abet.org/

Accreditation Board for Engineering and Technology. (2000). *Viewpoints – Vol. I: Issues of Accreditation in Higher Education*. https://www.abet.org/wp-content/uploads/2015/05/Viewpoints_Vol1.pdf

Accreditation Board for Engineering and Technology. (2006). *Engineering Change – A Study of the Impact of EC2000*. https://www.abet.org/wp-content/uploads/2015/04/EngineeringChange-executive-summary.pdf

Aedma, A., González Alen, D., Norbert, R., Petersen, N., & Rzeznik, P. (2014). *Reconfigurable LED Lamp* (EPS Project Wiki). Instituto Superior de Engenharia do Porto. http://www.eps2014-wiki4.dee.isep.ipp.pt/

Afroogh, S., Esmalian, A., Donaldson, J., & Mostafavi, A. (2021). Empathic Design in Engineering Education and Practice: An Approach for Achieving Inclusive and Effective Community Resilience. *Sustainability, 13*(7), 4060. doi:10.3390u13074060

Aghera, A., Emery, M., Bounds, R., Bush, C., Stansfield, R. B., Gillett, B., & Santen, S. A. (2018). A Randomized Trial of SMART Goal Enhanced Debriefing after Simulation to Promote Educational Actions. *The Western Journal of Emergency Medicine, 19*(1), 112–120. doi:10.5811/westjem.2017.11.36524 PMID:29383065

Agile Alliance. (2022). *What is Agile Alliance and What Does it Do?* https://www.agilealliance.org/the-alliance/

Ahmed, S. A., & Ali, M. (2004). Partnerships for solid waste management in developing countries: Linking theories to realities. *Habitat International, 28*(3), 467–479. doi:10.1016/S0197-3975(03)00044-4

AirShip Technologies Group. (2019). *Survival Air Stretcher*. https://www.airshiptg.com/air-stretcher

Akmajian, A., Demers, R. A., Farmer, A. K., & Harnish, R. M. (1990). *Linguistics. An Introduction to Language and Communication*. MIT Press.

Allport, G. W. (1954). *The Nature of Prejudice*. Perseus Books.

Alluwar, N. (2017). *What is IoT architecture?* https://www.quora.com/What-is-IoT-architecture

Alred, G., Byram, M., & Fleming, M. (Eds.). (2002). *Intercultural Experience and Education*. Multilingual Matters. doi:10.21832/9781853596087

American Society for Engineering Education. (2010). The Green Report – Engineering Education for a Changing World. Technical Report. American Society for Engineering Education.

Andersen, A. (2003). Collaborating Learning in Interdisciplinary and Intercultural Teams. *Proceedings of the Third International Conference on Engineering and Computer Education*, 1-3. https://copec.eu/congresses/icece2003/proc/pdf/T131.pdf

Andersen, A. (2012). The European Project Semester: A Useful Teaching Method in Engineering Education. In Project Approaches to Learning in Engineering Education. SensePublishers. doi:10.1007/978-94-6091-958-9_3

Andersen, A., & Hansen, J. (2002). Engineers of Tomorrow and Beyond, insight and skills needed to work across borders. In *Proceedings of the 2002 American Society for Engineering Education Annual Conference & Exposition* (pp. 7.500.1-7.500.7). 10.18260/1-2--10461

Andersen, A. (2000a). International Teamwork and Communication in Engineering Education (Discussion paper). *Workshop on Internationalisation, SEFI 28th Annual Conference.*

Andersen, A. (2001). Implementation of engineering product design using international student teamwork to comply with future needs. *European Journal of Engineering Education, 26*(2), 179–186. doi:10.1080/03043790110034438

Andersen, A. (2004). Preparing engineering students to work in a global environment to co-operate, to communicate and to compete. *European Journal of Engineering Education, 29*(4), 549–558. doi:10.1080/03043790410001711243

Andersen, A. (2004b). Student Based Learning in A Multicultural Environment. *Proceedings of the 2004 American Society for Engineering Education Annual Conference.* 10.18260/1-2--13893

Andersen, A. (2009). Keynote on Project Management and Teamwork. In *Proceedings of the First Ibero-American Symposium of Project Approaches in Engineering Education* (pp. 10-10). University of Minho.

Andersen, A. J. (2000b). Assessment Techniques used in Multidisciplinary and Cross-Cultural Student Teamwork. *Proceedings of the 2000 American Society for Engineering Education Annual Conference.* 10.18260/1-2--8174

Anguelov, N. (2016). *The dirty side of the garment industry*. CRC Press.

Ansems, T., & Ligthart, T. N. (2018). *ENVPACK: LCA-Based Tool for Assessing the Sustainability of Packaging Chains.* https://kidv.nl/media/verdiepingsbijeenkomsten/vd_oogst_wop_verslag.pdf?1.2.1

Archdaily. (2022). *Bus station projects*. https://www.archdaily.com/

Archdaily. (2022). *Sustainable architecture.* http://www.archdaily.com

Architecture Art Designs. (2019). *Five Most Renowned Green Architects.* https://www.architectureartdesigns.com/worlds-five-most-renowned-green-architects/

Architizer. (2022) *Bus station projects.* http://architizer.com

Arduino. (2019). *Arduino Uno.* https://www.arduino.cc/

Aritzeta, A., Swailes, S., & Senior, B. (2007). Belbin's Team Role Model: Development, Validity and Applications for Team Building. *Journal of Management Studies, 44*(1).

Armesto, L., Fuentes-Durá, P., & Perry, D. (2016). Low-cost Printable Robots in Education. *Journal of Intelligent & Robotic Systems*, *81*(1), 5–24. doi:10.100710846-015-0199-x

Arnó, E., Aguilar, M., & Tatzl, D. (2020). Engineering students' perceptions of the role of ESP courses in internationalized universities. *English for Specific Purposes*, *58*, 58–74. doi:10.1016/j.esp.2019.12.001

Arnó-Macià, E., & Mancho-Barés, G. (2015). The role of content and language in content and language integrated learning (CLIL) at university: Challenges and implications for ESP. *English for Specific Purposes*, *37*, 63–73. doi:10.1016/j.esp.2014.06.007

Ashworth, D. W. C. (2010). Project Based Learning in International Teams – Monitoring the Effectiveness of Teamwork. In *Key Engineering Materials* (Vol. 450, pp. 581–584). Trans Tech Publications, Ltd. doi:10.4028/www.scientific.net/KEM.450.581

Association of Educating Cities. (2012). *Educating Cities*. http://www.bcn.cat/edcities/aice/estatiques/espanyol/sec_educating.html

Asztalos, I., Giordano, R., Gunther, T., Navarrete Hill, M., & Rabone, G. J. (2015). *Autonomous Sailboat* (EPS Project Wiki). Instituto Superior de Engenharia do Porto. http://www.eps2015-wiki5.dee.isep.ipp.pt/

Augustyns, L., Pogoda, M., Milesi, M., Kang, M., Valls, P., Duarte, A. J., Malheiro, B., Ferreira, F., Ribeiro, M. C., Silva, M. F., Ferreira, P., & Guedes, P. (2017). Sustainable Desalinator – An EPS@ISEP 2016 Project. In *Proceedings of the 45th SEFI Annual Conference* (pp. 491-498). European Society for Engineering Education.

Balaji, K. V. A., & Somashekar, P. (2009). A comparative study of soft skills among engineers. *The IUP Journal of Soft Skills*, *3*(3/4), 50–57.

Balsa, C., Delrue, V., Hartl, P., Lambertus, A., Rohaert, S., & Scheldeman, M. (2019). *Sustainable market* (EPS Final report) [Unpublished bachelor thesis]. University of Antwerp, Belgium.

Barr, M., & Massa, A. (2006). *Programming Embedded Systems*. O'Reilly.

Bartosz, K., Bentin, C., Davies, E., Fernandez de Toro Ronda, P., Fountain, A., & Suits, K. (2015). *Aquaponic Shelter* (EPS Project Wiki). Instituto Superior de Engenharia do Porto. http://www.eps2015-wiki3.dee.isep.ipp.pt/

Batova, T. (2018). Global Technical Communication in 7.5 Weeks Online: Combining Industry and Academic Perspectives. *IEEE Transactions on Professional Communication*, *61*(3), 311–329. doi:10.1109/TPC.2018.2823598

Bauman, Z. (1993). *Postmodern Ethics*. Wiley-Blackwell.

Bazylinska, A., Jenei, A., Juht, T., Küttis, S., & Walczak, J. (2012). *UV-based Disinfection* (EPS Project Wiki). Instituto Superior de Engenharia do Porto. http://www.eps2012-wiki2.dee.isep.ipp.pt/

Beeline. (2022). *Bike Computer Beeline Velo | Beeline EU*. https://eu.beeline.co/pages/beeline-velo

Beilin, H. (1992). Piaget's enduring contribution to developmental psychology. *Developmental Psychology*, *28*(2), 191–204. doi:10.1037/0012-1649.28.2.191

Belbin Associates. (2014). *Method, Reliability & Validity, Statistics and Research, A Comprehensive Review of Belbin Team Roles*. https://www.belbin.com

Belbin, M. (2010). *Management Teams: Why They Succeed or Fail* (3rd ed.). Butterworth-Heinemann. doi:10.4324/9780080963594

Belbin, R. M. (2010). *Team roles at Work.* Elsevier Science.

Beleen, J., & Jones, E. (2015). Redefining Internationalization at Home. In A. Curaj, L. Matei, R. Pricopie, J. Salvi, & P. Scott (Eds.), *The European Higher Education Area* (pp. 59–72). Springer. doi:10.1007/978-3-319-20877-0_5

Bell, S. (2010). Project-Based Learning for the 21st Century: Skills for the Future. *The Clearing House: A Journal of Educational Strategies, Issues and Ideas, 83*(2), 39–43. doi:10.1080/00098650903505415

Belogolovsky, V. (2019). *Architecture should be about what it can do, not what it can look like: In Conversation with Michel Rojkind.* https://www.archdaily.com/910732/architecture-should-be-about-what-it-can-do-not-what-it-can-look-like-in-conversation-with-michel-rojkind

Bennett, J. M., & Bennett, M. J. (2004). Developing intercultural sensitivity: An integrative approach to global and domestic diversity. In D. Landis, J. Bennett, & M. Bennett (Eds.), *Handbook of Intercultural Training* (pp. 147–165). Sage. doi:10.4135/9781452231129.n6

Bennett, M. J. (2004). Becoming Interculturally Competent. In J. Wurzel (Ed.), *Toward multiculturalism: A reader in multicultural education* (pp. 62–77). Intercultural Resource Corporation.

Berényl, D., Brunke, F., de la Torre, A., Hood, L., & Nogal, H. (2015). *Solar Food Dehydrator* (EPS Project Wiki). Instituto Superior de Engenharia do Porto. http://www.eps2015-wiki1.dee.isep.ipp.pt/

Berger, T., Clarke, K., Koning, N., Kravchenko, D., & Runowski, B. (2020). Processing and development of ideas to solve existing problems in the local bus station of the city of Valencia (EPS Report). Valencia: Universitat Politècnica de València.

Bernhardt, B. C., & Singer, T. (2012). The neural basis of empathy. *Annual Review of Neuroscience, 35*(1), 1–23. doi:10.1146/annurev-neuro-062111-150536 PMID:22715878

Berry, J. W. (1990). Psychology of acculturation. In *Nebraska Symposium on Motivation: Cross-cultural perspectives. Current theory and research in motivation* (vol. 37, pp. 201–234). Lincoln, NE: University of Nebraska Press.

BessonL. (2014). *Lucy.* https://www.lucymovie.com

Best, K. (2019). *Design management: Managing design strategy, process and implementation.* Bloomsbury Visual Arts.

Bhardwaj, V., & Fairhurst, A. (2010). Fast fashion: Response to changes in the fashion industry. *International Review of Retail, Distribution and Consumer Research, 20*(1), 165–173. doi:10.1080/09593960903498300

Birnbaum, D. (2008). *Crisis in the 21st Century Garment Industry and Breakthrough Unified Strategy.* The Fashion Index Inc.

Björklund, T., & Keipi, T. (Eds.). (2019). *Design +: Organizational renewal and innovation through design.* Aalto University - Aalto Design Factory. Retrieved from https://aaltodoc.aalto.fi/handle/123456789/41115

Blazejewski, M., Heibeck, M., Moreno Ishii, A., & Nybjörk, R. G. (2014). *Bio-Inspired Swimming Robo*t (EPS Project Wiki). Instituto Superior de Engenharia do Porto. http://www.eps2014-wiki1.dee.isep.ipp.pt/

Blynk. (n.d.). *How Blynk Works.* http://docs.blynk.cc/#getting-started

Body, B. (2009). *Tips and observations of bird flight.* https://www.brendanbody.co.uk/flight_tutorial/index.html

Bombardó, C., Aguilar, M., & Barahona, C. (2007). *Technical Writing: A Guide for Effective Communication*. Edicions UPC.

Bonebright, D. A. (2010). 40 years of storming: A historical review of Tuckman's model of small group development. *Human Resource Development International, 13*(1), 111–120. doi:10.1080/13678861003589099

Borah, D., Malik, K., & Massini, S. (2021). Teaching-focused university–industry collaborations: Determinants and impact on graduates' employability competencies. *Research Policy, 50*(3), 104172. doi:10.1016/j.respol.2020.104172

Borja, M. B., & Valade-Amland, S. (2020). *Design: A business case: thinking, leading, and managing by design*. Business Expert Press.

Bournois, F., & Chevalier, F. (1998). Doing research with foreign colleagues: A project-life cycle approach. *Journal of Managerial Psychology, 13*(3/4), 206–212. doi:10.1108/02683949810215011

Boyne, G. A., & Chen, A. A. (2007). Performance targets and public service improvement. *Journal of Public Administration: Research and Theory, 17*(3), 455–477. doi:10.1093/jopart/mul007

Brahm, T., & Kunze, F. (2012). The role of trust climate in virtual teams. *Journal of Managerial Psychology, 27*(6), 595–614. doi:10.1108/02683941211252446

Brandenburg, U., Taboadela, O., & Vancea, M. (2016). Mobility Matters: the ERASMUS Impact Study. In Global and Local Internationalization. Transgressions: Cultural Studies and Education. SensePublishers. doi:10.1007/978-94-6300-301-8_16

Brassler, M., & Dettmers, J. (2017). How to Enhance Interdisciplinary Competence—Interdisciplinary Problem-Based Learning versus Interdisciplinary Project-Based Learning. *The Interdisciplinary Journal of Problem-Based Learning, 11*(2). Advance online publication. doi:10.7771/1541-5015.1686

Breymann, U. (2016). *C++ Eine Einführung*. Hanser Verlag. doi:10.3139/9783446449121

Brotherton, S. A., Fried, R. T., & Norman, E. S. (2008). Applying the work breakdown structure to the project management lifecycle. *2008 PMI Global Congress Proceedings*. https://www.pmi.org/learning/library/applying-work-breakdown-structure-project-lifecycle-6979

Brown, T. (2009). Change by design. How Design Thinking transforms organizations and inspires innovation. New York, NY: HarperCollins ebooks.

Brown, T. (2009). *Change by Design: How Design Thinking Transforms Organizations and Inspires Innovation*. HarperBusiness.

Brown, T., & Katz, B. (2019). *Change by design: How design thinking transforms organizations and inspires innovation*. Harper Business.

Brown, T., & Wyatt, J. (2010). Design Thinking for Social Innovation. *Stanford Social Innovation Review, 8*(1), 31–35. doi:10.48558/58Z7-3J85

Brundtland, G. (1987). *Report of the World Commission on Environment and Development: Our Common Future*. United Nations General Assembly, document A/42/427. https://sustainabledevelopment.un.org/content/documents/5987our-common-future.pdf

Brytting, T. (2005). *Företagsetik andra upplagan*. Liber Ekonomi.

Buckeridge, J. (2011). *4 Es Ethics, Engineering, Economics & Environment*. The Federation Press.

Budzinska, G., & Fuentes-Durá, P. (2014). Engineering Education for the Future. European Project Semester at Universitat Politècnica de València, Spain, and Lodz University of Technology, Poland. In *Proceedings of the Fourth Valencia Global Conference* (pp. 193-201). UPV.

Budzinska, G., & Fuentes-Durá, P. (2014). Engineering Education for the Future. European Project Semester at Universitat Politècnica de València, Spain, and Lodz University of Technology, Poland. In *Proceedings of the Fourth Valencia Global Conference* (pp.193-201). UPV.

Caramin, B. A., Dunn, I., Klawikowski, Y., & Ney, R. (2014). *Bio-inspired Flying Robot* (EPS Project Wiki). Instituto Superior de Engenharia do Porto. http://www.eps2014-wiki3.dee.isep.ipp.pt/

Caramin, B., Dunn, I., Ney, R., Klawikowski, Y., Duarte, A., Malheiro, B., Ribeiro, C., Ferreira, F., Silva, M., Ferreira, P., & Guedes, P. (2015). Design and Implementation of a Biologically Inspired Flying Robot – An EPS@ISEP 2014 Spring Edition Project. In *Proceedings of TEEM – 3rd Technological Ecosystems for Enhancing Multiculturality* (pp. 211-218). doi:10.1145/2808580.280861

Cărăşel, I.-S., Vandoorne-Feys, A., Nicoară, G.-G., & Michał, M. (2021). *FREE Desk, a reconfigurable and ergonomic smart desk.* Polytechnic of Porto, *Instituto Superior de Engenharia do Porto*, Porto. Retrieved from https://www.eps2021-wiki3.dee.isep.ipp.pt/doku.php?id=docs

Casado, M. G., Revert, C., Sales, V., & Veral, S. (2015). Smart Cities Trends: Tendencias en las Ciudades Inteligentes y oportunidades para los sectores del hábitat [Smart Cities Trends: Trends in smart cities and opportunities for sectors habitat]. ITC. AIDIMA y ASCER.

Casner-Lotto, J., & Barrington, L. (2006). *Are They Really Ready to Work? Employers' Perspectives on the Basic Knowledge and Applied Skills of New Entrants to the 21st Century US Workforce.* The Conference Board, Inc., the Partnership for 21st Century Skills, Corporate Voices for Working Families, and the Society for Human Resource Management. https://www.conference-board.org/pdf_free/BED-06-Workforce.pdf

Cateye. (2022). *Velo 9 | Products | CatEye.* https://www.cateye.com/intl/products/computers/CC-VL820/

Cech, E. A. (2014). Culture of Disengagement in Engineering Education? *Science, Technology & Human Values, 39*(1), 42–72. doi:10.1177/0162243913504305

Chakravorty, D. (2017). *STL File Format for 3D printing – Simply Explained.* all3dp. https://all3dp.com/what-is-stl-file-format-extension-3d-printing/

Chaney, L. H., & Martin, J. S. (2011). *Intercultural Business Communication* (5th ed.). Prentice Hall.

Chen, G.-m., & Starosta, W. J. (2005). *Foundations of Intercultural Communication* (2nd ed.). Allyn & Bacon.

Cheng, M., Adekola, O., Albia, J., & Cai, S. (2021). *Employability in higher education: a review of key stakeholders' perspectives.* Higher Education Evaluation and Development., doi:10.1108/HEED-03-2021-0025

Chen, J., Kolmos, A., & Du, X. (2020). Forms of implementation and challenges of PBL in engineering education: A review of literature. *European Journal of Engineering Education, 46*(1), 90–115. doi:10.1080/03043797.2020.1718615

City of Münster Presse- und Informationsamt. (2018). *Bicycles.* https://www.stadt-muenster.de/english/transport/bicycles.html

Claw, S., Folkertsma, L., Oliveira, H., Stoel, M., & Thor, L. (2014). *Be aware Mindset OpenBustop* (EPS Report). Universitat Politècnica de València.

Cleary, Y., Karreman, J., Closs, S., Drazek, Z., Engberg, J., Ghenghea, V., Meex, B., Minacori, P., Müller, J., & Straub, D. (2017). TecCOMFrame: A competence framework for technical communication. In *Proceedings of the 2017 IEEE International Professional Communication Conference (ProComm)* (pp. 1-5). 10.1109/IPCC.2017.8013937

CNI. (2018). *Ensino de engenharia: fortalecimento e modernização.* Propostas da indústria eleições 2018, Confederação Nacional da Indústria (CNI), Brasília. Retrieved from https://bucket-gw-cni-static-cms-si.s3.amazonaws.com/media/filer_public/5e/ec/5eec09ca-2b12-4880-8a0b-804411795ea7/ensino_de_engenharia_web.pdf

Cogo, A., & Jenkins, J. (2010). English as a lingua franca in Europe: A mismatch between policy and practice. *European Journal of Language Policy*, *2*(2), 271–294. doi:10.3828/ejlp.2010.16

Committee on the Engineer of. (2020). *Committee on Engineering Education, and National Academy of Engineering. In Educating the Engineer of 2020: Adapting Engineering Education to the New Century*. The National Academies Press.

Conway, B. E. (1999). *Electrochemical Supercapacitors: Scientific Fundamentals and Technological Applications.* Springer. doi:10.1007/978-1-4757-3058-6

Cooper, A. (1999). *The Inmates Are Running the Asylum: Why High Tech Products Drive Us Crazy and How to Restore the Sanity* (1st ed.). John Wiley & Sons. doi:10.1007/978-3-322-99786-9_1

Cooper, A., Reimann, R., Cronin, D., & Noessel, C. (2014). *About face: The essentials of interaction design* (4th ed.). John Wiley & Sons.

Covey, S. R. (2020). *7 Habits of Highly Effective People: Revised and Updated.* Simon Schuster.

Coward, C. (2015). *3D Printing*. DK Publishing.

Crane, A., Matten, D., Glozer, S., & Spence, L. (2019). *Business Ethics*. Oxford University Press.

Crawley, E. F., Malmqvist, J., Östlund, S., Brodeur, D. R., & Edström, K. (2014). *Rethinking Engineering Education.* Springer International Publishing., doi:10.1007/978-3-319-05561-9

Crome, L. (1939). The Stretcher Problem. *British Medical Journal*, *2*(4111), 819–820. doi:10.1136/bmj.2.4111.819 PMID:20782725

Cross, N. (1982). Designerly ways of knowing. *Design Studies*, *3*(4), 221–227. doi:10.1016/0142-694X(82)90040-0

Crowther, P., Joris, M., Otten, M., Nilsson, B., Teekens, H., & Wächter, B. (2000). *Internationalisation at Home. A position paper.* European Association for International Education.

Cuff, B. M. P., Brown, S. J., Taylor, I., & Howat, D. J. (2016). Empathy: A review of concept. *Emotion Review*, *8*(2), 144–153. doi:10.1177/1754073914558466

d.school. (2020). *Design Thinking Bootleg*. Retrieved from d.school - Stanford Design School: https://dschool.stanford.edu/resources/design-thinking-bootleg

Dalsgaard, A. M. (2020). *The human scale.* https://www.youtube.com/watch?v=tZcqJoaFIN0

Dandavate, U., Sanders, E. B. N., & Stuart, S. (1996). Emotions Matter: User Empathy in the Product Development Process. *Proceedings of the Human Factors and Ergonomics Society Annual Meeting*, *40*(7), 415–418. doi:10.1177/154193129604000709

Dannoritzer, C. (2010). *Comprar, tirar, comprar* [Buy, throw away, buy] (Documentary 28/12/2017). http://rtve.es/v/1382261

Davis, A., Dragos, S. V., Rozewski, J., Ruiz Ludeña, F. J., van Tendeloo, K., & von Bröckel, R. (2015). *Aquaponic Monitoring* (EPS Project Wiki). Instituto Superior de Engenharia do Porto. http://www.eps2015-wiki4.dee.isep.ipp.pt/

Davis, D. D. (2016). *Creative strategy and the business of design.* HowBooks.

de Certeau, M., & Mayol, P. (1998). *The Practice of Everyday Life: Living and Cooking* (Vol. 2). University of Minnesota Press.

de Couvreur, L., & Goossens, R. (2010). Design for (every) one: Co-creation as a bridge between universal design and rehabilitation engineering. *Proceedings of the Seventh International Conference on Design and Emotion.*

de Graaf, R. (2016). *Is iconicity good for architecture*? https://www.oma.com/lectures/is-iconicity-good-for-architecture

de Graaff, E., & Christensen, H. P. (2004). Editorial: Theme issue on active learning in engineering education. *European Journal of Engineering Education, 29*(4), 461–463. doi:10.1080/03043790410001716310

de los Ríos, I., Cazorla, A., Díaz-Puente, J. M., & Yagüe, J. L. (2010). Project–based learning in engineering higher education: Two decades of teaching competences in real environments. *Procedia: Social and Behavioral Sciences, 2*(2), 1368–1378. doi:10.1016/j.sbspro.2010.03.202

de Montaigne, M. (1580). *Essays of Michel de Montaigne* (C. Cotton, Trans. & W. C. Hazlitt, Ed.). https://www.gutenberg.org/files/3600/3600-h/3600-h.htm

De Vignemont, F., & Singer, T. (2006). The empathic brain: How, when, why? *Trends in Cognitive Sciences, 10*(10), 435–441. doi:10.1016/j.tics.2006.08.008 PMID:16949331

De Waal, F. B. M., & Preston, S D. (n.d.). Mammalian empathy: behavioral manifestations and neural basis. *Nature Reviews Neuroscience, 18*(8), 498–509. doi:10.1038/nrn.2017.72

Deardorff, D. K. (2006). Identification and assessment of intercultural competence as a student outcome of internationalization. *Journal of Studies in Intercultural Education, 10*(3), 241–266. doi:10.1177/1028315306287002

Decety, J., & Lamm, C. (2006). Human empathy through the lens of social neuroscience. *TheScientificWorldJournal, 6*, 1146–1163. doi:10.1100/tsw.2006.221 PMID:16998603

Decety, J., & Meyer, M. (2008). From emotion resonance to empathic understanding: A Social Developmental Neuroscience Account. *Development and Psychopathology, 20*(4), 1053–1080. doi:10.1017/S0954579408000503 PMID:18838031

Decety, J., & Moriguchi, Y. (2007). The empathic brain and its dysfunction in psychiatric populations: Implications for intervention across different clinical conditions. *BioPsychoSocial Medicine, 1*(1), 22. doi:10.1186/1751-0759-1-22 PMID:18021398

Deigh, J. (1995). Empathy and universalizability. *Ethics, 105*(4), 743–763. doi:10.1086/293751

Delgado, M. (2011). *El espacio público como ideología.* Los libros de la catarata.

Dezeen. (2022). *Sustainable architecture.* https://www.dezeen.com/

Diamond, A., Walkley, L., & Scott-Davies, S. (2011). *Global graduates into global leaders.* CIHE.

Diebäcker, M. (2000). Environmental and social benchmarking for industrial processes in developing countries: A pilot project for the textile industry in India, Indonesia and Zimbabwe. *Integrated Manufacturing Systems, 11*(7), 491–500. doi:10.1108/09576060010349794

Dispert, H., Boudin, C., Morgan, J. A., & McMahon, M. (2014). International study programs: The role of internships, R&D projects, and their evaluation. In *QScience Proceedings, Engineering Leaders World Congress on Engineering Education 2013* (pp. 1–10). 10.5339/qproc.2014.wcee2013.8

Docherty, A., Keane, S., Méry, G., Mesas Llauradó, A., & Sokolowska, N. (2014). *Aquaponics System* (EPS Project Wiki). Instituto Superior de Engenharia do Porto. http://www.eps2014-wiki5.dee.isep.ipp.pt/

Domingues, I. (Ed.). (2005). *Conhecimento e Transdisciplinaridade II: Aspectos metodológicos*. Humanitas.

Doran, G. (1981). There's a SMART way to write management's goals and objectives. *Management Review*, *70*(11), 35–36.

Dressen-Hammouda, D. (2008). Aligning EAP writing pedagogies across European universities: A case study from France. In I. Fortanet-Gómez & C. Räisänen (Eds.), *ESP in European Higher Education. Integrating Language and Content* (pp. 75–96). John Benjamins Publishing. doi:10.1075/aals.4.06dre

Duarte, A. J., Malheiro, B., Arno, E., Perrat, I., Fuentes, P., Silva, M. F., Guedes, P., & Ferreira, P. (2020). Engineering Education for Sustainable Development: The European Project Semester Approach. *IEEE Transactions on Education*, *63*(2), 108–117. doi:10.1109/TE.2019.2926944

Dudley-Evans, T., & St John, M. J. (1998). *Developments in English for specific purposes: A multidisciplinary approach*. Cambridge University Purposes.

Duhigg, C. (2016). *What Google Learned from its Quest to Build the Perfect Team*. https://www.nytimes.com/2016/02/28/

Dunn, I., Caramin, B., Klawikowski, Y., & Ney, R. (2014). *Bio-inspired Flying Robot*. https://drive.google.com/file/d/0B_4jquZNwPGmTjJUbVJpby1ob3M/edit?pli=1

Dziomdziora, A., Sin, D. N., Robertson, F., Mänysalo, M., & Pattiselano, N. (2016a). *Robotic Art – GraphBot*. http://www.eps2016-wiki1.dee.isep.ipp.pt/doku.php?id=report

Dziomdziora, A., Sin, D. N., Robertson, F., Mänysalo, M., Pattiselano, N., Duarte, A., Malheiro, B., Ribeiro, C., Ferreira, F., Silva, M. F., Ferreira, P., & Guedes, P. (2016b). Artistic Robot – An EPS@ISEP 2016 Project. In *Proceedings of ICL2016 – 19th International Conference on Collaborative Learning* (pp. 225-238). 10.1007/978-3-319-50337-0_20

EasyFairs. (2022). *Visit the future*. https://EasyFairs.com

Edmondson, A. (1999). Psychological Safety and Learning Behavior in Work Teams. *Administrative Science Quarterly*, *44*(2), 350–383. doi:10.2307/2666999

Efstratia, D. (2014). Experiential Education through Project Based Learning. *Procedia: Social and Behavioral Sciences*, *152*, 1256–1260. doi:10.1016/j.sbspro.2014.09.362

Eggleston, A. G., & Rabb, R. J. (2018) Technical Communication for Engineers: Improving Professional and Technical Skills. *Proceedings of the 2018 ASEE Annual Conference and Exposition*.

Eklund, H. J., & Meranius, M. S. (2021). Towards a consensus on the nature of empathy: A review of reviews. *Patient Education and Counseling*, *104*(2), 300–307. doi:10.1016/j.pec.2020.08.022 PMID:32888755

Elger, D. F., Beyerlein, S. W., & Budwig, R. S. (2000). Using design, build, and test projects to teach engineering. In *Proceedings of the 30th Annual Frontiers in Education Conference* (*vol. 2*, pp. F3C/9-F3C13). Institute of Electrical and Electronics Engineers. 10.1109/FIE.2000.896572

Endemol Nederland. (2012, April 2). *Blood, sweat and luxury problems* [Reality series]. BNN VARA.

ENoLL. (2013). *Tips & Tricks for Building a Sustainable Living Lab*. https://enoll.org/wp-content/uploads/2017/08/tips-tripcs-final-upload.pdf

ENoLL. (2021). *European Network of Living Labs*. https://www.enoll.org/

EOLA & Limitless. (2020). *SOCIA: Social Innovation Initiative Logbook.* https://www.socialenterprisebsr.net/wp-content/uploads/2020/10/SOCIA-Self-Assessment-Logbook-vFIN-pdf.pdf

EPS. (2021). *EPS - European Project Semester.* Retrieved from European Project Semester: http://europeanprojectsemester.eu/

EPS@ISEP. (2021). *Ergonomic Reconfigurable Home-Desk.* Retrieved from EPS@ISEP | The European Project Semester (EPS) at ISEP: https://www.eps2021-wiki3.dee.isep.ipp.pt/doku.php

Ertas, A. (2018). *Transdisciplinary engineering design process.* John Wiley & Sons.

Ertas, A., Gatchel, S., Rainey, V., & Tanik, M. M. (2007). Networked Approach to Transdisciplinary Research & Education. *The Academy of Transdisciplinary Learning & Advanced Studies*, *3*(2), 1–12.

Eskerod, P., & Riis, E. (2009). Project Management Models as Value Creators. *Project Management Journal*, *40*(1), 4–18. doi:10.1002/pmj.20098

European Commission. (2015). *Erasmus+ Programme Guide* (English Version 3). https://ec.europa.eu/programmes/erasmus-plus/documents/erasmus-plus-programme-guide_en.pdf

European Network for Accreditation of Engineering Education. (2015). *EUR-ACE® Framework Standards and Guidelines.* http://www.enaee.eu/wp-assets-enaee/uploads/2012/02/EAFSG_full_nov_voruebergehend.pdf

European Project Semester. (2021). *Concept | European Project Semester.* http://europeanprojectsemester.eu/concept

European Project Semester. (2022). *Concept | European Project Semester.* http://www.europeanprojectsemester.eu/concept

European Project Semester. (2022). http://europeanprojectsemester.eu/

European Project Semester. (2022). *Welcome | European Project Semester.* http://www.europeanprojectsemester.eu/

European Society for Engineering Education. (2016). *European Society for Engineering Education.* http://www.sefi.be

Eyskoot, M. (2018). *This is a Good Guide – for a Sustainable Lifestyle.* Laurence King Publishing.

Facca, C. A. (2020). A contribuição do pensamento do design na formação do engenheiro: o espaço do Fab Lab como experiência transversal. Doctoral Thesis, Universidade Anhembi Morumbi, PPG Design, São Paulo.

Facca, C. A., Alves, J. L., & Barbosa, A. M. (2019). Overview of Design Teaching on Engineering Courses: A Comparative Study between Brazil and Portugal. In *International Conference on Education and New Developments* (pp. 129-133). Porto: World Institute for Advanced Research and Science. doi:10.36315/2019v2end027

Fahrradbeleuchtung. (2012). *Shimano-Nabendynamos: Typen und ihre Unterschiede.* https://fahrradbeleuchtung-info.de/shimano-nabendynamos

Fathers, J. W. R. (2017). Engaged Design. *The International Journal of Design in Society*, *11*(2), 23–29. doi:10.18848/2325-1328/CGP/v11i02/23-29

FAZIT Communication GmbH. (2018, April 1). *Germany – land of bicycles.* https://www.deutschland.de/en/topic/life/sports-leisure/germany-land-of-bicycles

Felländer, A. (2019). Anna Felländer vill rädda oss från AI-undergång. *Digital.*

Fernández-Galiano, L. (Ed.). (2012). *Arquitectura: lo común. Fundación arquitectura y sociedad.* Fundación Arquitectura y Sociedad. http://arquitecturaysociedad.com/wp-content/uploads/2019/07/ARQUITECTURA_LO-COM%C3%9AN-comprimido.pdf

Ferreira, J., & Hignett, S. (2005). Reviewing ambulance design for clinical efficiency and paramedic safety. *Applied Ergonomics, 36*(1), 97–105. doi:10.1016/j.apergo.2004.07.003 PMID:15627427

Ferreira, P., Malheiro, B., Guedes, P., & Silva, M. (2015). Towards Active Course Marks for Autonomous Sailing Competitions. In F. Morgan & D. Tynan (Eds.), *Robotic Sailing 2014* (pp. 67–75). Springer International Publishing. doi:10.1007/978-3-319-10076-0_6

Ferreira, P., Malheiro, B., Silva, M. F., Duarte, A. J., & Guedes, P. (2019). Developing skills in engineering capstone projects with low-cost microcontroller solutions: the EPS@ISEP experience. In *Proceedings of the 11th International Conference on Education and New Learning Technologies* (pp. 5930-5933). IATED Academy. 10.21125/edulearn.2019.1430

Fielden, J. (2007). *Gobal Horizons for UK Universities*. CIHE.

Figas, M., Lejarza, A., Rappard, R., & Sarachagà, E. (2021). Students in the times of pandemic. European Project Semester, Lodz University of Technology.

Figueras, P. (2007). *La Ciudad Educadora*. Ajuntament de Barcelona.

Findeli, A. (2001). Rethinking Design Education for the 21st Century: Theoretical, Methodological, and Ethical Discussion. *Design Issues, 17*(1), 1–13. doi:10.1162/07479360152103796

Fohanno, B., Pires, B., Ionescu, C., Ladka, E., & Perek, M. (2021b). *Crowd Orchestration* (Project Wiki). https://www.eps2021-wiki1.dee.isep.ipp.pt/

Fohanno, B., Pires, B., Ionescu, C., Ladka, E., Perek, M., Malheiro, B., Ribeiro, C., Justo, J., Silva, M. F., Ferreira, P., & Guedes, P. (2021a). Crowd Orchestration – An EPS@ISEP 2021 Project. In *Proceedings of Ninth International Conference on Technological Ecosystems for Enhancing Multiculturality* (pp. 411-416). ACM. 10.1145/3486011.3486481

Ford, J. D., & Paretti, M. (2013). Introduction to special issue on international engineering communication. *Connexions: International Professional Communication Journal, 2*(1), 3–8.

Ford, J. D., Paretti, M., Kotys-Schwartz, D., Howe, S., & Ott, R. (2021). New engineers' transfer of communication activities from school to work. *IEEE Transactions on Professional Communication, 64*(2), 105–120. doi:10.1109/TPC.2021.3065854

Fountain, A., Kuron, B., Bentin, C., Davies, E., Suits, K., Toro, P., Duarte, A., Malheiro, B., Ribeiro, C., Ferreira, F., Lima, L., Ferreira, P., & Guedes, P. (2016). Learning Sustainability with EPS@ISEP – Development of an Insectarium. In *Proceedings of the Eight International Symposium on Project Approaches in Engineering Education* (pp. 109-116). Project Approaches in Engineering Education Association.

Frank, M., & Barzilai, A. (2004). Integrating alternative assessment in a project-based learning course for preservice science and technology teachers. *Assessment & Evaluation in Higher Education, 29*(1), 41–61. doi:10.1080/0260293042000160401

Freeman, M., Treleaven, L., Ramburuth, P., Leask, B., Caulfield, N., Ridings, S., & Sykes, C. (2009). *Embedding the Development of Intercultural Competence in Business Education*. Final Report to the Australian Learning and Teaching Council.

Friedman, M. S. (2002). *Martin Buber: The Life of Dialogue*. Routledge.

Friedman, V. J., & Berthoin Antal, A. (2005). Negotiating Reality. A Theory of Action Approach to Intercultural Competence. *Management Learning, 36*(1), 69–86. doi:10.1177/1350507605049904

Fuentes-Durá, P., Gaspar, F., & Abarca, J. M. (2015). *Innovación creativa: The NOMAD Travelbox* [Creative Innovation: The NOMAD Travelbox]. In *Proceedings of the Congreso de Innovación Educativa y Docencia en Red de la Universitat Politècnica de València*. Universitat Politécnica de València. 10.4995/INRED2015.2015.1601

Fuentes-Durá, P., Puyuelo Cazorla, M., Gasch Molina, M. I., & Perry, D. (2014). European Project Semester: Good practices for competence acquisition. In *Proceedings of the Fourth Valencia Global Conference* (pp.161-172). UPV.

Furstenberg, G. (2010). Making culture the core of the language class: Can it be done? *Modern Language Journal*, *94*(2), 329–332. doi:10.1111/j.1540-4781.2010.01027.x

Gantt, H. L. (1919). Organizing for Work. Harcourt, Brace, and Howe.

Garmin. (2022a). *Garmin Edge® 1030 | Bike GPS Computer*. https://www.garmin.com/en-US/p/567991/pn/010-01758-00

Garmin. (2022b). *Edge® 25 | Bike Computer | Garmin*. https://www.garmin.com/en-US/p/508489

Gasch, I., Fuentes-Durá, P., & Perry, D. (2012). European Project Semester. In Jornadas de Innovación Educativa 2012 (pp. 132-136). Academic Press.

Gasch, I., Fuentes-Durá, P., & Puyuelo, M. (2014). EPS, un proyecto multidisciplinar internacional. In *Jornadas de Innovación Educativa y Docencia en Red* (pp. 866–877). IN-RED.

Gazibara, S. (2013). Head, Heart and Hands Learning – A challenge for contemporary education. *Journal of Education Culture and Society*, *2013*(1), 71–82. doi:10.15503/jecs20131.71.82

Gehl people. (2022). *Masterplanning and Design projects*. https://gehlpeople.com/work/projects/

Gehl, J. (2011). *Cities for people – A lecture by Jan Gehl*. https://www.youtube.com/watch?v=KL_RYm8zs28

Generalitat Valenciana. (2022). *Visor cartogràfic de la Generalitat*. https://visor.gva.es/visor/

Gerdes, K. E., & Segal, E. A. (2009). A social work model of empathy. *Advances in Social Work*, *10*(2), 114–127. doi:10.18060/235

Gerdes, K. E., Segal, E. A., & Lietz, C. A. (2010). Conceptualising and measuring empathy. *British Journal of Social Work*, *40*(7), 2326–2343. doi:10.1093/bjsw/bcq048

Gerretsen, I., & John, T. (2019). *UK climate protest: Student walk out inspired by Greta Thunberg*. http://www.cnn.com

Gesteland, R. (2005). *Cross-Cultural Business Behavior*. Copenhagen Business School Press.

Glaser, E., Guilherme, M., Garcia, M. C. M., & Mughan, T. (2007). *Intercultural Competence for Professional Mobility*. Council of Europe.

Goleman, D. (1998). *Working with emotional intelligence*. Bloomsbury.

Gonzales-Liencres, C., Shamaay-Tsoory, S. G., & Brüne, M. (2013). Towards a neuroscience of empathy: Ontogeny, phylogeny, brain mechanism, context and psychopathology. *Neuroscience and Biobehavioral Reviews*, *37*(8), 1537–1538. doi:10.1016/j.neubiorev.2013.05.001 PMID:23680700

Google. (2021). *Technical Writing*. https://developers.google.com/tech-writing

Gopakumar, G. (2014). Public Leadership Framework: Studying Approaches to Diversify Engineering Education. *International Journal of Engineering Pedagogy*, *4*(1), 43–48. doi:10.3991/ijep.v4i1.3269

Gopsill, A., Hamish, C., McAlpine, B., & Hicks, B. J. (2015). Supporting engineering design communication using a custom-built social media tool – PartBook. *Advanced Engineering Informatics, 29*(3), 523–548. doi:10.1016/j.aei.2015.04.008

Graham, D. R. (2018). *The global state of the art in engineering education.* Massachusetts Institute of Technology, School of Engineering. Cambridge, MA: New Engineering Education Transformation. Retrieved from https://rhgraham.org/resources/Global-state-of-the-art-in-engineering-education---March-2018.pdf

Granet, K., & Hunt, H. (2021). *The business of design: Balancing creativity and profitability.* Princeton Architectural Press.

Grant, A. (2014). *Give and Take, Why Helping Others Drives our Success.* Penguin Books.

Grasso, D., & Burkins, M. (Eds.). (2010). *Holistic Engineering Education: Beyond Technology.* Springer. doi:10.1007/978-1-4419-1393-7

Grasso, D., & Martinelli, D. (2007). Holistic engineering. *The Chronicle of Higher Education, 53,* B8.

Grasso, D., & Martinelli, D. (2010). Holistic engineering education: Beyond technology. *Choice Reviews Online, 47*(10), 1–10. doi:10.5860/CHOICE.47-5674

Gray, C. F., & Larson, E. W. (2021). *Project Management, The Managerial Process* (8th ed.). McGraw – Hill Irwin.

Grenning, J. (2002). Planning Poker or How to avoid Analysis Paralysis while Release Planning. *Hawthorn Woods: Renaissance Software Consulting, 3,* 22-23.

Gustavsson, J., & Lagerquist, M. (2003). *MICUS Stretcher.* The Index Project. https://theindexproject.org/stories/micus-stretcher

Guzzo, F., & Facca, C. A. (2018). *O Design Thinking como Metodologia de Projeto Aplicada na Disciplina de Introdução à Engenharia. In Anais do 18º Congresso Nacional de Iniciação Científica - CONIC-SEMESP. 6.* Universidade Paulista. Retrieved from http://conic-semesp.org.br/anais/files/2018/trabalho-1000002444.pdf

Haboud, D. (2016). *IoT and The Three C's.* https://resources.altium.com/pcb-design-blog/iot-and-the-three-c-s

Handley, A. J., Koster, R., Monsieurs, K., Perkins, G. D., Davies, S., & Bossaert, L. (2005). European Resuscitation Council Guidelines for Resuscitation 2005: Section 2. Adult basic life support and use of automated external defibrillators. *Resuscitation, 67*(Suppl. 1), S7–S23. doi:10.1016/j.resuscitation.2005.10.007 PMID:16321717

Hansen, J. (2010). European Project Semester – how engineering students can achieve important competences. In *Proceedings of the 5th International Conference on Engineering Education* (pp. 1296–1298). Academic Press.

Hansen, J. (2010). European Project Semester – how engineering students can achieve important competences. In *Proceedings of the Fifth International Conference on Engineering Education* (pp. 1296-1298). Academic Press.

Hansen, J. (2010). European Project Semester – how Engineering Students Can Achieve Important Competences. In *Proceedings of the 5th International Conference on Engineering Education* (pp. 1296–1298). ICEE. https://www.ineer.org/Events/ICEE2010/papers/WS11/Paper_1107_1296.pdf

Hansen, J. (2012). An International Project/Internship Course. *Proceedings of the 3rd International Conference PRAXIS'2012 Open Discussion Forum.*

Hansen, J. (2012). Practical Elements in Danish Engineering Programmes, Including the European Project Semester. *Industry and Higher Education, 26*(4), 328–335. doi:10.5367/ihe.2012.0108

Hanzl, M., Młynarczyk, A., Tessier, A., Meiriño, M., Miciedes, I., & Noirot, F. (2014). Smart City - a Quest for Innovation within the EPS Framework. REAL CORP 2014 – PLAN IT SMART! Clever Solutions for Smart Cities. In *Proceedings of 19th International Conference on Urban Planning, Regional Development and Information Society* (pp. 563-571). Academic Press.

Harboe, G. (2015). *Material matters*. https://www.youtube.com/watch?v=JqIsZsBLkSkInternational

Harms, H., Janaszkiewicz, A., & Valauskaitė, J. (2012). *Context-aware Smart Object* (EPS Project Wiki). Instituto Superior de Engenharia do Porto. http://www.eps2012-wiki3.dee.isep.ipp.pt/

Harris, P. (2021). *Interpersonal analyses of bulimia: Normal-weight and obese* [Unpublished doctoral dissertation]. Universitat Politècnica de València.

HAS. (2022). *HAS University of Applied Sciences*. https://www.hasuniversity.nl

Heath, C., & Heath, D. (2010). *Made to Stick, Why Some Ideas Survive and Others Die*. Random House Publishing.

Heintze, H.-J., Kirch, L., Küppers, B., Mann, H., Mischo, F., Mucke, P., Pazdzierny, T., Prütz, R., Radtke, K., Strube, F., & Weller, D. (2018). *The WorldRiskReport 2018: Child Protection and Children's Rights*. Bündnis Entwicklung Hilft and Institute for International Law of Peace and Armed Conflict. https://www.WorldRiskReport.org

Heitmann, G. (2005). Challenges of engineering education and curriculum development in the context of the Bologna Process. *European Journal of Engineering Education*, *30*(4), 447–458. doi:10.1080/03043790500213136

Hemsing, R. (2018). *SmartHalo Review – Bring Your Bike to Life | Momentum Mag*. https://momentummag.com/smarthalo-product-review-bring-bike-life/

Hengsadeekul, T., Zevert, D., Dau, M., & Feron, A. (2019). Survival Unit (EPS Final Report). Nottingham Trent University.

Henry, R. K., Yongsheng, Z., & Jun, D. (2006). Municipal solid waste management challenges in developing countries—Kenyan case study. *Waste Management (New York, N.Y.)*, *26*(1), 92–100. doi:10.1016/j.wasman.2005.03.007 PMID:16006111

Heylighen, A., & Dong, A. (2019). To empathise or not to empathise? Empathy and its limits in design. *Design Studies*, *65*, 107–124. doi:10.1016/j.destud.2019.10.007

Hignett, S., Crumpton, E., & Coleman, R. (2009). Designing emergency ambulances for the 21st century. *Emergency Medicine Journal*, *26*(2), 135–140. doi:10.1136/emj.2007.056580 PMID:19164631

Hoffman, K. (2015). *Co-Design, Not Redesign* (Event Apart Video). https://aneventapart.com/news/post/co-design-not-redesign-by-kevin-m.-hoffman-an-event-apart-video

Hoffman, M. L. (2000). *Empathy and moral development: Implications for caring and justice*. Cambridge University Press. doi:10.1017/CBO9780511805851

Hofstede, G. (1980). *Culture's Consequences: International Differences in Work-Related Values*. Sage Publications.

Hofstede, G. (1991). *Cultures and Organizations: Software of the Mind: Intercultural Cooperation and its Importance for Survival*. McGraw Hill.

Hofstede, G., Hofstede, G. J., & Minkov, M. (2010). *Cultures and Organizations: Software of the Mind* (3rd ed.). McGraw-Hill Professional.

Holt, M., Kosterna, K., Pechero Villota, D., Nyby, E. C. V., & Selçuk, O. (2014). *Travel Logging System* (EPS Project Wiki). Instituto Superior de Engenharia do Porto. http://www.eps2014-wiki2.dee.isep.ipp.pt/

Huczynski, A. A., & Buchanan, D. A. (2019). *Organizational Behaviour*. Pearsons.

Hulls, C. C. W., & Rennick, C. (2020). Use of a Cornerstone Project to Teach Ill-Structured Software Design in First Year. *IEEE Transactions on Education, 63*(2), 98–107. doi:10.1109/TE.2019.2959591

Hume, D. (1978). *(1739–40 [1978]). A Treatise of Human Nature*. Oxford University Press.

Hutton, M. (1980). Learning from action: a conceptual framework. In S. Warner Weil & M. McGill (Eds.), *Making Sense of Experiential Learning* (pp. 50–59). SRHE/Open University Press.

HvA. (2022). *Amsterdam University of Applied Sciences*. http://www.hva.nl

Hyland, K. (2016). General and specific EAP. In K. Hyland & P. Shaw (Eds.), *The Routledge Handbook of English for Academic Purposes* (pp. 17–29). Routledge. doi:10.4324/9781315657455

Iacoboni, M. (2008). *Mirroring people: The science of empathy and how we connect with others*. Picador.

IAEC. (2020). *Charter of Educating Cities*. https://www.edcities.org/wp-content/uploads/2020/11/ENG_Carta.pdf

IDEO. (2020). *Design Thinking*. Retrieved from IDEO Design Thinking: https://designthinking.ideo.com/

ideo.org. (2022). *IDEO.org is a nonprofit design studio | IDEO*. https://www.ideo.org/

IEEE. (2021). *IEEE Learning Network*. https://iln.ieee.org/public/trainingcatalog.aspx

Igloo Magazine. (2012). Interview with Eduardo Souto de Moura. *Igloo Habitat & Arhitectură, 126*. https://www.igloo.ro/interviu/

Imbesi, L. (2016). *Design for Social Innovation in Canada*. Common Ground Publishing. doi:10.18848/978-1-61229-811-5/CGP

ISEP. (2021). *European Project Semester at ISEP (EPS@ISEP)*. Retrieved from ISEP - Instituto Superior de Engenharia do Porto: https://www.isep.ipp.pt/Course/Course/44

Ishii, A., Heibeck, M., Blazejewski, M., & Nybjörk, R. (2014). *Design and Implementation of a Biologically Inspired Swimming Robot – Final Report*. http://www.eps2014-wiki1.dee.isep.ipp.pt/lib/exe/fetch.php?media=final_report_team1.pdf

Ishii, A., Heibeck, M., Blazejewski, M., Nybjörk, R., Duarte, A., Malheiro, B., Ribeiro, C., Ferreira, F., Silva, M., Ferreira, P., & Guedes, P. (2015). Design and Implementation of a Biologically Inspired Swimming Robot – An EPS@ISEP 2014 Spring Edition Project. In *Proceedings of TEEM – 3rd Technological Ecosystems for Enhancing Multiculturality* (pp. 219-226). 10.1145/2808580.2808613

ISO. (1994). *ISO 8402:1994. Quality management and quality assurance – Vocabulary*. https://www.iso.org/standard/20115.html

ISO. (2012). *ISO 21500:2012. Guidance on project management*. https://www.iso.org/standard/50003.html

ISO. (2018). *ISO 21511:2018. Work breakdown structures for project and programme management*. https://www.iso.org/standard/69702.html

Itin, C. M. (1999). Reasserting the Philosophy of Experiential Education as a Vehicle for Change in the 21st Century. *Journal of Experiential Education*, *22*(2), 91–98. doi:10.1177/105382599902200206

Jameson, D. (1993, March). Using a Simulation to Teach Intercultural Communication in Business Communication Courses. *The Bulletin*, 3–11.

Janssen, D., & Carradini, S. (2021). Generation Z Workplace Communication Habits and Expectations. *IEEE Transactions on Professional Communication, 64*(2), 137–153. doi:10.1109/TPC.2021.3069288

Jazar, P. R. (2006). *Theory of Applied Robotics*. Springer.

Johnson, D. W., Johnson, R. T., & Smith, K. A. (1991). *Cooperative Learning: Increasing College Faculty Instructional Productivity*. ASHE-ERIC Higher Education Report No. 4 1991. Education Resources Information Center. https://files.eric.ed.gov/fulltext/ED343465.pdf

Joint Quality Initiative Informal Group. (2004). *Shared 'Dublin' descriptors for Short Cycle, First Cycle, Second Cycle and Third Cycle Awards*. https://www.aqu.cat/doc/doc_24496811_1.pdf

Jones, E. (2013). *Internationalization and employability: the role of intercultural experiences in the development of transferable skills. Public Money and Management*. March Issue.

Jones, J. C. (1970). *Design Methods: Seeds of Human Futures*. Wiley-Interscience.

Jones, S. A., Michelfelder, D., & Nair, I. (2017). Engineering Managers and Sustainable Systems: The Need for and Challenges of Using an Ethical Framework for Transformative Leadership. *Journal of Cleaner Production, 140*, 205–212. doi:10.1016/j.jclepro.2015.02.009

Joshi, A., & Lazarova, M. (2005). Do "Global" Teams Need "Global" Leaders? Identifying Leadership Competencies in Multinational Teams. In Managing Multinational Teams: Global Perspectives (Advances in International Management) (vol. 18, pp. 281 – 302). Emerald Group Publishing Limited. doi:10.1016/S0747-7929(05)18011-1

Kans, M. (2021). Engineering education development – a business modelling approach. *Higher Education Evaluation and Development, 15*(1), 53–77. doi:10.1108/HEED-02-2020-0003

Kedrowicz, A. A., & Taylor, J. L. (2013). Engineering communication and the global workplace: Preparing professionals and global citizens. *Connexions: International Professional Communication Journal, 1*(2), 81–105.

Kelley, T., & Kelley, D. (2013). *Creative confidence*. Crown Business.

Kerzner, H. (2017). *Project Management: A System Approach on Planning, Scheduling, and Controlling* (12th ed.). John Wiley & Sons.

Keskeys, P. (2017). *Epic Overview: Benjamin Grant on the Sky-High Art of Aerial photography*. Architizer. https://architizer.com/blog/practice/materials/epic-overview/#.YTnIGa6JnAM.link

KIDV. (2019). *Overview Scientific Research Programme 2015-2019 – Interdisciplinary approach to sustainable packaging* [Brochure]. KIDV.

KIDV. (2022). *Getting started with Sustainable Packaging*. https://www.kidv.nl

King, R., & Ruiz-Gelices, E. (2003). International student migration and the European 'Year Abroad': Effects on European identity and subsequent migration behaviour. *International Journal of Population Geography, 9*(3), 229–252. doi:10.1002/ijpg.280

Klein, A., & Dytham, M. (2003). *PechaKucha — Klein Dytham architecture*. http://www.klein-dytham.com/pechakucha

Kohut, H. (1959). Introspection, empathy, and psychoanalysis: An examination of the relationship between the mode of observation and therapy. *Journal of the American Psychoanalytic Association, 7*(3), 459–483. doi:10.1177/000306515900700304 PMID:13672863

Kolb, D. (1984). *Experiential Learning: Experience as the Source of Learning and Development*. Prentice Hall.

Kolb, D. A. (2014). *Experiential learning: Experience as the source of learning and development* (2nd ed.). Pearson FT Press.

Kolb, D. A., & Fry, R. (1975). Toward an applied theory of experiential learning. In C. Cooper (Ed.), *Theories of Group Process*. John Wiley.

Kolko, J. (2012). *Wicked Problems: Problems Worth Solving*. Austin Center for Design.

Kolmos, A., Bertel, L. B., Holgaard, J. E., & Routhe, H. W. (2020). Project Types and Complex Problem-Solving Competencies: Towards a Conceptual Framework. In A. Guerra, A. Kolmos, M. Winther, J. Chen (Eds.). *Educate for the future: PBL, Sustainability and Digitalisation 2020* (pp. 56-65). Aalborg Universitetsforlag. International Research Symposium on PBL.

König, L. S., & Ribarić, H. M. (2019). Is there a mismatch between employers' and university teachers' perceptions on graduate employability in Croatia? *Journal of Managerial Issues, 24*(1), 87–102. https://hrcak.srce.hr/221552

Koolhaas, R. (2010). *Navigating modernization*. https://www.oma.com/lectures/navigating-modernization

KoolhaasR. (2001). *Junkspace*. https://www.readingdesign.org/junkspace

Kouprie, M., & Visser, F. S. (2009). A framework for empathy in design: Stepping into and out of the user's life. *Journal of Engineering Design, 20*(5), 437–448. doi:10.1080/09544820902875033

Krajcik, J. S., & Blumenfeld, P. (2006). Project-based learning. In R. K. Sawyer (Ed.), The Cambridge Handbook of the Learning Sciences (pp. 317–333). Cambridge.

Kristensson, P., Gusyafsson, A., & Archer, T. (2004). Harnessing the creative Potential among Users. *The Journal of Innovation Management, 21*(1), 4–14. doi:10.1111/j.0737-6782.2004.00050.x

Kullman, K. (2016). Prototyping bodies: A post-phenomenology of wearable simulations. *Design Studies, 47*, 73–90. doi:10.1016/j.destud.2016.08.004

La Città dei Bambini. (2022). https://www.lacittadeibambini.org

Ladkin, D. (2016). *Mastering the Ethical Dimension of Organizations*. Elgar.

Lane, L. (2021). Interstitial design processes: how design thinking and social design processes bridge theory and practice in TPC pedagogy. In M. J. Klein (Ed.), *Effective Teaching of Technical Communication Theory, Practice, and Application* (pp. 29–43). WAC Clearinghouse.

Lannon, J. (2000). *Technical Communication* (8th ed.). Longman.

Lannon, J., & Gurak, L. (2021). *Technical Communication, EBook* (Global Edition). Pearson Education.

Lathem, S. A., Neumann, M. D., & Hayden, N. (2011). The Socially Responsible Engineer: Assessing Student Attitudes of Roles and Responsibilities. *Journal of Engineering Education, 100*(3), 444–474. doi:10.1002/j.2168-9830.2011.tb00022.x

Laurel, B. (2003). *Design research: Methods and perspectives*. MIT Press.

Laut, J., Bartolini, T., & Porfiri, M. (2015). Bioinspiring an Interest in STEM. *IEEE Transactions on Education, 58*(1), 48–55. doi:10.1109/TE.2014.2324533

Leask, B. (2015). *Internationalizing the Curriculum*. Routledge. doi:10.4324/9781315716954

Lehmann, M., Christensen, P., Du, X., & Thrane, M. (2008). Problem-oriented and project-based learning (POPBL) as an innovative learning strategy for sustainable development in engineering education. *European Journal of Engineering Education, 33*(3), 283–295. doi:10.1080/03043790802088566

Leiberg, S., & Anders, S. (2006). The multiple facets of empathy: A survey of theory and evidence. *Progress in Brain Research, 156*, 419–440. doi:10.1016/S0079-6123(06)56023-6 PMID:17015094

Leigh, B., & MacGregor, J. T. (1992). What is Collaborative Learning? In A. S. Goodsell, M. R. Maher, V. Tinto, B. Leigh Smith, & J. MacGregor (Eds.), *Collaborative Learning: A Sourcebook for Higher Education* (pp. 10–29). National Center of Postsecondary Teaching, Learning and Assessment.

Levenson, R., & Ruef, A. (1992). Empathy: A physiological substrate. *Journal of Personality and Social Psychology, 63*(2), 234–246. doi:10.1037/0022-3514.63.2.234 PMID:1403614

Levinas, E. (1969). *Totality and Infinity*. Duquesne University Press.

Levy, O., Beechler, S., Taylor, S., & Boyacigiller, N. (2015). Global Mindset. Wiley Encyclopedia of Management, 1-4. doi:10.1002/9781118785317.weom060094

Lewis, R. D. (2015). *Introduction to Cross-Culture*. Richard Lewis Communications. https://www.crossculture.com/services/cross-culture/introduction-to-cross-culture/

Lewis, R. D. (2006). *When cultures collide: leading across cultures*. Nicholas Brealey Publishing.

Lidwell, W., Butler, J., & Holden, K. (2010). *Universal principles of design: 125 ways to enhance usability, influence perception, increase appeal, make better design decisions, and teach through design*. Rockport Publishers.

Ligthart, T. N., & Ansems, T. M. M. (2019b). EnvPack, an LCA-based tool for environmental assessment of packaging chains: Part 2: influence of assessment method on ranking of alternatives. *The International Journal of Life Cycle Assessment, 24*(5), 915–925. doi:10.100711367-018-1531-z

Ligthart, T. N., Thoden van Velzen, E. U., & Brouwer, M. (2019a). EnvPack an LCA-based tool for environmental assessment of packaging chains. Part 1: Scope, methods and inventory of tool. *The International Journal of Life Cycle Assessment, 24*(5), 900–914. doi:10.100711367-018-1530-0

Lindsey, C. C. (1978). Form, Function, and Locomotory Habits in Fish. In W.S. Hoar & D.J. Randall (Eds.), Fish Physiology. Academic Press. doi:10.1016/S1546-5098(08)60163-6

Lipps, T. (1903). Einfühlung, Innere Nachahmung und Organempfindung [Empathy, Inner Imitation and Sense-Feelings]. *Archiv fur die Gesamte Psychologie, 1*, 465–519.

Llauradó, A. M., Docherty, A., Méry, G., Sokolowska, N., Keane, S., Duarte, A. J., Malheiro, B., Ribeiro, C., Ferreira, F., Silva, M., Ferreira, P., & Guedes, P. (2015). Aquaponics System – An EPS@ISEP 2014 Spring Project. In *Proceedings of the Third International Conference on Technological Ecosystems for Enhancing Multiculturality* (pp. 537-544). Association for Computing Machinery. 10.1145/2808580.2808662

Loch, W. (2020). Critical Thinking in Problem Exploration in Design and Technology Design Project. *Design and Technology Education: An International Journal, 25*(1), 35–58.

Lockwood, J. (2019). What do we mean by 'workplace English'? A multilayered syllabus framework for course design and assessment. In K. Hyland & L. L. C. Wong (Eds.), *Specialised English. New Directions in ESP and EAP Research and Practice* (pp. 22–35). Routledge. doi:10.4324/9780429492082-3

Lubin, G. (2013). The Lewis Model Explains Every Culture in The World. *Business Insider*. https://www.businessinsider.com/the-lewis-model-2013-9

Lucas, B., Hanson, J., & Claxton, G. (2014). Thinking like an engineer: Implications for the education system. Technical Report. Royal Academy of Engineering.

Lyons, J. (1977). *Semantics* (Vol. 2). Cambridge University Press.

Magistretti, S., Bianchi, M., Calabretta, G., Candi, M., Dell'Era, C., Stigliani, I., & Verganti, R. (2021, October). Framing the multifaceted nature of design thinking in addressing different innovation purposes. *Long Range Planning*, 102163. Advance online publication. doi:10.1016/j.lrp.2021.102163

Malheiro, B., Silva, M., Ferreira, P., & Guedes, P. (2015). CDIO and the European Project Semester: a match for capstone projects? In *Proceedings of the 11th International CDIO Conference* (pp. 1-14). http://rocketship.cdio.org/knowledge-library/documents/cdio-and-european-project-semester-match-capstone-projects

Malheiro, B., Guedes, P., Silva, M. F., & Ferreira, P. (2019a). Fostering Professional Competencies in Engineering Undergraduates with EPS@ISEP. *Education Sciences*, *9*(2), 119. doi:10.3390/educsci9020119

Malheiro, B., Silva, M. F., Ferreira, P., & Guedes, P. (2019b). Learning Engineering with EPS@ISEP: Developing Projects for Smart Sustainable Cities. *International Journal of Engineering Pedagogy*, *9*(4), 33–49. doi:10.3991/ijep.v9i4.10259

Malheiro, B., Silva, M., Duarte, P., & Guedes, P. (2015). CDIO and the European Project Semester: a Match for Capstone Projects? In *Proceedings of the 11th International CDIO Conference* (pp. 1-14). 10.13140/RG.2.1.3621.9042

Malheiro, B., Silva, M., Ferreira, P., & Guedes, P. (2015a). CDIO and the European Project Semester: a match for capstone projects? In *Proceedings of the 11th CDIO International Conference* (pp. 1–14). CDIO.

Malheiro, B., Silva, M., Ribeiro, M. C., Guedes, P., & Ferreira, P. (2013). The European Project Semester at ISEP – Learning to Learn Engineering. In *Proceedings of the 1st International Conference of the Portuguese Society for Engineering Education* (pp. 1–8). 10.1109/CISPEE.2013.6701977

Malheiro, B., Silva, M., Ribeiro, M. C., Guedes, P., & Ferreira, P. (2015b). The European Project Semester at ISEP: The Challenge of Educating Global Engineers. *European Journal of Engineering Education*, *40*(3), 328–346. doi:10.1080/03043797.2014.960509

Manzini, E. (2015). *Design, When Everybody Designs: An Introduction to Design for Social Innovation*. The MIT Press. doi:10.7551/mitpress/9873.001.0001

Markham, T., Larmer, J., & Ravitz, J. (2003). *Project based learning handbook: A guide to standards-focused project based learning* (2nd ed.). Buck Institute for Education.

Martin, B., & Hanington, B. M. (2012). *Universal methods of design: 100 ways to research complex problems, develop innovative ideas, and design effective solutions*. Rockport Publishers.

Martínez-López, J. I., & González, C. M. A. (2020). Evaluation of Decision-making Quality using Multi-attribute Decision Matrix and Design Thinking. In *2020 IEEE Global Engineering Education Conference (EDUCON)* (pp. 622-629). 10.1109/EDUCON45650.2020.9125174

Martin, R. L. (2009). *The design of business: Why design thinking is the next competitive advantage*. Harvard Business Press.

Marx, R. W., Blumenfeld, P. C., Krajcik, J. S., & Soloway, E. (1997). Enacting project-based science: Challenges for practice and policy. *The Elementary School Journal*, *97*(4), 341–358. doi:10.1086/461870

Mathieau, J. E., Heffner, T. S., Goodwin, G. F., Salas, E., & Cannon-Bowers, J. A. (2000). The Influence of Shared Mental Models on Team Process and Performance. *The Journal of Applied Psychology, 85*(2), 273–283. doi:10.1037/0021-9010.85.2.273 PMID:10783543

Matos, S., & Lopes, E. (2013). Prince2 or PMBOK – A Question of Choice. *Procedia Technology, 9*, 787–794. doi:10.1016/j.protcy.2013.12.087

Mawson, M., & Haworth, A. C. (2018). Supporting the employability agenda in university libraries: A case study from the University of Sheffield. *Information and Learning Science, 119*(1/2), 101–108. doi:10.1108/ILS-04-2017-0027

Maxeon Solar Technologies. (2021). *Maxeon Solar Panels*. https://maxeon.com/us/node/1163

Maylath, B., King, T., & Arnó, E. (2013). Linking engineering students in Spain and technical writing students in the U.S. as co-authors: The challenge and outcomes of subject-matter experts and language specialists collaborating internationally. *Connexions: International Professional Communication Journal, 2*(1), 150–185.

Maylor, H. (2010). *Project Management* (4th ed.). Pearson Education Limited.

McChrystal, S. A., Silverman, D., Fussell, C., & Collins, T. (2015). *Team of Teams: New Rules of Engagement for a Complex World.* Portfolio/Penguin.

Melo, A., & Abelheira, R. (2015). *Design Thinking & Thinking Design.* Novatec.

Mendes, A., Tatuc, E., Joos, F., Wyka, J., & Petrevski, K. (2021b). *Sustainable Food Production Through Vermicomposting* (Project Wiki). https://www.eps2021-wiki2.dee.isep.ipp.pt/

Mendes, A., Tatuc, E., Joos, F., Wyka, J., Petrevski, K., Malheiro, B., Ribeiro, C., Justo, J., Silva, M. F., Ferreira, P., & Guedes, P. (2021a). Sustainable Food Production Through Vermicomposting – An EPS@ISEP 2021 Project. In *Proceedings of Ninth International Conference on Technological Ecosystems for Enhancing Multiculturality* (pp. 553-559). ACM. 10.1145/3486011.3486515

Merino, L., & Puyuelo, M. (2021). El espacio público como el contexto del diseño y la innovación [Public space as a context of design and innovation]. In M. Puyuelo (Ed.), *Todo sobre Diseño* [All about Design] (pp. 47–60). Editorial Universitat Politècnica de València. https://riunet.upv.es/handle/10251/166499

Meyer, M. W., & Norman, D. (2020). Changing Design Education for the 21st Century. *She Ji: The Journal of Design, Economics, and Innovation, 6*(1), 13–49. doi:10.1016/j.sheji.2019.12.002

Michael, J. (2006). Where's the evidence that active learning works? *Advances in Physiology Education, 30*(4), 159–167. doi:10.1152/advan.00053.2006 PMID:17108243

Mikhieieva, O., & Waidmann, M. (2017). Communication Management Tools for Managing Projects in an Intercultural Environment. *PM World Journal, 6*(8), 1–15.

Mikkelsen, H., & Riis, J. O. (2017). *Project Management - A Multi-Perspective Leadership Framework.* Emerald Publishing Limited. doi:10.1108/9781787148291

Minney, S. (2016). *Slow fashion: Aesthetics Meets Ethics.* New Internationalist Publications Ltd.

MIRO. (2021). *Miro Help Center.* Retrieved from MIRO: https://help.miro.com/hc/en-us/articles/360017730533-What-is-Miro-

Mittal, H., Mathew, L., & Gupta, A. (2015). Design and development of an infant incubator for controlling multiple parameters. *International Journal of Emerging Trends in Electrical and Electronics, 11*(5), 65–72.

Möller, B., Tarkowski, M., van den Bleeken, T., & Wlazlo, E. K. (2013). *Autonomous Environmental Buoy* (EPS Project Wiki). Instituto Superior de Engenharia do Porto. http://www.eps2013-wiki3.dee.isep.ipp.pt/

Moore, P., & Conn, C. P. (1985). *Disguised: A true story.* Word Books.

Morgan, A. (Director). (2015). *The True Cost* [Documentary]. An Untold Production in association with Life is My Movie Entertainment.

Morris, L., Ma, M., & Wu, P. C. (2014). *Agile innovation: The revolutionary approach to accelerate success, inspire engagement and ignite creativity.* John Wiley & Sons.

Mulder, K. F., Ferrer, D., Segalas, J., Kordas, O., Nikiforovich, E., & Pereverza, K. (2015). Motivating students and lecturers for education in sustainable development. *International Journal of Sustainability in Higher Education, 16*(3), 385–401. doi:10.1108/IJSHE-03-2014-0033

Mulder-Nijkamp, M., de Koeijer, B., & Torn, R.-J. (2018a). Synthesizing Sustainability Considerations through Educational Interventions. *Sustainability, 11*(1), 21. doi:10.3390u11010021

Mulder-Nijkamp, M., de Koeijer, B., & Torn, R.-J. (2018b). Implementing Sustainability Considerations in Packaging Design Curricula. 1-8. In *Proceedings of the 21st IAPRI World Conference on Packaging* (pp. 834-841). Curran Associates, Inc. 10.12783/iapri2018/24465

Mungazi, D. A. (1991). *Colonial Policy and Conflict in Zimbabwe: A study of cultures in collision 1890-1979.* Taylor & Francis.

Mungazi, D. A. (1996). *The Mind of Black Africa.* Wetsport, Greenwood Publishing Group.

National Academy of Engineering. (2012). *Infusing Real World Experiences into Engineering Education.* The National Academies Press.

Nieto, S. (1999). *The Light in Their Eyes: Creating Multicultural Learning Communities. Multicultural Education Series.* Teachers College Press.

Nimon, W., & Beghin, J. (1999). Are eco-labels valuable? Evidence from the apparel industry. *American Journal of Agricultural Economics, 81*(4), 801–811. doi:10.2307/1244325

Nordquist, R. (2015). *Glossary of Grammatical and Rhetorical Terms.* About.com. http://grammar.about.com/od/c/g/communicaterm.htm

Nottingham University Hospitals NHS Trust. (2019). *Our organization | NUH.* https://www.nuh.nhs.uk/our-organisation

NXP. (2016). *LPC176x/5x User manual.* https://www.nxp.com/docs/en/user-guide/UM10360.pdf

Nylund, R. (2019). *Unga ingenjörers psykologiska kontrakt inom teknologi-industrin.* Acta Wasaensia 433, University of Vaasa.

OECD. (2015). Skills for Social Progress: The Power of Social and Emotional Skills. OECD Skills Studies. Paris: OECD Publishing. doi:10.1787/9789264226159-en

OECD. (2018). *The Future of Education and Skills - Education 2030.* Organisation for Economic Co-operation and Development (OECD). Paris, France: OECD Publishing. Retrieved from https://www.oecd.org/education/2030/E2030%20Position%20Paper%20(05.04.2018).pdf

Omata. (2022). *Omata | Smart GPS Bicycle Speedometer.* https://omata.com/

Onono, E. (Executive Producer). (2018, October 9). *Fashion's dirty secrets* [Stacey Dooley investigates]. BBC.

openpsychometrics.org. (2019). *Big Five Personality Test*. https://openpsychometrics.org/tests/IPIP-BFFM/

Organisation for Economic Co-operation and Development. (2014). OECD Science. *Technology and Industry Outlook, 2014*. Advance online publication. doi:10.1787ti_outlook-2014-en

Organisation for Economic Co-operation and Development. (2018). The future of education and skills. *Education, 2030*. https://www.oecd.org/education/2030/oecd-education-2030-position-paper.pdf

Osterman, K. F., & Kottkamp, R. B. (1993). *Reflective practice for educators: Improving schooling through professional development*. Corwin Press Inc.

Ovacen. (2021). *Cómo crear mapas personalizados, sin calles, elegantes y en 1 minuto*. https://ovacen.com/mapas-personalizados/

Pallasmaa, J. (1996). *The eyes of skin*. https://arts.berkeley.edu/wpcontent/uploads/2016/01/Pallasmaa_The-Eyes-of-the-Skin.pdf

Paltridge, B. (2014). Genre and second-language academic writing. *Language Teaching*, *47*(3), 303–318. doi:10.1017/S0261444814000068

Panitz, T. (1999). *Collaborative versus Cooperative Learning: A Comparison of the Two Concepts Which Will Help Us Understand the Underlying Nature of Interactive Learning*. Education Resources Information Center. https://files.eric.ed.gov/fulltext/ED448443.pdf

Panke, S. (2019). Design Thinking in Education: Perspectives, Opportunities and Challenges. *Open Education Studies*, *1*(1), 281–306. doi:10.1515/edu-2019-0022

Paretti, M. C., Eriksson, A., & Gustafsson, M. (2019). Faculty and Student Perceptions of the Impacts of Communication in the Disciplines (CID) on Students' Development as Engineers. *IEEE Transactions on Professional Communication*, *62*(1), 27–42. doi:10.1109/TPC.2019.2893393

Paretti, M. C., McNair, L. D., & Holloway-Attaway, L. (2007). Teaching technical communication in an era of distributed work: A case study of collaboration between U.S. and Swedish students. *Technical Communication Quarterly*, *16*(3), 327–352. doi:10.1080/10572250701291087

Paul, R., & Elder, L. (2008). *The miniature guide to critical thinking: Concepts and tools*. Foundation for Critical Thinking.

Penzenstadler, B., Haller, G., Schlosser, T., & Frenzel, G. (2009). Soft skills required: A practical approach for empowering soft skills in the engineering world. *Proceedings of the 2009 Collaboration and Intercultural Issues on Requirements: Communication, Understanding and Softskills*. 10.1109/CIRCUS.2009.5

Peppler, K. (2017). *The SAGE Encyclopedia of Out-of-School Learning*. doi:10.4135/9781483385198

Peragine, J., & Hudgins, G. (2017). *365 Low or No Cost Workplace Teambuilding Activities*. Atlantic Publishing Group.

Pérez Arroyo, S. (1993). *Escritos de arquitectura*. Pronaos.

Pinto, J. K. (2007). *Project Management, Achieving Competitive Advantage*. Pearson Education.

PMI. (2017). PMBOK® Guide (Sixth Edition). Project Management Institute.

Ponsa, P., Roman, J., Arno, E., & Pérez, J. (2015). Professional skills in international multidisciplinary teams. *International Journal of Engineering Education*, *31*(4), 998–1006.

Poole, H. H. (1989). Fundamentals of Robotics Engineering. Van Nostrand Reinhold. doi:10.1007/978-94-011-7050-5

Poppendieck, M., & Poppendieck, T. (2003). *Lean Software Development: An Agile Toolkit for Software Development Managers*. Addison Wesley.

Prechtl, E., & Davidson-Lund, A. (2007). Intercultural competence and assessment: perspectives from the INCA project. In H. Kotthoff & H. Spencer-Oatey (Eds.), *Handbook of Intercultural Communication* (pp. 467–490). Mouton de Gruyter. doi:10.1515/9783110198584.5.467

Preckel, K., Kanske, P., & Singer, T. (2018). On the interaction of social affect and cognition: Empathy, compassion and theory of mind. *Current Opinion in Behavioral Sciences*, *19*, 1–6. doi:10.1016/j.cobeha.2017.07.010

Publications Office of the European Union. (2021). *EU law – EUR-Lex*. https://eur-lex.europa.eu/homepage.html

Putnam, R. D. (2001). *Making democracy work: Civic traditions in modern Italy*. Princeton University Press.

Puyuelo, M., & Merino, L. (2021). The public space as a design observatory. The SmartTuria Hub project in Valencia (Spain). In *Proceedings International Conference on Built Environment, Science and Technology* (pp. 52-43). SRM Institute of Science and Technology.

Puyuelo, M., Fuentes-Durá, P., & Ballester, E. (2016). The opportunity in Industrial Design Engineering. A profile that matches the challenges of contemporary design. *Temes de Disseny*, *32*, 110-117. https://riunet.upv.es/handle/10251/94457

Puyuelo, M., Fuentes-Dura, P., & Ballester, E. (2016). The opportunity in Industrial Design Engineering. A profile that matches the challenges of contemporary design. *Temes de Disseny*, *32*, 110–117.

Puyuelo, M., Merino, M. D., & Fernández, N. (2011). Design Challenges Today: An Overview of Its Sustainable Principles. *Design Principles and Practices an International Journal*, *5*(4), 63–78. doi:10.18848/1833-1874/CGP/v05i04/57896

Quintrell, N., & Westwood, M. (1994). The influence of a peer-pairing programme on international students' first year experience and the use of student services. *Higher Education Research & Development*, *13*(1), 49–57. doi:10.1080/0729436940130105

Rasoal, C., Danielsson, H., & Tomas, J. (2012). Empathy among students in engineering programmes. *European Journal of Engineering Education*, *37*(5), 427–435. doi:10.1080/03043797.2012.708720

Razzouk, R., & Shute, V. (2012). What Is Design Thinking and Why Is It Important? *Review of Educational Research*, *83*(3), 330–348. doi:10.3102/0034654312457429

Reddy, M. J. (1979). The Conduit Metaphor: A Case of Frame Conflict in our Language about Language. In A. Ortony (Ed.), *Metaphor and Thought*. Cambridge University Press.

Reinhardt, A., Chousa Esteban, A., Urbanska, J., McPhee, M., & Greene, T. (2016a). *Didactic Robotic Fish Kit – Final Report*. http://www.eps2016-wiki3.dee.isep.ipp.pt/lib/exe/fetch.php?media=final_report.pdf

Reinhardt, A., Chousa Esteban, A., Urbanska, J., McPhee, M., Greene, T., Duarte, A., Malheiro, B., Ribeiro, C., Ferreira, F., Silva, M. F., Ferreira, P., & Guedes, P. (2016b). Didactic Robotic Fish – An EPS@ISEP 2016 Project. In *Proceedings of ICL2016 – 19th International Conference on Collaborative Learning* (pp. 239-253). 10.1007/978-3-319-50337-0_21

Reniers, R. L. E. P., Corcoran, R., Drake, R., Shryane, N. M., & Vollm, B. A. (2011). The QCAE: A Questionnaire of Cognitive and Affective Empathy. *Journal of Personality Assessment*, *93*(1), 84–95. doi:10.1080/00223891.2010.528484 PMID:21184334

Robinson, C. (2011). *5 cities, 5 places, one day*. https://fundacion.arquia.com/mediateca/filmoteca/explora-documental/

Rogers, C. (1957). The necessary and sufficient conditions for therapeutic personality change. *Journal of Consulting Psychology*, *21*(2), 95–103. doi:10.1037/h0045357 PMID:13416422

Rohaert, S., Baelus, C., & Lacko, D. (2012). Project work on wellbeing in multi-disciplinary student teams: a triple testimonial on EPS at Artesis. In *Proceedings of the International Conference on Engineering and Product Design Education* (pp. 1296–1298). The Design Society, Institution of Engineering Designers.

Rohaert, S., Baelus, C., & Lacko, D. (2012). Project work on wellbeing in multi-disciplinary student teams: a triple testimonial on EPS at Artesis. In *Proceedings of the International Conference on Engineering and Product Design Education* (pp. 1296–1298). The Design Society. https://www.designsociety.org/download-publication/33244/Project+Work+on+Wellbeing+in+Multidisciplinary+Student+Teams%3A+A+Triple+Testimonial+on+EPS+at+Artesis

Romero, G., Mesías, R., Enet, M., Oliveras, R., García, L., Coipel, M., & Osorio, D. (2012). *La participación en el diseño urbano y arquitectónico en la producción social del hábitat*. CYTED. http://cdam.unsis.edu.mx/files/Desarrollo%20Urbano%20y%20Ordenamiento%20Territorial/Otras%20disposiciones/Participaci%C3%B3n_dise%C3%B1o_urbano.pdf

Rosenblatt, L. (1994). The transactional theory of reading and writing. In R. B. Ruddell, M. R. Ruddell, & H. Singer (Eds.), *Theoretical models and processes of reading* (4th ed., pp. 1057–1092). International Reading Association.

Rosenthal, E. (2007, Jan. 25). Can polyester save the world? *The New York Times*.

Roth, L., & Altman, T. (2021). The self-other agreement of multiple informants on empathy measures and its relation to empathic accuracy. *Personality and Individual Differences*, *171*, 110499. doi:10.1016/j.paid.2020.110499

Royal Academy of Engineering. (2016). *Thinking like an engineer – Implications for the education system*. https://www.raeng.org.uk/publications/reports/thinking-like-an-engineer-implications-full-report

RungeK. (2013). *Philosophy*. https://afilii.com/en/philosophy/

Ryan, R. M., & Deci, E. L. (2000). Self-Determination Theory and the Facilitation of Intrinsic Motivation, Social Development, and Well-Being. *The American Psychologist*, *55*(1), 68–78. doi:10.1037/0003-066X.55.1.68 PMID:11392867

Sachs, J. D. (2005). *The end of Poverty*. Penguin.

Saénz de Oiza, F. (n.d.). *No te mueras sin ir a Ronchamp*. https://www.rtve.es/alacarta/videos/imprescindibles/imprescindibles-no-mueras-sin-ir-ronchamp-saenz-oiza/4881972/

Sánchez de la Guía, L. (2019). *Modelos y Métodos de Participación del Usuario en el Diseño de Productos y Servicios* [Models and Methods of User Participation in the Design of Products and Services] [Unpublished doctoral dissertation]. Universitat Politècnica de València.

Sánchez De la Guía, L., & Puyuelo, M. (2019). Dealing with Users Participation in Design of Products and Services. *The International Journal of Design Management and Professional Practice*, *13*(1), 21–29. doi:10.18848/2325-162X/CGP/v13i01/21-28

Sandberg, S., & Grant, A. (2017). *Option B: Facing Adversity, Building Resilience, and Finding Joy*. Barnes & Noble.

Schein, E., & Schein, P. (2017). *Organizational Culture and Leadership* (5th ed.). John Wiley Sons inc.

Schjær-Jacobsen, H., Abou-Hayt, I., Ashworth, D., Jensen, M. P., & Schreiber, M. P. (2012). Industrial design as an innovative element in engineering education. In *Proceedings of the International Conference on Engineering Education* (pp. 281-288). Academic Press.

Schön, D. A. (1987). *Educating the reflective practitioner: Towards a new design for teaching and learning in the professions*. Jossey-Bass.

Schön, D. A. (2017). *The Reflective Practitioner: How Professionals Think in Action*. Routledge., doi:10.4324/9781315237473

Schwaber, K., & Sutherland, J. (2020). *The Scrum Guide. The Definitive Guide to Scrum: The Rules of the Game.* https://scrumguides.org/docs/scrumguide/v2020/2020-Scrum-Guide-US.pdf

Schwartz, S. H. (2012). An Overview of the Schwartz Theory of Basic Values. *Online Readings in Psychology and Culture, 2*(1). Advance online publication. doi:10.9707/2307-0919.1116

Segalàs Coral, J., & Benson, P. (2012). International design project semester. In *Proceedings of the 14th International Conference on Engineering and Product Design Education* (pp. 585–590). The Design Society. https://www.designsociety.org/publication/33251/International+Design+Project+Semester

Segalàs, J. (2012). The EPS Experience at UPC-Barcelona Tech. In L. C. Campos, E. A. T. Dirani, A. L. Manrique, & N. Hattum-Janssen (Eds.), *Project Approaches to Learning in Engineering Education* (pp. 109–124). SensePublishers. doi:10.1007/978-94-6091-958-9_8

Segalàs, J., Esbrí, M. E., & Benson, P. (2011). European project semester: 30 ECTS of PBL in sustainability with multicultural and multidisciplinary bachelor student groups. *Proceedings of 17th International Conference on Engineering Education.*

Segal, E. A. (2011). Social empathy: A model built on empathy, contextual understanding, and social responsibility that promotes social justice. *Journal of Social Service Research, 37*(3), 266–277. doi:10.1080/01488376.2011.564040

Segal, L. D., & Suri, J. F. (1997). The Empathic Practitioner: Measurement and Interpretation of User Experience. *Proceedings of the Human Factors and Ergonomics Society Annual Meeting, 41*(1), 451–454. doi:10.1177/107118139704100199

Sfakiotakis, M., Lane, D. M., & Davies, J. B. C. (1999). Review of fish swimming modes for aquatic locomotion. *IEEE Journal of Oceanic Engineering, 24*(2), 237–252. doi:10.1109/48.757275

Shalamova, N. (2016). Blending engineering content with design thinking and UX to maximize student engagement in a technical communication class. In *Proceedings of the IEEE International Professional Communication Conference (IPCC)* (pp. 1-5). 10.1109/IPCC.2016.7740493

Shannon, C. E., & Weaver, W. (1949). *A Mathematical Model of Communication.* University of Illinois Press.

Sheng, C.-W., Tian, Y.-F., & Chen, M.-C. (2010). Relationships Among Teamwork Behavior, Trust, Perceived Team Support, and Team Commitment. *Social Behavior and Personality, 38*(10), 1297–1306. doi:10.2224bp.2010.38.10.1297

Sheppard, K., Dominick, P., & Aronson, Z. (2004). Preparing engineering students for the new business paradigm of international teamwork and global orientation. *International Journal of Engineering Education, 20*(3), 475–483. https://www.ijee.ie/articles/Vol20-3/IJEE1486.pdf

Shtub, A., Bard, J. F., & Globerson, S. (1994). *Project Management, Engineering, Technology and Implementation.* Prentice-Hall.

Silva, J. A. M., Silva, M. F., & Barbosa, R. S. (2014). Design and implementation of a biological inspired swimming robot. *Advances in Robotics Research International Journal, 1*(3), 239–250. doi:10.12989/arr.2015.1.3.239

Silva, M. F., Malheiro, B., Ferreira, P., & Guedes, P. (2019). Promoting Ethical Practices in Engineering Undergraduates. In *Proceedings of the Fourth International Conference on Robot Ethics and Standards* (pp. 55-63). CLAWAR Association. 10.13180/icres.2019.29-30.07.002

Silva, M. F., Malheiro, B., Guedes, P., Ferreira, P., Ribeiro, C., Ferreira, F., & Duarte, A. J. (2015). Development of Biomimetic Robots in the EPS Engineering Programme Capstone Project. In *Proceedings of Third International Conference on Technological Ecosystems for Enhancing Multiculturality* (pp. 227-234). Association for Computing Machinery. 10.1145/2808580.2808614

Silva, M. F., Malheiro, B., Ribeiro, C., Guedes, P., Duarte, A. J., & Ferreira, P. (2018). Collaborative Learning with Sustainability-driven Projects: A Summary of the EPS@ISEP Programme. *International Journal of Engineering Pedagogy, 8*(4), 106–130. doi:10.3991/ijep.v8i4.8260

Silva, M., Malheiro, B., Guedes, P., Ferreira, P., Ribeiro, C., Ferreira, F., & Duarte, A. (2017). The EPS@ISEP – European Project Semester Program at ISEP - Implementation results and ideas for improvement. In *Proceedings of the 45th SEFI Annual Conference* (pp. 129-136). SEFI.

Sinek, S. (2011). *Start with Why*. Penguin Books.

Skates, G. W. (2003). Interdisciplinary project working in engineering education. *European Journal of Engineering Education, 28*(2), 187–201. doi:10.1080/0304379031000079022

Smart, D., Volet, S., & Ang, G. (2000). *Fostering social cohesion in universities: Bridging the cultural divide*. Australian Education International Department of Education, Training and Youth Affairs.

Smith, C. (2016). *Lezyne Super GPS Enhanced Review | GearLab*. https://www.outdoorgearlab.com/reviews/biking/bike-computer/lezyne-super-gps-enhanced

Smith, K. A. (1995). Cooperative Learning: Effective Teamwork for Engineering Classrooms. *IEEE Education Society/ASEE Electrical Engineering Division Newsletter*, 1-6.

Smith, J. (2017). What is empathy for? *Synthese, 194*(3), 709–722. doi:10.100711229-015-0771-8 PMID:32226174

Smith, K. E., Porges, E. C., Norman, G. J., Connelly, J. J., & Decety, J. (2014). Oxytocin receptor gene variation predicts empathic concern and autonomic arousal while perceiving harm to others. *Social Neuroscience, 9*(1), 1–9. doi:10.1080/17470919.2013.863223 PMID:24295535

Snook, R. (1972). Medical aspects of ambulance design. *British Medical Journal, 3*(5826), 574–578. doi:10.1136/bmj.3.5826.574 PMID:5069643

Society for Technical Communication. (2021). *Defining Technical Communication*. https://www.stc.org/about-stc/defining-technical-communication/

Spencer-Oatey, H. (2011). *Achieving mutual understanding for effective intercultural management*. Working Paper, Centre for Applied Linguistics, University of Warwick. https://www2.warwick.ac.uk/fac/cross_fac/globalpeople/resourcebank/researchpapers

Sperber, D., & Mercier, H. (2018). *The Enigma of Reason*. Penguin Books.

Spielwarenmesse. (2014). *Toys go green*. https://www.spielwarenmesse.de/uploads/media/Toys_go_green_Wissensbroschuere_de.pdf

Staab, H. (2017). *Das Salutogenesekonzept nach Aaron Antonovsky*. GRIN Verlag.

Stanicka, N., Zwinkels, J., Darmody, C., Larke, C., Ball, E., & van Nerum, P. (2016). *Hortus (EPS Final Report)*. Universitat Politècnica de València.

Steen, K., & van Bueren, E. (2017). *Urban Living Labs: A Living Lab Way of Working*. AMS Institute for Advanced Metropolitan Solutions. Delft University of Technology. https://www.ams-institute.org/news/urban-living-labs-living-lab-way-working/

Steerman, S., Tuominen, A., Gregersen, M. S., Mickiewicz, K., Linsen, M., Schuler, L. M., & Grabowska, K. (2019). Cities, also for children. Strategies to rethink where we live (EPS Report). Valencia: Universitat Politècnica de València.

Stewart, V. (2007). Becoming citizens of the world. *Educational Leadership, 64*(7), 8–14.

Struik. (2022). *Struik Foods Europe*. https://struik.com

Stueber, K. (2019). Empathy. In E. N. Zalta (Ed.), *The Stanford Encyclopedia of Philosophy* (Fall 2019 Edition). https://plato.stanford.edu/archives/fall2019/entries/empathy/

Swales, J. (1985). *Episodes in ESP*. Pergamon Press.

Swales, J. (1990). *Genre Analysis. English in Academic and Research Settings*. Cambridge University Press.

Talbot, D., Geelhoed, R., & Ninggal, M. (1999). A qualitative study of Asian international students' attitudes toward African Americans. *NASPA Journal, 36*(3), 210–221. doi:10.2202/1949-6605.1081

Tectónica. (2022). *Constructive materials and techniques*. https://tectonica.archi/

Thackara, J. (2015). *How to Thrive in the Next Economy: Designing Tomorrow's World Today*. Thames & Hudson.

The Myers-Briggs Company. (2021), *Myers-Briggs Type Indicator® (MBTI®) | Official Myers Briggs Personality Test*. https://www.themyersbriggs.com/en-US/Products-and-Services/Myers-Briggs

The National Association of City Transportation Officials. (2022). *Bus station design guide*. https://nacto.org/

Thomas, J. W., & Mergendoller, J. R. (2000). *Managing project based learning: Principles from the field*. Paper presented at the Annual Meetings of the American Educational Research Association, New Orleans, LA.

Thomas-Hope, E. (Ed.). (1998). *Solid Waste Management: Critical Issues for Developing Countries*. Canoe Press.

THUAS. (2022). *The Hague University of Applied Sciences, Industrial Design Engineering (IPO)*. https://www.dehaagsehogeschool.nl

TiFN. (2022). *Way of working*. https://www.tifn.nl

Ting-Toomey, S. (1999). *Communicating across Cultures*. The Guilford Press.

Titchener, E. B. (1909). *Lectures on the Experimental Psychology of Thought-Processes*. Macmillan. doi:10.1037/10877-000

TNO. (2022). *Innovation for life*. https://www.tno.nl

Toivakka, S., & Maasola, M. (2011). *Itsetunto kohdalleen, harjoituksia itsetuntemuksen ja vuorovaikutustaitojen oppimiseen*. PS-Kustannus.

Tonucci, F. (2009). Ciudades a escala humana: La Ciudad de los Niños. *Revista de Educación, número extraordinario 2009*, 147-168.

Tonucci, F. (2004). *La ciudad de los niños: Un modo nuevo de pensar la ciudad*. Fundación Germán Sánchez Ruipérez.

Trompenaars, F., & Hampden-Turner, C. (2020). *Riding on the Waves of Culture: Understanding Diversity in Global Business* (4th ed.). McGraw-Hill Professional.

Tuckman, B. W. (1965). Developmental sequence in small groups. *Psychological Bulletin, 63*(6), 384–399. doi:10.1037/h0022100 PMID:14314073

U.S. Census Bureau. (1995). *Celebrating Our Nation's Diversity*. CBET 95/01.

UG. (2022). *University of Groningen*. https://www.rug.nl

UNESCO. (2010). *Engineering: Issues, Challenges and Opportunities for Development.* UNESCO. https://unesdoc.unesco.org/ark:/48223/pf0000189753

United Nations Educational, Scientific and Cultural Organization. (2010). *Engineering: Issues, Challenges and Opportunities for Development.* United Nations.

United Nations Educational, Scientific and Cultural Organization. (2017). *Engineering.* http://www.unesco.org/new/en/natural-sciences/science-technology/engineering/

United Nations. (2021). United Nations 17 *Sustainable Development Goals to Transform Our World.* https://www.un.org/sustainabledevelopment/

UT. (2022). *University of Twente.* https://www.utwente.nl

Van Doorsselaer, K., & Koopmans, R. J. (2020). *Ecodesign: A Life Cycle Approach for a Sustainable Future.* Hanser-Fachbuch. doi:10.3139/9781569908624

Van Hoof, H. B., & Verbeeten, M. J. (2005). Wine Is for Drinking, Water Is for Washing: Student Opinions About International Exchange Programs. *Journal of Studies in International Education, 9*(1), 42–61. doi:10.1177/1028315304271480

Vandelvelde, X., Caesar, N., Suys, M., Kirchhöfer, V., & Suppes, D. (2018). Positive Energy Building Design (EPS final report). École Nationale d'Ingénieurs de Tarbes, France.

Vanhoucke, M. (2018). *The Data-Driven Project Manager. A Statistical Battle Against Project Obstacles.* Springer. doi:10.1007/978-1-4842-3498-3

Varney, S. (2021). *Leadership in Complexity and Change.* Walter de Gruyter GmbH. doi:10.1515/9783110713343

Vendoorne-Feys, A., Nicoara, G.-G., Carasel, I.-S., Karpiak, M., & Koceski, N. (2021b). *Ergonomic Reconfigurable Home-Desk* (Project Wiki). https://www.eps2021-wiki3.dee.isep.ipp.pt/

Vendoorne-Feys, A., Nicoara, G.-G., Carasel, I.-S., Karpiak, M., Koceski, N., Malheiro, B., Ribeiro, C., Justo, J., Silva, M. F., Ferreira, P., & Guedes, P. (2021a). Reconfigurable and Ergonomic Smart Desk – An EPS@ISEP 2021 Project. In *Proceedings of Ninth International Conference on Technological Ecosystems for Enhancing Multiculturality* (pp. 464-470). ACM. 10.1145/3486011.3486566

Vento, E., Kurz, M., Sarv, M., Mutlu, E., Evans, R., & Christensen, N. (2011). *WAW what a waste* (EPS Final Report). Universitat Politècnica de València.

Verein Deutscher Ingenieure. (2021). *Ethische Grundsätze des Ingenieurberufs* [Ethical Principles of the Engineering Profession]. https://www.vdi.de/themen/ethische-grundsaetze

Verganti, R. (2009). *Design-driven innovation: Changing the rules of competition by radically innovating what things mean.* Harvard Business Press.

Verhulst, E., Rohaert, S., & Doorsselaer, K. V. (2015). How to incorporate Sustainable Design in the International European Project Semester programme: Insights from practice. In *Proceedings of the International Conference on Engineering and Product Design Education* (pp. 1-6). The Design Society, Institution of Engineering Designers.

Verschueren, J. (2008). Intercultural Communication and the Challenges of Migration. *Language and Intercultural Communication, 8*(1), 21–35. doi:10.2167/laic298.0

Vianna, M., Vianna, Y., Adler, I., Lucena, B., & Russo, B. (2012). *Design Thinking: inovação em negócios* (2nd ed.). MJV Press.

von Hippel, E. (2006). *Democratizing Innovation*. The MIT Press.

Vos, H., & de Graaff, E. (2004). Developing metacognition: A basis for active learning. *European Journal of Engineering Education, 29*(4), 543–548. doi:10.1080/03043790410001716257

Wahoo Fitness. (2022). *Instructions | Elemnt Bolt GPS Bike Computer | Wahoo Fitness EU*. https://eu.wahoofitness.com/instructions/elemnt-bolt

Walther, J., Brewer, M., Sochacka, N., & Miller, S. (2019). Empathy and engineering formation. *Journal of Engineering Education, 109*(1), 1–23. doi:10.1002/jee.20301

Ward, C. (2001). *The impact of international students on domestic students and host institutions. Export Education Policy Project*. New Zealand Ministry of Education.

Washington Accord. (2022). *International Engineering Alliance*. International Engineering Alliance. https://www.ieagreements.org/accords/washington/

Weigand, E. (1999). Misunderstanding: The Standard Case. *Journal of Pragmatics, 31*(6), 763–785. doi:10.1016/S0378-2166(98)00068-X

Weinberg, U. (2015). *Network Thinking: Beyond Brockhaus Thinking*. Murmann Publishers.

Wharton, C. Y., & Horrocks, A. J. (2015). Students' perceptions of employability within their degree programme: highlighting the disparity between what academics believe is included and the student experience. *Proceedings of the 2nd International Conference on Enhancement and Innovation in Higher Education.*

Wheelan, S. A., Åkerlund, M., & Jacobsson, C. (2020). *Creating Effective Teams – Guide for Members and Leaders* (5th ed.). Sage.

Widdig, B., & Breslow, L. (2005). *Communicating Across Cultures 21G.019*. Spring 2005. Massachusetts Institute of Technology. https://ocw.mit.edu

Wingate, U. (2018). Academic literacy across the curriculum: Towards a collaborative instructional approach. *Language Teaching, 51*(3), 349–364. doi:10.1017/S0261444816000264

Woodley, A. W., Chabris, C. F., Pentland, A., Hashmi, N., & Malone, T. W. (2010). Evidence for a collective intelligence factor in the performance of human groups. *Science, 330*(6004), 686–688. doi:10.1126cience.1193147 PMID:20929725

Woolever, K. R. (2005). *Writing for the Technical Professions* (3rd ed.). Pearson Longman.

WUR. (2022). *Wageningen University & Research*. https://www.wur.nl

Yeager, D., & Dweck, C. (2020). What can be learned from growth mindset controversies? *The American Psychologist, 75*(9), 1269–1284. doi:10.1037/amp0000794 PMID:33382294

Yiu, J. (2010). *The Definitive Guide to the ARM Cortex-M3*. Newnes.

Yong, E., & Ashman, P. J. (2019). Integration of the Structured Development of Communication Skills within a Chemical Engineering Curriculum at the University of Adelaide. *Education for Chemical Engineers, 27*, 20–27. doi:10.1016/j.ece.2018.12.002

Zegarac, V. (2008). Culture and communication. In H. Spencer-Oatey (Ed.), *Culturally Speaking. Culture, Communication and Politeness Theory* (pp. 48–70). Continuum.

Zhang, Z., & Chang, E. (2017). Editorial: Current research and practice in teaching disciplinary literacies. *ESP Today, 5*(2), 132–147. doi:10.18485/esptoday.2017.5.2.1

Zwikael, O., & Chih, Y.-Y. (2014). Project benefit management: formulation and appraisal of target benefits. In *Proceedings of the Project Management Institute Research and Education Conference*. Newtown Square, PA: Project Management Institute.

About the Contributors

Benedita Malheiro holds a five-year degree in Electrical Engineering followed by an M.Sc. and a Ph.D. in Electrical and Computers Engineering, all from the University of Porto. She is a Coordinator Professor at the Electrical Engineering Department of Instituto Superior de Engenharia do Porto, the School of Engineering of the Polytechnic Institute of Porto, and director of the European Project Semester. She is specialized in engineering education and, as a senior researcher of the Centre of Robotics and Autonomous Systems of INESC TEC, in solving distributed, dynamic, and decentralized problems with the help of artificial intelligence (AI) and distributed computing. She is a member of AAAI, ACM, APPIA (Portuguese Association for AI) and OE, the Portuguese Engineers Association.

Pedro Fuentes-Durá received a Ph.D. degree in environmental engineering from the Universitat Politècnica de València, Spain, participating in several research projects. He trained as an Industrial Chemist with the University of Valencia. He is a Professor at the Nuclear and Chemical Engineering Department, Universitat Politècnica de València, where he has been the Head of International relations and the Head of Business relationships with the School of Design Engineering. He has participated in several innovation projects in higher education and as a committee member in several international events in the framework of higher education. He is the Founder and the Co-Director of Valencia Global and the International Coordinator of the European Project Semester. He is the Director of Open Innovation chair UPV-Istobal.

* * *

Duane Abata has been in academia for over forty years. He has been a member of the faculty at Michigan Technological University in Houghton, Michigan, at Northern Arizona University in Flagstaff, Arizona, and currently at South Dakota School of Mines and Technology in Rapid City, South Dakota. He has also served on the administration as Dean in the engineering colleges at each of these schools. In 2001-2003 Dr. Abata was elected to serve as President of the American Society for Engineering Education. He has spent considerable time in Europe, primarily in Denmark, Norway, and Italy, and served as the International Coordinator for EPS in the United States.

Elisabet Arnó-Macià is associate professor of English for Specific Purposes (ESP) at the Polytechnic University of Catalonia (Barcelona, Spain), where she teaches technical communication courses to engineering students. Her research interests include technical and academic communication, the use of technology in languages for specific purposes, and internationalisation and multilingualism in higher education, especially English-Medium Instruction (EMI).

David Ashworth is Associate Professor (retired), MSc Computer Aided Engineering (Nottingham Trent University), BA (Open University), Cert.Ed (The University of Birmingham).

Joe Barr has taught Technical Communication as part of the EPS programme at the Universitat Politècnica de Catalunya since 2013. He teaches Technical Writing and academic skills to engineering undergraduates at the same university. Moreover, he teaches Basic Concepts in Experimental Research in the Health Sciences faculty of the Universitat Internacional de Catalunya to undergraduates studying medicine, dentistry, and physical therapy as well as on Academic English pre-sessional courses at Keele University in the UK.

Philip Breedon is Professor of Smart Technologies at Nottingham Trent University and leads the Medical Engineering Design Research Group (MEDRG). He is a Chartered Engineer and Chartered IT professional. As PI and Co-I he and a multidisciplinary research team have recently delivered a number of funded projects at a value of over £2M. His research interests and latest projects centre on new and emerging technologies and materials. This includes wearable technologies, 3D/4D printing, robotics, extended reality systems, additive and subtractive manufacturing for medical applications, the surgical pathway and investigative research related to the utilisation of smart materials for medical applications.

Grazyna Budzińska graduated from the University of Lodz in English philology and in Executive EMBA, joint program with the University of Maryland, USA. Works as a senior lecturer at the Lodz University of Technology where for over 10 years has coordinated an international program European Project Semester as well as conducts Capstone projects in collaboration with companies at Clark University, Poland. The main areas of her academic interests are team dynamics and team management, business communication, innovative project, and problem-based learning methodologies. Conducted team management and team building courses at universities in France, The Netherlands, Spain, Germany, Switzerland, and other European countries.

Wander Colenbrander (1967) is senior lecturer Sustainable Packaging Design & Innovation and researcher of Lectorate Circular Business at The Hague University of Applied Sciences. His teaching and research activities focus upon sustainability aspects of packaging design education in close relation to business practice.

Nuria Matarredona Desantes is an architect, teacher and researcher. PhD from the Polytechnic University of Valencia (2015) where she has been teaching since 2016. She has been a visiting scholar at several research centers as University of Pennsylvania, Harvard University, the Universidad Nacional Autónoma of Mexico and the Polytechnic University of Madrid, specializing herself in architectural heritage preservation and sustainable development of local communities. She has participated in several specialized publications and forums. Mainly interested in collaborative design and innovative solutions to enhance the places where we live in.

Abel J. Duarte has a BSc and a 5-year degree in chemical engineering from the Porto Polytechnic followed by a M.Sc. and a PhD in chemistry from the University of Porto. He is Adjunct Professor at the Department of Chemical Engineering of ISEP and a researcher of REQUIMTE/LAQV. His research interests include sustain-able development, biochemical systems, and engineering education.

Claudia Facca is a Designer, educator and researcher. She has a Doctorate in Design (Universidade Anhembi Morumbi, SP/Brazil - 2020), "Sandwich" Doctorate (Faculty of Engineering of the University of Porto, Porto/Portugal - 2019), Master in Design (Universidade Anhembi Morumbi, SP/Brazil - 2008), Specialization in Communication and Arts and Didactics of Higher Education (Universidade Presbiteriana Mackenzie, SP/Brazil - 1992 and 1993), and a Bachelor Degree in Industrial Design with qualification in Product Design (Universidade Presbiteriana Mackenzie, SP/Brazil - 1991). She is a full professor and Coordinator of the Undergraduate Program in Design at Centro Universitário do Instituto Mauá de Tecnologia (São Paulo / Brazil). She has more than 30 years of teaching and professional experience in Design, having worked mainly in the areas of product development.

Paulo Ferreira graduated and received the MSc. degree in electrical and comput-er engineering from the Faculty of Engineering of the University of Porto, Portugal, in 1988 and 1994, respectively. He is Adjunct Professor at the Department of Informatics Engineering of ISEP. His research focuses on embedded systems. He is a member of the IEEE.

Philippe Fillatreau graduated from the ENSEEIHT Engineering School in 1988. He obtained his PhD in perception for mobile robotics in 1994 at the LAAS-CNRS Laboratory, Toulouse, France. His PhD works, funded by the Framatome Company, dealt with perception, 3D environment modelling and localization for an autonomous all-terrain mobile robot, in the framework of joint projects dealing with public safety or planetary exploration. In 1991–92 he spent one year at the Matra Marconi Space (now Airbus Group Defence & Space) Company, where he was involved in the european Eureka-AMR project. In 1994–95, he was a post-doctoral visiting researcher at the Department of Artificial Intelligence of the University of Edinburgh, Scotland; funded by a Human Capital and Mobility fellowship of the European Community, his works dealt with non polyhedral objects recognition and model update for public safety mobile robots. In 1996, Philippe Fillatreau joined the Delta Technologies Sud-Ouest Company (Toulouse) where he was technical head of the computer vision activities. There, he lead industrial projects often related to quality control on production lines and of industrial processes, and Research and Development projects in the fields of Computer Vision and Algorithm Architecture Adequacy, aiming at integrating advanced image processing algorithms in smart cameras based on FPGAs for very high throughput applications (e.g. transports safety, on-board satellite images processing or quality control of industrial processes). In 2009, he joined the National Engineering School of Tarbes (ENIT) as a researcher and lecturer. His current research interests are related to Virtual Reality (interactive task and path planning for the validation of complex tasks / scenarios), Computer Vision, Mobile Robotics and HW/SW co-design. Targeted applications include Industry 4.0 (validation of complex tasks / scenarios from design stage on, digital twin) or Advanced Driver Assistance Systems (ADAS). At the ENIT, Philippe Fillatreau teaches software and real-time programming, computer vision and critical embedded systems, and Virtual Reality. He also teaches project management (theory and project based learning, at the ENIT, the Paul Sabatier University and in several European Project Semester providers universities in France, Spain, Belgium, Poland, Finland, and Scotland), and quality engineering (Paul Sabatier University, Toulouse).

Isabel Gasch is an Associate Professor from the Department of Continuous Medium Mechanics and Theory of Structures at the Universitat Politècnica de València (UPV). Since 2007 she has taught 7 different subjects in the Degree of Mechanical Engineering and the Degree of Automatic and Industrial Electronic Engineering. Since the 2009-2010 academic year she participated as a supervisor in the European Project Semester (EPS). She has supervised 30 final degree projects and have tutored more than three hundred students in company internships. She is the co-author of 4 educational books, one of them published by the Limusa publishing house with diffusion in Central and South America. In order to improve teaching quality, she has taken more than 40 courses at the Institute of Education Sciences (ICE) of the UPV. Regarding Educational Research and Innovation. She has participated in 6 Educational Innovation and Improvement Projects (PIME), as well as in different Teaching Innovation Groups. In relation to these innovations, she has participated in 15 National and International Educational Congresses.

Nichiket Goyal is a student of Chitkara university.

Anca Greculescu (born in 1978) earned her PhD in Education Sciences from University of Bucharest and pursues a career in technical higher education as a Lecturer for the Department of Communication in Modern Languages, University Politehnica of Bucharest and as a communication advisor for EPS (European Project Semester):http://www.europeanprojectsemester.eu/providers/Bucharest. Main research areas: linguistic/professional communication, ESP (English for Specific Purposes); EST (English for Science and Technology); quality assurance in technical higher education, organizational communication. Amongst the latest works published: The Quality Paradigm. The Study of English in Technical Higher Education, Transversal Publishing House, Bucharest; Mechanical Transmissions Based on One Step Reducer (Design Guidebook) – in collaboration, Politehnica Press, Bucharest, 2013; European Project Semester – University Politehnica of Bucharest – Guidebook, Politehnica Press, Bucureşti 2017; Teacher Dimensions in Technical Higher Education –A Student Perspective – in collaboration, Revista Românească pentru Educaţie Multidimensională, 2019.

Pedro Guedes holds a five-year degree in electrical and computers engineering from the Faculty of Engineering of the University of Porto, Portugal and a M.Sc. in Systems Control from the University of Technology of Compiègne, France. He is Adjunct Professor at the Mathematics Department of ISEP and Researcher with the Robotics and Autonomous Systems Centre, INESC TEC, Porto. His research interests include engineering education, modeling and simulation of autonomous marine vehicles.

Jørgen Hansen, born in Odense 1945, graduated as a chemical engineer from DTU in 1969, retired associate professor (1985-2013), worked with EPS from 1998 to 2013, head of EPS from 2003 to 2013. Has also graduated as a singer from Sjællands Kirkemusikskole.

Martin Higginson taught Materials Science & Technology to Product Design and Engineering students at Nottingham Trent University for 32 years retiring from full-time employment at the end of 2020. In October 2021, he accepted a post back at NTU as an hourly paid lecturer in Engineering Science. Martin was responsible for the European Project Semester programme within the Product Design subject area at NTU between 2014 – 2020.

Anna Karczemska is an Assistant Professor in the Institute of Turbomachinery, Lodz University of Technology. Since the year 2000 she is part of the Division of Medical Apparatus where she works in the area of biomedical engineering, material science, allotropic forms of carbon (nanodiamonds), meteoritics (carbon materials in the Space), antibacterial coatings etc. She is a coordinator of the international project M-ERA.NET2/2019/2020 AnBaCo "Antibacterial Coatings Containing Carbon Nanoparticles Obtained by Sol-Gel Method". She has been involved in different students projects such as Problem Based Learning or European Project Semester, for many years. Higher education: Lodz University of Technology, Faculty of Technical Physics, Applied Mathematics and Computer Science of Lodz University of Technology, specialty - Solid State Physics.1997, MSc. PhD Studies: Imperial College of Science, Technology and Medicine, London, 1997-2000 PhD degree: Warsaw University of Technology, Materials Science Faculty, 2002. ORCID iD: https://orcid.org/0000-0003-4635-8019.

Malgorzata Klimka-Kolysko, psychologist, doctor of management sciences, works at the Department of Work and Organizational Psychology at the Faculty of Psychology and Law at the SWPS University in Poznań. He specializes in the psychology of career counseling and organizational psychology. Teaching and practical interests focus on career planning issues, both in an individual and organizational perspective.

Aurora Martínez-Corral, Ph.D. in Architecture (2012), is a part-time professor in the Department of Architectural Constructions of the UPV (2005), combining teaching with the practice of the profession as an architect of the Legal Services and Urban Infractions of the City Council of Bétera (Valencia) (2006). Her research and publications focused on railway architecture, its constructive-material analysis, and the protection of this heritage, together with the investigation of materials focused on reuse and sustainability.

Roger Nylund is a Senior lecturer in Management, Leadership and Engineering Ethics.

Andrei Opris has been working for the last 3 years as an Automation Engineer, developing graphical interfaces that facilitate the communication between the human (the user) and the industrial machine. Until now, he has worked in 2 major industries: Oil & Gas and Metallurgy. The project developed at EPS was a good starting point, it was an automation for a robotic arm and it had also a graphical-interface part in order to control, monitor and interact with it. Participant in the EPS Program at University Politehnica of Bucharest, where he worked on a hands-on project on IoT Capabilities, a valuable experience that helped him broaden his horizons and gave him the opportunity to start studying different technologies and subjects.

Ignasi Perat is an Associate Professor with the Department of Electrical Engineering, Polytechnic University of Catalonia. Since 1989, he has been lecturing with the School of Engineering of Vilanova i la Geltrú, where he has been a member of the GAECE Research Group (Electronically Commutated Drives). He is currently taking part in numerous activities related to technology transfer and lifelong learning. He has published numerous specialized articles and has participated in different national and international conferences. He is currently the Director of the Master's degree in Rail Systems and Electric Drive.

David Perry was born in London. He worked in Italy and Croatia before settling in Spain where he obtained his doctorate in general linguistics from the University of Valencia. Since joining the School of Design Engineering at the Universitat Politècnica de València he has travelled extensively, giving courses and lectures in a number of European and Asian countries. He has also been a member of several international projects, and thus has first-hand experience of the pitfalls and the benefits to be found when working or socializing in international contexts. His involvement with the European Project Semester at the UPV goes back to the very first edition. Besides giving short courses on aspects of communication and human resources, he has supervised several projects and been coordinator of the final evaluation panel.

Kim Poldner (1978) is Professor Circular Business at The Hague University of Applied Sciences and her research interests evolve at the crossroads of entrepreneurship, aesthetics and sustainability. She has written award winning case studies on sustainable fashion pioneers such as Veja and Osklen and publishes in journals such as Organization, Journal of Business Venturing, Organization Studies, Organization & Environment, Business & Society and Journal of Cleaner Production.

Marina Puyuelo has a background in Fine Arts in the specialty of sculpture, and a doctorate from the School of Architecture of the Universtat Politècnica de València. She has led several research projects with public and private funding, on inclusive design and accessibility in products and environments for public spaces and collective use. Lately, she has been working in the areas of informal education and the cultural heritage interpretation. She is the author of many articles and communications published in international journals and congresses.

Sarah Rohaert, after her Master in Product Development, gained international experience in product development, technical support & sales in the Portuguese development centre and the Belgian hub of a Scandinavian company, coordinating design, prototype testing & serial production in production plants worldwide. Inspired by the translation process of customer requirements to production specifications and complete product-service solution development for clients, she completed a PhD on a cross functional knowledge management approach for New Product Development for SMEs in transition. As a European Project Semester (EPS) Program coordinator, Product-Service-System design coach and guest professor on design sciences didactics at the University of Antwerp, she shares her research insights with multi-disciplinary student teams to address design and education innovation challenges in collaboration with industry, social organisations, museums, and STEM education pioneering schools. She has coordinated professionalisation courses for practice-based research staff, summer schools, masterclasses, events and campaigns about sustainability and globalisation as a Global Minds collaborator.

Amanda Abós Sanchez was born in Tarragona (Spain) in 1996. She studied Industrial Design Engeenering in UPC Vilanova i la Geltrú, Barcelona.

Francesco Luke Siena holds a PhD focussed on medical device design, in addition to an MSc in Smart Design, a BSc in Computer Aided Product Design and a PgDip in Professional Research Practice, all from Nottingham Trent University. Dr Siena is a Lecturer in Product Design, a Creative Short Course and Summer School Tutor within the School of Architecture Design & The Built Environment at Nottingham Trent University. Dr Siena primarily teaches across both BA/BSc and MA/MSc Product Design courses in addition to supporting research projects and PhD supervision alongside academics and members of the Medical Engineering Design Research Group. Dr Siena's research interests include medical device design and engineering, 3D printing, emerging technologies and materials, and the accurate assessment and measurement of health outcomes.

Manuel F. Silva was born on the 11th April 1970. He obtained his BSc, MSc and PhD degrees in Electrical and Computer Engineering from the Faculty of Engineering of the University of Porto, in 1993, 1997 and 2005, respectively. He is currently Coordinating Professor at the Department of Electrical Engineering of the Instituto Superior de Engenharia do Porto (having supervised or co-supervised more than 60 MSc theses and 80 graduation projects/internships) and Senior Researcher at the Centre for Robotics in Industry and Intelligent Systems of INESC TEC (https://criis.inesctec.pt/). He is the author of more than 150 publications in international journals and conferences (https://www.researchgate.net/profile/Manuel-Silva-8) and has been involved in several R&D projects. He has also been actively involved in the organization of several international conferences, being part of the Management Team of CLAWAR Association (https://clawar.org/) and the General Assembly Board of the Portuguese Robotics Society (of which he has been the President of the Executive Board). His research interests focus on modelling, simulation, robotics, bio-inspired robotics, control and education in engineering, robotics and control.

Gabriel Songel is the Chair Professor at Politechnic University of Valencia. Docent since 1988. Doctorate in Design Methodology at Engineering School at UPV. Master in Industrial Design at IED Milan. Founder of the Design Research an Management Group UPV. Founder partner of Innoarea Strategic Design Consulting S.L. TBE spinoff from UPV. Coordinator of the exhibitions Design Vision of Innovation, Valencia Design Innovation and 30 years of Industrial Design at UPV. 4 six-year tracks of Research and 1 of Transference on Design recognised by the Ministry of Education.

Gina Florica Stoica (born in 1955). Ph.D Professor at the Faculty of Mechanical Engineering and Mechatronics, University Politehnica of Bucharest. Research interests in Mechanical Engineering and Industrial Management, Electro discharge machining, Tribology, Materials, Design of Machine Parts. An EPS coordinator since 2014 (http://www.europeanprojectsemester.eu/providers/Bucharest). Publications: "Tribological behavior of new titanium alloys – a comparative study", "Applied Mechanics and Materials"; "Tests in biological environment for wear resistance" and in collaboration with Anca Greculescu, "Mechanical Transmissions Based on one step reducer".

Michael Vloeberghs qualified as a general practitioner in 1985 at the Free University Brussels, trained in general surgery, Neurology and Neurosurgery and finished training in the UK in 1993. After being a Consultant in Neurosurgery at the Free University of Brussels, Michael took up a post at NUH from 1995 to date. Michael's clinical activity consists of Children's Neurosurgery, in particular movement disorder and its treatment with Selective Dorsal Rhizotomy (SDR) and Deep Brain Stimulation (DBS) and Vagus Nerve Stimulation (VNS) in children with intractable epilepsy. Michael is involved in general Neurosurgical emergencies both adult and paediatric and is a UK National Pathway provider for adult Neurosurgical treatment of depression with Vagus Nerve Stimulators. Academically Michael has an interest in the application of engineering in medicine and led a number of project regarding simulation of child injury, circulation of cerebro-spinal fluid (CSF) as well as clinical research in the treatment of movement disorder.Michael was a Special Scientific Adviser at the Rutherford Appleton Laboratories with a comparable role in applying the RAL high end technology in medical applications.His same role later moved to NTU where Michael is a Visiting Professor involved in medical applications of technology as part of student and post graduate projects.

Dariusz Witkowski, Lodz University of Technology, Institute of Turbomachinery, Division of Medical Apparatus, wrote his master's thesis at the Ecole Centrale de Lyon, France during a 4-month internship, the title of thesis: "Cooling of gas turbine blades". Mr. Witkowski defended his doctoral thesis in 2006 at Lodz University of Technology, Title: Analysis of the influence of the mechanical disc heart valve geometry and its angular position in the aorta on flow haemodynamic characteristics". Since 1999, he has been employed at the Institute of Turbomachinery Lodz University of Technology, Mr. Witkowski conducts classes with students (lectures, tutorials, projects) including European Project Semester, Problem Based Learning and others. Mr. Witkowski participated in many research projects in the fields of Mechanical Engineering and Biomedical Engineering, Mr. Witkowski is involved in the research of flow through artificial heart valves, Ventricular Assist Devices, microflows, and research on antibacterial layers.

Index

K

L

M

N

O

P

Q

R

S

T

U

V

W

Recommended Reference Books

IGI Global's reference books are available in three unique pricing formats:
Print Only, E-Book Only, or Print + E-Book.

Shipping fees may apply.

www.igi-global.com

ISBN: 978-1-5225-9276-1
© 2020; 612 pp.
List Price: US$ **345**

ISBN: 978-1-5225-8325-7
© 2019; 169 pp.
List Price: US$ **165**

ISBN: 978-1-4666-8693-9
© 2015; 337 pp.
List Price: US$ **215**

ISBN: 978-1-4666-8823-0
© 2016; 824 pp.
List Price: US$ **420**

ISBN: 978-1-4666-8150-7
© 2015; 380 pp.
List Price: US$ **205**

ISBN: 978-1-4666-6379-4
© 2015; 602 pp.
List Price: US$ **225**

Do you want to stay current on the latest research trends, product announcements, news, and special offers?
Join IGI Global's mailing list to receive customized recommendations, exclusive discounts, and more.
Sign up at: **www.igi-global.com/newsletters.**

Publisher of Peer-Reviewed, Timely, and Innovative Academic Research

IGI Global
DISSEMINATOR OF KNOWLEDGE

www.igi-global.com Sign up at www.igi-global.com/newsletters facebook.com/igiglobal twitter.com/igiglobal linkedin.com/igiglobal

Ensure Quality Research is Introduced to the Academic Community

Become an Evaluator for IGI Global Authored Book Projects

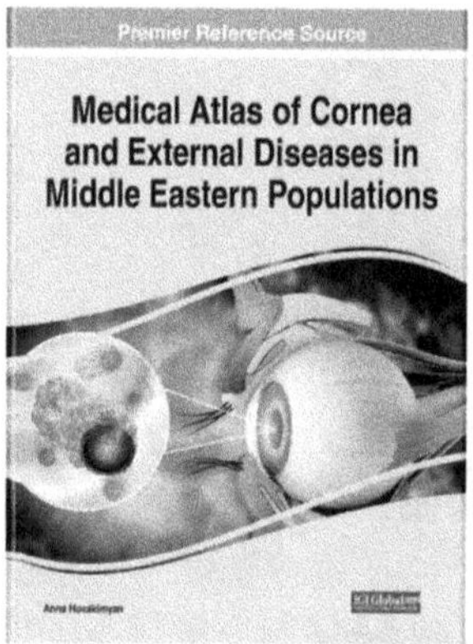

The overall success of an authored book project is dependent on quality and timely manuscript evaluations.

Applications and Inquiries may be sent to:
development@igi-global.com

Applicants must have a doctorate (or equivalent degree) as well as publishing, research, and reviewing experience. Authored Book Evaluators are appointed for one-year terms and are expected to complete at least three evaluations per term. Upon successful completion of this term, evaluators can be considered for an additional term.

If you have a colleague that may be interested in this opportunity, we encourage you to share this information with them.

Increase Your Manuscript's Chance of Acceptance

IGI Global Author Services

Copy Editing & Proofreading

Professional, native English language copy editors improve your manuscript's grammar, spelling, punctuation, terminology, semantics, consistency, flow, formatting, and more.

Scientific & Scholarly Editing

A Ph.D. level review for qualities such as originality and significance, interest to researchers, level of methodology and analysis, coverage of literature, organization, quality of writing, and strengths and weaknesses.

Figure, Table, Chart & Equation Conversions

Work with IGI Global's graphic designers before submission to enhance and design all figures and charts to IGI Global's specific standards for clarity.

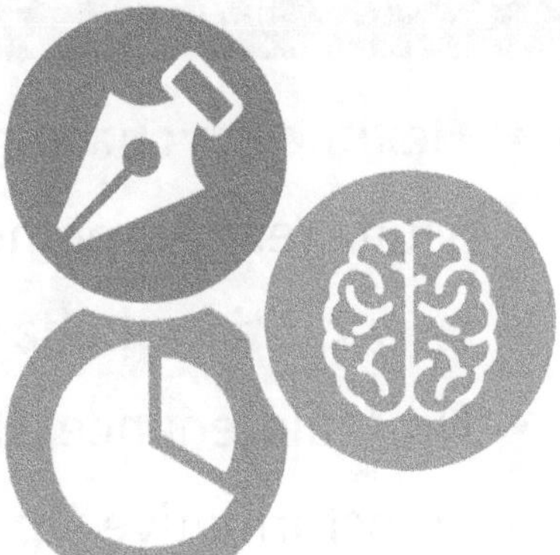

- **Professional Service**
- **Quality Guarantee & Certificate**
- **Timeliness**
- **Affordable Pricing**

What Makes IGI Global Author Services Stand Apart?

Services/Offerings	IGI Global Author Services	Editage	Enago
Turnaround Time of Projects	*3-5 Business Days*	6-7 Busines Days	6-7 Busines Days
Pricing	Fraction of our Competitors' Cost	Up to **2x** Higher	Up to **3x** Higher

Learn More or Get Started Here:

For Questions, Contact IGI Global's Customer Service Team at cust@igi-global.com or 717-533-8845

IGI Global
PUBLISHER of TIMELY KNOWLEDGE
www.igi-global.com

6,600+ E-BOOKS. ADVANCED RESEARCH. INCLUSIVE & ACCESSIBLE.

IGI Global
PUBLISHER of TIMELY KNOWLEDGE

IGI Global e-Book Collection

- **Flexible Purchasing Options** (Perpetual, Subscription, EBA, etc.)
- Multi-Year Agreements with **No Price Increases** Guaranteed
- **No Additional Charge** for Multi-User Licensing
- No Maintenance, Hosting, or Archiving Fees
- Transformative **Open Access Options** Available

Request More Information, or Recommend the IGI Global e-Book Collection to Your Institution's Librarian

Among Titles Included in the IGI Global e-Book Collection

Research Anthology on Racial Equity, Identity, and Privilege (3 Vols.)
EISBN: 9781668445082
Price: US$ 895

Handbook of Research on Remote Work and Worker Well-Being in the Post-COVID-19 Era
EISBN: 9781799867562
Price: US$ 265

Research Anthology on Big Data Analytics, Architectures, and Applications (4 Vols.)
EISBN: 9781668436639
Price: US$ 1,950

Handbook of Research on Challenging Deficit Thinking for Exceptional Education Improvement
EISBN: 9781799888628
Price: US$ 265

Acquire & Open

When your library acquires an IGI Global e-Book and/or e-Journal Collection, your faculty's published work will be considered for immediate conversion to Open Access *(CC BY License)*, at no additional cost to the library or its faculty *(cost only applies to the e-Collection content being acquired)*, through our popular **Transformative Open Access (Read & Publish) Initiative**.

For More Information or to Request a Free Trial, Contact IGI Global's e-Collections Team: eresources@igi-global.com | 1-866-342-6657 ext. 100 | 717-533-8845 ext. 100

Have Your Work Published and Freely Accessible

Open Access Publishing

With the industry shifting from the more traditional publication models to an open access (OA) publication model, publishers are finding that OA publishing has many benefits that are awarded to authors and editors of published work.

Freely Share Your Research

Higher Discoverability & Citation Impact

Rigorous & Expedited Publishing Process

Increased Advancement & Collaboration

Acquire & Open

When your library acquires an IGI Global e-Book and/or e-Journal Collection, your faculty's published work will be considered for immediate conversion to Open Access *(CC BY License)*, at no additional cost to the library or its faculty *(cost only applies to the e-Collection content being acquired)*, through our popular **Transformative Open Access (Read & Publish) Initiative**.

Provide Up To **100%** OA APC or CPC Funding

Funding to Convert or Start a Journal to **Platinum OA**

Support for Funding an **OA Reference Book**

IGI Global publications are found in a number of prestigious indices, including Web of Science™, Scopus®, Compendex, and PsycINFO®. The selection criteria is very strict and to ensure that journals and books are accepted into the major indexes, IGI Global closely monitors publications against the criteria that the indexes provide to publishers.

WEB OF SCIENCE™ Compendex Scopus®

PsycINFO® IET Inspec

Learn More Here:

For Questions, Contact IGI Global's Open Access Team at openaccessadmin@igi-global.com

IGI Global PUBLISHER of TIMELY KNOWLEDGE www.igi-global.com

Are You Ready to

Publish Your Research ?

IGI Global
PUBLISHER of TIMELY KNOWLEDGE

IGI Global offers book authorship and editorship opportunities across 11 subject areas, including business, computer science, education, science and engineering, social sciences, and more!

Benefits of Publishing with IGI Global:

- Free one-on-one editorial and promotional support.
- Expedited publishing timelines that can take your book from start to finish in less than one (1) year.
- Choose from a variety of formats, including Edited and Authored References, Handbooks of Research, Encyclopedias, and Research Insights.
- Utilize IGI Global's eEditorial Discovery® submission system in support of conducting the submission and double-blind peer review process.
- IGI Global maintains a strict adherence to ethical practices due in part to our full membership with the Committee on Publication Ethics (COPE).
- Indexing potential in prestigious indices such as Scopus®, Web of Science™, PsycINFO®, and ERIC – Education Resources Information Center.
- Ability to connect your ORCID iD to your IGI Global publications.
- Earn honorariums and royalties on your full book publications as well as complimentary copies and exclusive discounts.

Join Your Colleagues from Prestigious Institutions, Including:

Australian National University
Massachusetts Institute of Technology
Johns Hopkins University
Tsinghua University
Harvard University
Columbia University in the City of New York

Learn More at: www.igi-global.com/publish

or Contact IGI Global's Aquisitions Team at: acquisition@igi-global.com

Easily Identify, Acquire, and Utilize Published Peer-Reviewed Findings in Support of Your Current Research

IGI Global OnDemand

Purchase Individual IGI Global OnDemand Book Chapters and Journal Articles

For More Information:

www.igi-global.com/e-resources/ondemand/

Browse through 150,000+ Articles and Chapters!

Find specific research related to your current studies and projects that have been contributed by international researchers from prestigious institutions, including:

Massachusetts Institute of Technology

HARVARD UNIVERSITY

COLUMBIA UNIVERSITY IN THE CITY OF NEW YORK

Australian National University

- Accurate and Advanced Search
- Affordably Acquire Research
- Instantly Access Your Content
- Benefit from the InfoSci Platform Features

"It really provides an excellent entry into the research literature of the field. It presents a manageable number of highly relevant sources on topics of interest to a wide range of researchers. The sources are scholarly, but also accessible to 'practitioners'."

- Ms. Lisa Stimatz, MLS, University of North Carolina at Chapel Hill, USA

Interested in Additional Savings?

Subscribe to

IGI Global OnDemand *Plus*

Learn More

Acquire content from over 128,000+ research-focused book chapters and 33,000+ scholarly journal articles for as low as US$ 5 per article/chapter (original retail price for an article/chapter: US$ 37.50).

www.ingramcontent.com/pod-product-compliance
Lightning Source LLC
LaVergne TN
LVHW062114300926
849161LV00010B/2461
9781668423097